AF581672

Entrepreneurial Education Strengthening Resilience, Societal Change and Sustainability

Entrepreneurial Education Strengthening Resilience, Societal Change and Sustainability

Editors

Jaana Seikkula-Leino
Priti Verma
Maria Salomaa
Mats Westerberg

MDPI • Basel • Beijing • Wuhan • Barcelona • Belgrade • Manchester • Tokyo • Cluj • Tianjin

Editors
Jaana Seikkula-Leino
Mid Sweden University
Sweden

Priti Verma
Higher Colleges of Technology
United Arab Emirates

Maria Salomaa
Tampere University of Applied Sciences
Finland

Mats Westerberg
Luleå University of
Technology
Sweden

Editorial Office
MDPI
St. Alban-Anlage 66
4052 Basel, Switzerland

This is a reprint of articles from the Special Issue published online in the open access journal *Sustainability* (ISSN 2071-1050) (available at: https://www.mdpi.com/journal/sustainability/special_issues/entrepreneurial_education).

For citation purposes, cite each article independently as indicated on the article page online and as indicated below:

LastName, A.A.; LastName, B.B.; LastName, C.C. Article Title. *Journal Name* **Year**, *Volume Number*, Page Range.

ISBN 978-3-0365-6241-4 (Hbk)
ISBN 978-3-0365-6242-1 (PDF)

Contents

About the Editors

Jaana Seikkula-Leino

Dr. Jaana Seikkula-Leino works as a Professor of Pedagogy at Mid Sweden University. Furthermore, she is an Associate Professor (docent) of entrepreneurship education at the University of Turku, Finland. Dr. Seikkula-Leino has a research record specializing in entrepreneurship education, entrepreneurial organization, change management, curriculum, strategy, leadership, psychology, teacher education, assessment, and digitalization. She has over 150 publications. Over the years, she has worked as an expert for governments, the EU, the United Nations, and OECD. Her current projects are, for example, SKILLOON, www.skilloon.com; Entrepreneurship Competence in The Vocational Education and Training, Europe, Cedefop. Dr. Prof. Jaana Seikkula-Leino is a selected member of the Academy of Finland's scientific committee. Orcid: https://orcid.org/0000-0001-8624-6930.

Priti Verma

Dr. Priti Verma serves as the Associate Executive Dean and Assistant Professor, Bachelor of Education (Early Childhood) program, at the Faculty of Education, Higher Colleges of Technology. She received her PhD in Child Development (2003) from the Chaudhary Charan Singh Haryana Agricultural University, India. She recently completed a research project in the UAE funded by the Sheikh Saud bin Saqr Al Qasimi Foundation for Policy Research where she studied the quality of privately owned Early Childhood Education Centers and nurseries based in Ras Al Khaimah. Her main research interests are in the area of child development, early childhood education, entrepreneurship education, early childhood teacher education, innovative and entrepreneurial society, micronutrient interventions and their impact on young children growth and development. She is a member of the National Association for Education of Young Children and the Society for Research in Child Development. Orcid: https://orcid.org/0000-0003-2029-6526.

Maria Salomaa

Dr. Maria Salomaa works as an RDI specialist at Tampere University of Applied Sciences and she is a visiting research fellow at the University of Lincoln (UK). She holds MA degrees in theatre and drama research and higher education administration, and a PhD in management. Her main areas of expertise are project management, research impact, regional development and planning, and national and international higher education policies, and currently, her research interests include regional development policies, structural funds, higher education institutions and rural regions. Orcid: https://orcid.org/0000-0002-4415-0161.

Mats Westerberg

Dr. Mats Westerberg works as a professor at Luleå University of Technology in Sweden. He is mainly doing research and development work in the entrepreneurship education field and there with a special interest in how entrepreneurship can be brought to other subject fields in education on all levels in the education system as a tool for strengthening motivation and equipping the students with entrepreneurial competences so they better can make use of their subject knowledge. Dr. Westerberg has published extensively in both education and entrepreneurship journals and anthologies. Currently he is involved in projects investigating entrepreneurship in vocational education, instilling innovation in municipalities and train-the-trainer programs in both entrepreneurship and sustainability.

Preface to "Entrepreneurial Education Strengthening Resilience, Societal Change and Sustainability"

The societal role of higher education extends beyond the traditional academic core missions of education and research. This so-called 'third mission' is often linked with various engagement and knowledge transfer actions with external stakeholders, such as businesses, public actors, and non-governmental organisations (e.g., [1]). However, the capability to balance the needs and challenges of both regional stakeholders and university systems to reinforce their strategic capacity [2], [3], continuing the development of mechanisms and entrepreneurial skills within the university as well as individual capacities of both staff and students to adapt and thrive in ever-changing circumstances should be a priority. Furthermore, environmental, societal, and economic problems challenge all human activities, including higher education and its surrounding communities, networks, and ecosystems.

Entrepreneurial thinking, behavior, and actions are not only related to traditional educational institutions, but entrepreneurial behavior is also linked to the psychological and affective factors that steer human behavior [4]. Therefore, we must also consider the existing human capability in businesses and working life. How could we develop education and training in this arena? How could we develop higher education to support the development of these competences?

Our book guides the reader to how political documents steer the implementation of entrepreneurship education and the global Sustainable Development Goals. Secondly, we present studies on how different pedagogical solutions and models can meaningfully support the development of entrepreneurial competencies in higher education. The training organizer and teacher can thus also get ideas for developing their teaching systems, curriculum, and pedagogy. However, we will also show the complexity of this phenomenon and its related concepts, thus aiming to create a deeper understanding of the educational phenomenon. Thirdly, we will focus on how teacher training could contribute to promoting the introduction of entrepreneurship education to different fields of education. Teacher education has a far-reaching impact on the future because it is responsible for developing the knowledge and skills of teachers and trainers, whose actions can be easily seen fifty years ahead: through their students, effects can be seen hundreds of years ahead. Fourth, we open up about the importance of ecosystems and networks. What could they be, and how could they be developed, even during a time of crisis? Finally, we present the training in the business world and how this sector can contribute to the strengthening of entrepreneurial activities. Finally, we offer an analysis of the entrepreneurial ecosystem and how it could be developed. Our book consists of 13 articles from experts in their field, not only from science fields—many authors also have extensive knowledge of education and the business world behind them. The authors have contributed to designing, researching, and developing entrepreneurial ecosystems promoting resilience and sustainability in their countries and globally. We believe that the contribution of this book is beneficial to researchers, policy-makers, higher education providers, stakeholders, educators, and trainers. Next, we will describe the book's content in more detail by presenting articles grouped in relation to the main themes shown above.

Policies supporting entrepreneurial and sustainable education development

First, we focus on the European Framework for Entrepreneurial Competences (EntreComp). This study aims to discover how European policies can drive the development of entrepreneurial competencies. This study, conducted by Seikkula-Leino, Salomaa, Jónsdóttir, McCallum, and Israel [5], focused on examining the case of EntreComp and its integration into education and training development in different countries. The outcomes suggest that EntreComp has been widely recognized as a relevant driver of competence in entrepreneurship education. However, a lack of shared vision and development of practice in the use of EntreComp might hinder the implementation of the framework. Therefore, further support and guidance are needed to promote the learning acquisition of policy-driven frameworks, such as EntreComp.

How to implement entrepreneurship education, and what could be the pre-conditions and outcomes, even during a time of crisis?

Entrepreneurship is increasingly seen as a set of competencies needed in many areas. As a result, it needs to be integrated into higher education even in seemingly distant areas, such as, e.g., the public sector. Thus, there is a need for further research-based guidance to introduce and develop entrepreneurship education as an enabler of multidisciplinary approach towards the transition in higher education. Therefore, Klucznik-Töro [6] aims in her study to support that transition and address related challenges by presenting a novel progression model with guidelines for the development of courses at the higher education level with an entrepreneurial university approach.

Next, we present how entrepreneurial competencies could be developed by leveraging a community of practice (CoP) during a crisis. Mead, Pietsch, Matthew, Lipkin-Moore, Metzger, Avdeev, and Ruzycki [7] discuss the role of the community of practice in sustaining and supporting the faculty. Furthermore, faculty-level case studies reveal how sustainable design and social responsibility can be integrated into teaching activities and experiences. The main contribution of this research is suggesting that the learning framework can play an essential role in informing how to best optimize the CoP format and approach in a way that it leverages and addresses strengths, challenges, and experiences, thus supporting the needs of CoP members even during challenging times.

The investigation of the crisis period continues in the study of Lee and Jung [8], in which they state that the entrepreneurial mindset has not been examined concerning career adaptability, especially with the increased uncertainty in this world. Therefore, it explored the relations among intolerance of uncertainty—specifically its sub-factors, prospective anxiety and inhibitory anxiety—career adaptability, and entrepreneurial mindset students facing school-to-work transition during COVID-19. In the study, it is stressed that entrepreneurial mindset and career adaptability show a significantly positive relation. Furthermore, the study also supported the mediating role of an entrepreneurial mindset between intolerance of uncertainty and career adaptability.

Team Learning as a model for facilitating entrepreneurship competencies in higher education, the case of Proakatemia by Nevalainen, Seikkula-Leino, and Salomaa [9] will be presented as strengthening our understanding of how an entrepreneurial approach based on curriculum and pedagogy development could lead to more meaningful results in higher education institutions. The study investigated university students' perception of the development of their entrepreneurial competencies (Proakatemia, Tampere University of Applied Sciences, Finland). The results indicate that the Team Learning concept of Proakatemia facilitates learning entrepreneurial competencies, thus providing further ideas and insights for education providers and educators to develop their curricula, programmes, and teaching practices.

Ashari, Abbas, Abdul-Talib, and Zamani [10] apply the Theory of Planned Behavior (TPB) in their research to study the effect of an entrepreneurship course on the entrepreneurial intentions (engineering students at Universiti Teknologi Petronas) as the entrepreneurial intention is effective in predicting behavior. The results of the study stress that the TPB explains and predicts entrepreneurial intention. However, the outcomes also highlight that participating in entrepreneurship studies do not empower the relationship between the exogenous and endogenous construct (compared to the students who do not participate in these kinds of studies). The results of this research raise a favorable implication for improving the curricula and pedagogy, thus highlighting the enhancement of the implementation of the SDGs through entrepreneurship studies.

We finalize this section of the book with an analysis of the complexity of this educational phenomenon, and the topics of our book, thus giving some directions for future education design. In the study of Pascucci, Hernández-Sánchez, and Sánchez-García [11], they conclude with their review in which they analyze the state of entrepreneurial education as it applies to business resilience. They search records over the last 20 years about entrepreneurial resilience that consider their social impact and focus on sustainability. The target of the study was to determine whether an enterprise that stresses social impact and sustainability rather than profits could empower entrepreneurial resilience. The study offers a more complex description of the entrepreneurial resilience by connecting social and environmental sensitivity with a profit-oriented logic. The authors state that there are three clusters, "education and sustainability", "entrepreneurship and social impact," and "innovation", and these three clusters are related to superior entrepreneurial resilience. The authors suggest this approach could be adopted in real time to be able to adapt to socio-economic crises. Furthermore, the awareness of complexity is an asset, for example, for policymakers, education providers, and educators to consider more meaningfully the previously mentioned pro-conditions in designing educational frameworks, programmes, and activities accordingly.

How can teacher education enhance entrepreneurship education and the development of a sustainable society?

Next, we gaze again at the documents that steer education and training. However, at this stage, we concentrate on teacher education. Seikkula-Leino, Jónsdóttir, Håkansson-Lindqvist, Westerberg, and Eriksson-Bergström [12] studied how entrepreneurial, sustainable, and pro-environmental education has been developed in Nordic teacher education curricula. In their research, the authors analyzed the B.Ed. Curricula of three academic teacher organizations in Nordic countries, Sweden, Finland, and Iceland. According to their results, teacher education curricula incorporated entrepreneurship education and sustainable development to some extent, although very explicitly. Given the urgency of challenges such as global climate change, the educational goals and contents in these curricula related to entrepreneurial education and sustainable development are minimal. The idea of integrating environmental/sustainable and entrepreneurship education could be enhanced more explicitly, with these interdisciplinary educational themes stressed more strongly.

To continue with teacher education research, we have an example of teacher education in Wales Weicht and Jónsdóttir [13]. This study aims to learn how creativity, innovation, and an enterprising mindset of learners are developed in teacher education. Documentary evidence, such as module and assignment handbooks, explored how teacher educators deliver entrepreneurial education for social change. Their finding indicates that entrepreneurial education in teacher training has enabled constructive learning, cultivating creativity and action competence, thus promoting social change. In their research, they also provide examples that display how the intentions of the Curriculum for Wales and entrepreneurial education approaches of The University of Wales Trinity Saint David (UWTSD) emerge in practice.

In the final teacher education article, we concentrate on how education for sustainable development and entrepreneurial education has been designed and developed in Finnish vocational teacher education by Asikainen and Tapani [14]. The case study explores teacher students' process of sense-making of sustainable development and how becoming a teacher who practices Education for Sustainable Development connects with entrepreneurship. The qualitative content analysis of students' writings focused on signs of transformative learning and was guided by the approaches of sustainable development and learning goals for teachers' sustainability competencies in the vocational teacher education curricula. The results indicate that transformative learning is possible. Moreover, it highlights the importance of specific entrepreneurial capabilities in actualizing change agency in vocational teacher education.

Networks, entrepreneurial businesses, and ecosystems promoting sustainability and social change

The latter part of this book concentrates on how higher education external networks could support the design of training programmes and curricula. Furthermore, we concentrate on businesses and their entrepreneurial actions, e.g., talent management and cooperation activities, stressing environmental responsibility and entrepreneurial resilience. Finally, we offer an insight into the entrepreneurial ecosystem. What could it be and how does it develop in higher education?

Tehseen and Haider [15] start this section of the book with their study, which explored the impact of universities' partnerships on students' sustainable entrepreneurship intentions. The authors investigated the impact of entrepreneurial attitude and perceived desirability and feasibility concerning sustainable entrepreneurship intentions regarding the moderating impact of entrepreneurial passion. The students participated in their undergraduate studies in Malaysia. This quantitative and comparative study highlights, for example, that universities having a partnership with other overseas' universities may offer high-quality entrepreneurship courses or other activities due to which these students have high entrepreneurial passion and attitudes and are more willing and able to of starting their businesses as compared to students of other local universities that have no partnership with overseas universities.

Next, we present Ferreiro-Seoane's, Miguéns-Refojo's, and Atrio-Lema's [16] study to analyze the characteristics of the most attractive companies in the labor market, which each year maintained their position in the ranking. The Spanish business magazine 'Actualidad Económica' (AE) created the ranking system. In this research, it is concluded that the permanence in the ranking significantly increases the total value and training, thus leading businesses to excellence, along with the fact that they are in the capital of the country and they concentrate on commerce, professional, scientific and technical, and finance and insurance sectors. However, the evaluation of training is explained by the study's employee valuation, working culture, and talent management. However, factors such as gender variable in the business direction, nationality, size, and stock market membership do not significantly influence the overall valuation. The study offers ideas on developing entrepreneurial human capital, labor market, talent management, and corporate governance.

Finally, we conclude our book with Liu, Kulturel-Konak, and Konak [17] study, in which they examine two core issues of the university-based entrepreneurship education ecosystem by presenting the critical elements of the ecosystem, their roles, and the development process and sustainable construction strategy of the ecosystem. In the study, it is highlighted that the main elements of the higher-education based entrepreneurship education ecosystem consist of six units (colleges/universities, students, educators, government, industry, and community) acting as initiators and seven factors (entrepreneurship curricula, entrepreneurial activities and practices, organizational structure, resources, leadership vision, core faculty, and operating mechanism) acting as the intermediaries. Furthermore, these key elements constitute three independent functional subsystems: teaching and innovation, support, and operation interconnected by the universities. The development process of a higher-education based entrepreneurship education ecosystem involves seven steps: preparation, germination, growth, equilibrium, stagnation, recession, and collapse. For sustainability, suggestions on a solid foundation, continuous investment, and constant monitoring are provided to university administrators and policymakers to advance higher education's contribution to social and economic development.

[1] P. Laredo, "Revisiting the Third Mission of Universities: Toward a Renewed Categorization of University Activities?," Higher Education Policy, vol. 20, no. 4, pp. 441–456, Dec. 2007, doi: 10.1057/palgrave.hep.8300169.

[2] D. Charles, F. Kitagawa, and E. Uyarra, "Universities in crisis?–new challenges and strategies in two English city-regions," Cambridge Journal of Regions, Economy and Society, vol. 7, no. 2, pp. 327–348, Jul. 2014, doi: 10.1093/cjres/rst029.

[3] M. Salomaa and D. Charles, "The university third mission and the European Structural Funds in peripheral regions: Insights from Finland," Sci Public Policy, vol. 48, no. 3, pp. 352–363, Jul. 2021, doi: 10.1093/scipol/scab003.

[4] J. Seikkula-Leino and M. Salomaa, "Bridging the Research Gap—A Framework for Assessing Entrepreneurial Competencies Based on Self-Esteem and Self-Efficacy," Educ Sci (Basel), vol. 11, no. 10, p. 572, Sep. 2021, doi: 10.3390/educsci11100572.

[5] J. Seikkula-Leino, M. Salomaa, S. R. Jónsdóttir, E. McCallum, and H. Israel, "EU Policies Driving Entrepreneurial Competences—Reflections from the Case of EntreComp," Sustainability, vol. 13, no. 15, p. 8178, Jul. 2021, doi: 10.3390/su13158178.

[6] A. Klucznik-Törő, "The New Progression Model of Entrepreneurial Education—Guideline for the Development of an Entrepreneurial University with a Sustainability Approach," Sustainability, vol. 13, no. 20, p. 11243, Oct. 2021, doi: 10.3390/su132011243.

[7] T. Mead et al., "Leveraging a Community of Practice to Build Faculty Resilience and Support Innovations in Teaching during a Time of Crisis," Sustainability, vol. 13, no. 18, p. 10172, Sep. 2021, doi: 10.3390/su131810172.

[8] A. Lee and E. Jung, "The Mediating Role of Entrepreneurial Mindset between Intolerance of Uncertainty and Career Adaptability," Sustainability, vol. 13, no. 13, p. 7099, Jun. 2021, doi: 10.3390/su13137099.

[9] T. Nevalainen, J. Seikkula-Leino, and M. Salomaa, "Team Learning as a Model for Facilitating Entrepreneurial Competences in Higher Education: The Case of Proakatemia," Sustainability, vol. 13, no. 13, p. 7373, Jul. 2021, doi: 10.3390/su13137373.

[10] H. Ashari, I. Abbas, A.-N. Abdul-Talib, and S. N. Mohd Zamani, "Entrepreneurship and Sustainable Development Goals: A Multigroup Analysis of the Moderating Effects of Entrepreneurship Education on Entrepreneurial Intention," Sustainability, vol. 14, no. 1, p.

431, Dec. 2021, doi: 10.3390/su14010431.

[11] T. Pascucci, B. R. Hernández-Sánchez, and J. C. Sánchez-García, "Cooperation and Environmental Responsibility as Positive Factors for Entrepreneurial Resilience," Sustainability, vol. 14, no. 1, p. 424, Dec. 2021, doi: 10.3390/su14010424.

[12] J. Seikkula-Leino, S. R. Jónsdóttir, M. Håkansson-Lindqvist, M. Westerberg, and S. Eriksson-Bergström, "Responding to Global Challenges through Education: Entrepreneurial, Sustainable, and Pro-Environmental Education in Nordic Teacher Education Curricula," Sustainability, vol. 13, no. 22, p. 12808, Nov. 2021, doi: 10.3390/su132212808.

[13] R. Weicht and S. R. Jónsdóttir, "Education for Social Change: The Case of Teacher Education in Wales," Sustainability, vol. 13, no. 15, p. 8574, Jul. 2021, doi: 10.3390/su13158574.

[14] E. Asikainen and A. Tapani, "Exploring the Connections of Education for Sustainable Development and Entrepreneurial Education—A Case Study of Vocational Teacher Education in Finland," Sustainability, vol. 13, no. 21, p. 11887, Oct. 2021, doi: 10.3390/su132111887.

[15] S. Tehseen and S. A. Haider, "Impact of Universities' Partnerships on Students' Sustainable Entrepreneurship Intentions: A Comparative Study," Sustainability, vol. 13, no. 9, p. 5025, Apr. 2021, doi: 10.3390/su13095025.

[16] F. J. Ferreiro-Seoane, V. Miguéns-Refojo, and Y. Atrio-Lema, "Can Talent Management Improve Training, Sustainability and Excellence in the Labor Market?," Sustainability, vol. 13, no. 12, p. 6645, Jun. 2021, doi: 10.3390/su13126645.

[17] H. Liu, S. Kulturel-Konak, and A. Konak, "Key Elements and Their Roles in Entrepreneurship Education Ecosystem: Comparative Review and Suggestions for Sustainability," Sustainability, vol. 13, no. 19, p. 10648, Sep. 2021, doi: 10.3390/su131910648.

Conclusion

There is a definite need to re-assess our traditional higher education from the perspective of promoting entrepreneurial competencies to promote the needed changes in societies. This book provides an overview of how entrepreneurship education and training can strengthen the activities that promote social resilience and sustainability in higher education. We provide an overview of how the global framework for entrepreneurial learning is implemented, enhancing the development of entrepreneurial competencies and what we could learn from the experience. Moreover, we provide insights into the phenomenon's complexity, entrepreneurial pedagogy, and models, which we think are helpful to know and understand in designing, e.g., entrepreneurial curricula and their activities not only in 'the normal life' but also in times of crises. We also highlight the essence of developing teacher education in this context, thus providing sustainability for meaningful education development over decades further.

Furthermore, we stress that entrepreneurial competencies are also needed in working life. Therefore, we also offer an insight into talent management from this point of view. Additionally, students' quick transitions from higher education to working life are relevant, as well as the combinations of studies and working life. Therefore, it would be desirable for the development of entrepreneurial competencies to be looked at on a larger scale in the surrounding community. Lastly, we conclude our book by presenting some interesting key points in developing entrepreneurial higher education from the entrepreneurial ecosystem point of view. We hope our book provides new knowledge, ideas, and practices to researchers, policy-makers, higher education providers, stakeholders, educators, and trainers.

Jaana Seikkula-Leino, Priti Verma, Maria Salomaa, and Mats Westerberg
Editors

Article

EU Policies Driving Entrepreneurial Competences—Reflections from the Case of EntreComp

Jaana Seikkula-Leino [1,2,*], Maria Salomaa [1,3], Svanborg Rannveig Jónsdóttir [4], Elin McCallum [5] and Hazel Israel [5]

1 RDI and Business Operations, Tampere University of Applied Sciences, 33520 Tampere, Finland; maria.e.salomaa@gmail.com
2 Department of Education, Faculty of Human Sciences, Mid Sweden University, 852 30 Sundsvall, Sweden
3 International Business School, University of Lincoln, Lincoln LN6 7TS, UK
4 School of Education, University of Iceland, 102 Reykjavík, Iceland; svanjons@hi.is
5 Bantani Education, 3080 Tervuren, Belgium; elin@bantani.com (E.M.); hazel@bantani.com (H.I.)
* Correspondence: jaana.seikkula-leino@tuni.fi

Abstract: The United Nations' Sustainable Development Goals promote entrepreneurial competences as a means of supporting young people to innovate, start businesses, and create jobs. Furthermore, the European Union considers entrepreneurial skills to be essential in creating welfare and economic sustainability. Empowering individuals with entrepreneurship education, an entrepreneurial mindset and behaviors, are tools to develop human capital. This article explores how European policies can drive development of entrepreneurial competences by examining the case of the European Framework for Entrepreneurial Competences (EntreComp) and its integration into education and training development in different countries. With this research, we contribute education development from the practical point of view by analyzing how a cross section of actors, using EntreComp as a European framework for entrepreneurial competences, see that entrepreneurial learning has been realized and could be further supported in transnational education contexts. We will also expand the theoretical discussion of entrepreneurship education from the perspective of education sciences, as we have not previously obtained clarifying results or conclusions on how, for example, the educational change related to the development of entrepreneurship education should be implemented. The research data was collected through a case study, for which an online survey including both quantitative and qualitative approaches was conducted in 2020. Responses from 348 respondents from 47 countries were analyzed through an extended model for learning originally conceptualized by Shulman and Shulman (2004). The findings suggest that EntreComp has been widely recognized as a critical driver of competence in entrepreneurial education. However, a lack of shared vision and development of practice in the use of EntreComp can hinder the effective implementation of the framework. Thus, further support and guidance are needed in promoting the learning process of policymakers, educators, trainers, and other stakeholders, on both micro- and macro-level education design, to support successful adoption and adaptation of the policy-driven frameworks.

Keywords: entrepreneurship education; entrepreneurial competences; learning community; EntreComp; EU policy

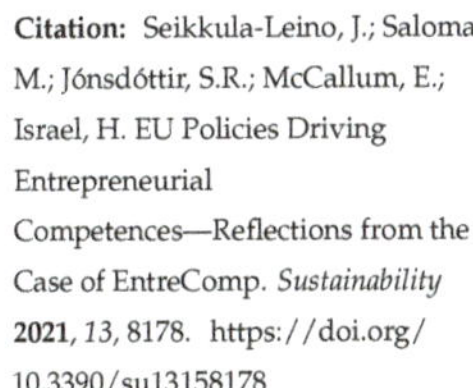

Citation: Seikkula-Leino, J.; Salomaa, M.; Jónsdóttir, S.R.; McCallum, E.; Israel, H. EU Policies Driving Entrepreneurial Competences—Reflections from the Case of EntreComp. *Sustainability* **2021**, *13*, 8178. https://doi.org/10.3390/su13158178

Academic Editor: Andrea Pérez

Received: 25 May 2021
Accepted: 18 June 2021
Published: 22 July 2021

1. Introduction

Sustainable Development Goals (SDGs), adopted by the United Nations [1] Member States in 2015, have increasingly focused on entrepreneurial learning interventions to support ambition in young people to start their own businesses and generate their own employment opportunities (SDG4.4 and SDG8.3). Furthermore, the European Union (EU) considers entrepreneurial skills to be an essential factor in creating social and economic sustainability. Entrepreneurship education (EE) seeks to empower individuals with sufficient formal education and training to support entrepreneurial behavior and thinking. According to European policy documents [2], we need skills and competences that support personal development, social inclusion, active citizenship, and employment. The key competences believed to enhance human capital, welfare, and competitiveness include literacy, numeracy, science, and foreign languages as well as transversal skills, e.g., digital competences, entrepreneurship, critical thinking, and problem solving [3]. Entrepreneurship education can cultivate such competences throughout life in different contexts and for different purposes, such as using these competences to create entrepreneurial actions impacting on sustainable development goals (e.g., [1,4], in particular SDG4.7). Furthermore, entrepreneurial culture enables, e.g., SMEs to turn environmental challenges into opportunities [5].

The European Entrepreneurship Competence Framework (EntreComp) is one of the EU's responses to support common understanding and widespread integration of entrepreneurship, within and across education systems, promoting entrepreneurial learning towards social, cultural, or financial value creation. EntreComp is suggested as a tool for supporting development of the entrepreneurial capacities of European citizens and organizations by establishing a consistent reference point to support development of shared concepts of entrepreneurship competences, goal setting, and evaluation. Previous studies on EntreComp indicate that it is possible to build education programs on EntreComp [6–10] by using the framework as a basis for curricula development and learning activities. As Bacigalupo et al. [3] stress, the use of the framework is claimed to foster entrepreneurship as a transversal and holistic competence, applicable to diverse purposes and contexts. The framework can also be used to describe and differentiate outcomes and attainment in the assessment of entrepreneurial competences [3]. For example, in Finland EntreComp has been adapted into higher education teaching practices with good results. Recent empirical evidence indicates that the EntreComp competences are attributes of success aligned with the expectations of the future of working life, and are therefore meaningful for transforming education [11].

Although entrepreneurship education appears to be gaining ground [12], with widespread policy recognition of the need for educational reforms, educators and other education professionals have struggled to identify the content and methods with which to implement EE. It has also been acknowledged that educators' learning processes have been missing from education policy implementation [13]). However, consideration of educators' reflections and learning processes are a crucial feature of successful educational reforms, such as the implementation of the EU strategies and frameworks that form a vision for the future of education. Additionally, motivation for change, sufficient substance knowledge, and understanding of how to implement these reforms in practice are essential [14,15]. This is also supported by Kelchtermans' [16] claim that educators' sense of identity and their individual roles are key drivers of active engagement with educational review and renewal processes. Strengthening this, the Eurydice report [12] states that educators' attitudes and behaviors in EE may be even more important than substance knowledge.

Previous studies imply that the successful development of entrepreneurship education by engaging educators is a complex issue. In response, many scholars have introduced new approaches to widen the discussion using learning communities [17–21] and extending resource bases (e.g., society) for EE activities [20,22]. These approaches offer new possibilities for learners to gain experiences and insight through a participatory approach, which involves, e.g., policy-makers, representatives from private and public sectors, and parents [13]. To cultivate a supportive ecosystem for EE, teachers, educators, policy-makers and communities may need to collaboratively reconsider their values and assumptions about school and education [23,24]. As Perrotta [25] argues, educational cultures, policies, rationality, and emotional dimensions have high significance when adopting new approaches to education. This complexity is also highlighted in curriculum reform research: promoting an enterprising culture through curriculum reform requires meaningful, well-designed partnerships, securing wide participation of internal and external stakeholders [26]. Such partnerships need to take into account local curricula, specific characteristics of different communities, and a focus on practice-oriented reforms [23,26].

However, there can be a significant gap between policy document recommendations and the design and implementation of entrepreneurship education in practice. In this paper, we anticipate that the development of entrepreneurial competences, driven by EU policies, requires the engagement and involvement of the whole learning community, acting as co-creators in the policy implementation through practice-oriented reforms. This approach is also highlighted by the chosen case study of the EntreComp framework. Therefore, this study investigates how the EntreComp framework has been implemented in transnational education contexts, looking particularly at how EntreComp guides the creation of a future vision for integrating entrepreneurship education to educational practices. Empirical data was collected from educators to examine how it is implemented and understood in practice: how can the framework motivate educational change and reform? By addressing these issues, we highlight how a policy-driven entrepreneurial competence framework can provoke, support, and drive educational change in Europe, and describe what kind of activities might further strengthen entrepreneurial competences and the sustainability of education in practice. Our case study also opens up opportunities for further research. For example, Cohen et al. [27] argue the generalizability of such single experiments (e.g., case and pilot studies) can be further extended through replication or multiple experiment strategies, allowing single case studies to contribute to the development of a growing pool of data. In this context, we may consider case studies methodologically, both quantitative and qualitative, that we adapt for our purpose in studying the case of EntreComp.

A case study of the EntreComp360 project [28] allowed the gathering of a wide transnational data set. The project's main goal was to develop opportunities to connect the wider community and provide guidance and resources to support those who are inspired by, or are already using, EntreComp across lifelong learning. To achieve this, it was essential to obtain an overview of European entrepreneurship education and how EntreComp is, and might be, integrated into it. Therefore, the EntreComp online survey was issued in 2020 targeting policymakers, lifelong learning organizations, educators, and other stakeholders. A comparative approach reveals: what are the learning communities' key reflections and learning processes related to development of entrepreneurial competences? What is their vision for the future implementation of EntreComp? What is their motivation for change? How do they perceive and utilize EntreComp? Both quantitative and qualitative data were collected as part of the study. In total, there were 348 respondents from 47 countries. The country comparison is shown between the U.K., Finland, Spain, Germany, Italy, and Iceland to identify varying levels of integration of the framework in different national contexts.

The paper is structured as follows: firstly, we highlight the key points of the theoretical debate around EU policies driving entrepreneurship education and learning communities to create a framework for analysis by adapting the Shulman and Shulman framework [14]. Secondly, we present the case of EntreComp, methodological choices, and data collection methods. Thirdly, we summarize the key results from the survey. Fourthly, we summarize the results gained in terms of what gaps can be identified, and what leverage is needed to support coherent educational reform to promote entrepreneurial competencies across Europe. Finally, we conclude that a lack of shared vision, motivation, practices, and understanding of entrepreneurship education can hinder the effective implementation of the framework. We propose further research on the integration of individual and community learning processes into the designing of the policy tools.

2. Literature Review

2.1. European Policies Driving Entrepreneurship Education

Since 2003, there has been a concerted effort by the European Commission to drive entrepreneurship education forward. The inclusion of entrepreneurship as one of the eight European Key Competences in 2006 led to the agreement by all EU Member States to embed these key competences into their education and training systems; however, the 2016 Eurydice study [12] demonstrated that the approach is fragmented and not mainstreamed for the entrepreneurship key competence. This visibility has increased since the launch of the Europe 2020 strategy through education and training focused policy documents culminating in the 2016 and 2020 Skills Agendas [2]. These reaffirmed the importance of the entrepreneurship key competence with three key work strands: improving the quality and relevance of skills formation; making skills and qualifications more visible and comparable; and advancing skills intelligence, documentation, and informed career choice. The 2020 European Skills Agenda is now setting the framework for policy and reform at EU and national levels, stressing the importance of developing an entrepreneurial mindset and giving all learners at least one entrepreneurial hands-on experience during compulsory education, as well as strengthening VET and STEM education through entrepreneurial work-based learning. The importance of transversal skills was reaffirmed by the European Council in the 2018 conclusions and emphasized within the European Pillar of Social Rights [29].

In constantly developing societies, in which technological solutions innovate at pace, and social and environmental challenges relentlessly evolve and emerge, versatile and action-orientated competences are indeed essential for taking part in and contributing to society. Education is the primary lever to develop human consciousness and mentally self-conscious individuals to secure long-term sustainability and ensure a sustainable future [30]. General education is expected to provide a broad and balanced range of knowledge and skills in modern societies. Creative thinking, social skills, and the capacity to solve relevant and meaningful problems, are considered to be fundamental competences in promoting sustainable societies [31]. Entrepreneurial education, also referred to as entrepreneurship education, offers a way to nurture such competences. Entrepreneurship education is integrative in nature, with applications across different knowledge areas and purposes [26,32]. The core pedagogy of entrepreneurship education has been analyzed as emancipatory pedagogy, aiming to give the student agency and freedom to develop creative, independent, and action competences [23,32,33].

The purpose of EE is to educate students to take more responsibility for themselves and their learning; to achieve their goals, to become creative, active and critical citizens; to discover existing opportunities and create new ones; and to cope and thrive in a complicated society [32,34]. An essential aim is that students take an active role in the labor market, considering entrepreneurship as a natural career choice [15]. Entrepreneurial education involves developing behaviors, skills, and attributes, applied both individually and collectively, to help individuals and organizations of all kinds, to create, cope, enjoy, change, and innovate (e.g., [20]). It is worth emphasizing that academics have warned that EE should not be too simplistic or too focused on market or business creation in order to offer viable solutions to complex social, cultural, and economic issues [35,36].

Since this study explores how European policies can drive development of entrepreneurial competences by examining the case of the European Framework for Entrepreneurial Competences (EntreComp) and its integration into education and training development in different countries, we will next focus on the key elements of the EntreComp framework.

2.2. *EntreComp Framework Strengthening Entrepreneurial Competences*

The EntreComp framework is comprehensive and detailed, creating a construct of entrepreneurial competences that can be developed through entrepreneurial education. Its primary purposes are to create a common understanding of entrepreneurship as a key competence and to establish a common language for entrepreneurial competences to bridge education and work [3]. EntreComp identifies fifteen entrepreneurial competences that can be developed through learning within a progression model that maps across developing autonomy of use and complexity of application, and emphasizes the broad-based potential of these competences to create value for others across the social, cultural, and economic spheres. The framework consists of three interrelated and interconnected competence areas: 'Ideas and opportunities', 'Resources', and 'Into action'. Each of these areas comprises five competences, which, together, constitute the building blocks, or threads, of entrepreneurship as competences. The framework describes development of the 15 competences along a progression model, expressed in learning outcomes, over eight levels (see Figure 1.)

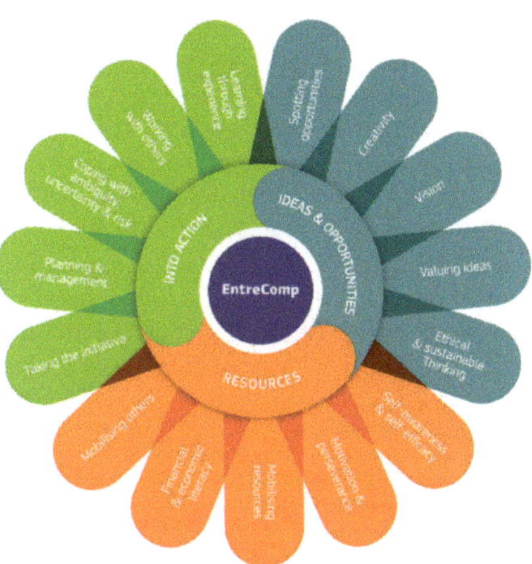

Figure 1. EntreComp Framework [29].

A developing body of studies and research evidences the use of EntreComp across Europe and globally. Stakeholders have translated the original report and framework into thirteen languages to date, with partial translations available in a further three languages. The European Commission has developed a suite of publications driving implementation of the framework into youth-work, education, employment, and enterprise sectors. The most policy-relevant is the 2018 Revision of the European Key Competence Framework, where all European member States agreed to EntreComp as the basis of the entrepreneurship key competence within education and training systems based on the original report published in 2016.

A number of EC documents provide guidance and case examples to promote use of the framework [29,37]. The utilization of the EntreComp framework has been broken down into five specific goals: mobilize interest and inspire action; create value by adapting the framework to specific contexts; appraise or assess levels of entrepreneurship competence; implement entrepreneurial ideas and projects; and recognize entrepreneurship skills [29]. The EU has further directed funding towards raising awareness and implementation of EntreComp (e.g., targeted calls via COSME and Horizon2020 with a specific focus on using EntreComp). On a larger scale, the Erasmus+ 2014–2020 program includes development of key competences as a program priority including entrepreneurship and 107 current Erasmus+ projects explicitly mention EntreComp in their application project summary. These initiatives encourage the use of EntreComp to underpin projects related to entrepreneurship key competences. However,, EntreComp as a framework remains less developed and exploited than other European key competence frameworks, such as DigComp, where the first framework version was published in 2013 and development has been supported by well-established European stakeholder networks such as AllDigital. The EntreComp community platform, initiated by the Erasmus+ funded EntreComp360 partnership, will be launched as a community of practice in 2021 to support those using or inspired by the EntreComp framework. The implementation of EE is related to the reflection of both individuals and communities. Next, we will present learning communities in the context of driving changes in education.

2.3. Learning Communities of EE Driving Educational Change

Communities of practice, including communities of learning, have been acknowledged as a powerful method of continuing professional development in education, implementing changes, and sustaining changes [38,39]. Learning is not seen as something that only takes place within the individual, but rather, something that happens through participation in communities of practice [40]. Thus, knowledge is understood as situated in the social context, and its influenced by what is valued in the community. The development of knowledge in a community of learning involves participants collaborating to learn from various resources. [23,39]. We may consider that knowledge management is the process of creating, sharing, using, and managing the knowledge and other information. Therefore, it is also about creating such changes in an organization supporting the most effective knowledge development. [41]. To implement and sustain change, the whole social ecology involved in the development of EE, e.g., stakeholders and influencers of the community of the practice, must participate in the landscapes of learning that educators may belong to [23,39].

Community-based learning refers to a wide variety of instructional methods and programs that can be used to connect classroom teaching to the surrounding community. It is also a pedagogical strategy that intentionally integrates services to the community through classroom learning [42]. The literature describes numerous and diverse benefits of student engagement with community-based learning across multiple domains, including academic development, socio-personal development and civic engagement (e.g., [42–47]. It also supports active citizenship through sustainable development [48], enables and encourages people to take direct actions to tackle the challenges of a rapidly changing, increasingly global world by offering knowledge and skills to improve citizens' lives in sustainable ways (e.g., eco-friendly farming and addressing social and economic inequalities). Greater shared ownership empowers people to shape their community's future and address global issues. This is also acknowledged in international policies, e.g., the UNESCO Institute for Lifelong Learning [49] emphasizes the reciprocal relationship between community-based learning practices and national and local public policies, as both should be guided by a shared vision.

At the same time as we talk about developing learning communities, our discussion has expanded to what we understand by 'real' learning. The analysis of learning has shifted towards the processes of individual educators and learners, as a means of creating an understanding of a learner's reflection, within the broader context of community, institution, policy, and profession. This has laid the foundation for more recent discussion on the broader goals of student learning in a variety of settings, including that of learning professions and communities ([14], see Figure 2.) This learning is most effective when accompanied by metacognitive awareness and analysis of one's own learning processes, supported by membership of a learning community. Indeed, this approach applies to student learning processes, as well as to the learning processes of educators, practitioners, parents, policy makers, and stakeholders [14,21]. Despite recent gains in the profile and reach of entrepreneurship education, the community learning process has been less represented [21]. Therefore, in this study we focus on how the European Union framework for entrepreneurial learning drives key competence development in Europe within learning communities by extending the original framework by Shulman and Shulman ([14], see Figure 2) consisting of four core modules —vision, motivation, practice, and understanding—with integration of the EntreComp framework's goals as described in the following section.

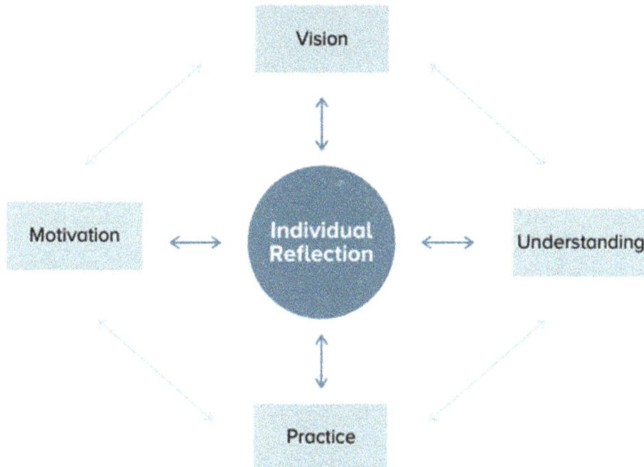

Figure 2. Teacher's learning. Adapted with permission from ref. [14]. Copyright© 2021 Informa UK Limited.

According to the Shulman and Shulman [14] model, an accomplished teacher or other educational practitioner, should be a member of a professional community, and be ready, willing and able to learn from their teaching or other educational experiences. An educator's vision generates the readiness and willingness that induces motivation. Educators should be ready to pursue a vision of a classroom, a school, or other learning unit as a 'learning community' in which teachers and other educators deepen their understanding and are motivated to further develop the forms of pedagogical and organizational practices needed in transforming their visions, motives, and understandings into a functioning, pragmatic reality. When they form learning communities, and work as members of them, the educators are capable of learning from their own and others' experiences through active reflection. Shulman and Shulman [14] also suggest that accomplished educators can smoothly integrate their vision, motivation, understanding, and practice into their teaching or other educational activities through active reflection, which is key to learning and development. Furthermore, Shulman and Shulman highlight the concept of the learning community from a broad perspective, e.g., expanding the learning communities from the classroom to the level of society. Thus, policy-makers and stakeholders can have an essential role in advancing the learning processes of themselves and their communities, expanding the learning communities from micro to macro level in society.

By examining the case of EntreComp employing the Shulman and Shulman [14] approach, we investigate how the framework has been adapted in different national contexts. A comparative approach reveals: what are the learning communities' key reflections and learning processes? What is their vision for the future by implementing EntreComp? What is their motivation for change? How do they perceive and utilize EntreComp?

To conclude, despite all possibilities and efforts in developing education and competencies, the state-of-the-art literature on entrepreneurial behavior promoting sustainable societies still lacks understanding of how entrepreneurial frameworks and strategies strengthen the learning process of communities. With our study, we aim to increase both theoretical and practical understanding of how global strategies, which address entrepreneurial behavior and sustainability, could be operationalized meaningfully in a real-life setting. Therefore, this study contributes to the education development from a practical point of view by investigating how different educational actors, using EntreComp as a European framework for entrepreneurial competences, perceive entrepreneurial learning, how it has been strengthened by the policy framework and how it could be further supported in transnational education contexts. We also expand the theoretical discussion related to entrepreneurship education from the perspective of education sciences, as there are no unambiguous results or conclusions from previous studies explaining how the educational change related to the development of entrepreneurship education can be efficiently implemented. In this context, we stress the importance of the learning process of communities involving different educational actors. These starting points naturally form our research questions and methodology described in the next section.

3. Methodology

3.1. Data Collection

An online survey was created in 2020 for policymakers, educators, and other stakeholders. Both quantitative and qualitative data were collected from respondents. The quantitative survey was mainly based on Likert statements in which we used four-point scales. In some cases, respondents had the opportunity to answer, e.g., yes/no. In addition, a few statements were rather practice orientated (e.g., "Do you have an existing example of practice that could be shared with others?"). Often in this respect, the respondents further supplemented their answer with a descriptive, qualitative response. Applying purposive sampling [27], the survey was shared across EntreComp360 project's partner communication channels, through direct mail to existing databases, social media, and broad dissemination across online media to engage actors and networks from youth work, formal education, non-formal education, employment, and enterprise practice and policy organizations. We chose purposive sampling because we wanted to target our survey selectively to the kind of respondents we considered having some previous experience, understanding, or knowledge of entrepreneurship education. Furthermore, we describe that our methodology is based on a case study approach to examine the practical implications of the EntreComp framework. A survey was administered to gather data related to this particular case, being the implementation of the EntreComp, to collect first-hand data from either a small sample or a larger entire population of individuals to describe different aspects and characteristics of the respondents related to the phenomenon. Case studies based on surveys do not involve any experimental manipulation of the conditions, and thus, there was no activity of this type in our study. ([27].)

There were a total of 348 respondents from 47 countries. Since most of the countries had only a few (less than 10) respondents, the comparison between different countries was made using data from only 6 countries (U.K. (n = 52), Finland (n = 37), Spain (n = 36), Germany (n = 35), Italy (n = 30), and Iceland (n = 27). There were 12 NAs). However, in bar plots, we mostly used all the answers when no comparison between different countries was done. The different fields of work the respondents are involved in are described in 4.1. Background Variables. One respondent may be involved in several fields, that is, in many questions one respondent may provide several answers to the same question.

3.2. The Assessment Frame and Research Questions

The framework used in this survey, based on the work of Shuman and Shulman [14], Seikkula-Leino [26], and Seikkula-Leino et al. [21,50,51], has been widely used. It has proven effective in qualitatively assessing educators' learning and reflection within education reform. The frame consists of four core modules which are: vision, motivation, practice, and understanding, as described in the literature review. In our study, 'vision' refers to, e.g., these questions: "What kind of outcomes and impact may I proceed with EntreComp?". 'Practice' refers to questions such as: "How could I implement EntreComp in education and training?" while 'motivation' refers to: "Would you like to integrate EntreComp into your daily work?". One example of questions referring to 'understanding' is: "Does your work link to some or all of the competences highlighted in EntreComp?"

The survey answers were analyzed by each research question as follows:

1. What is the understanding of EntreComp?
2. What is the motivation for using EntreComp?
3. How is EntreComp implemented in practice?

4. What is the vision for the future developed by using EntreComp?

Figure 3, derived from previous work by Shuman and Shulman [14] and Seikkula-Leino [26], and complemented with EntreComp goals, illustrates how individual reflections within learning communities play a crucial part in educational reform related to entrepreneurship education.

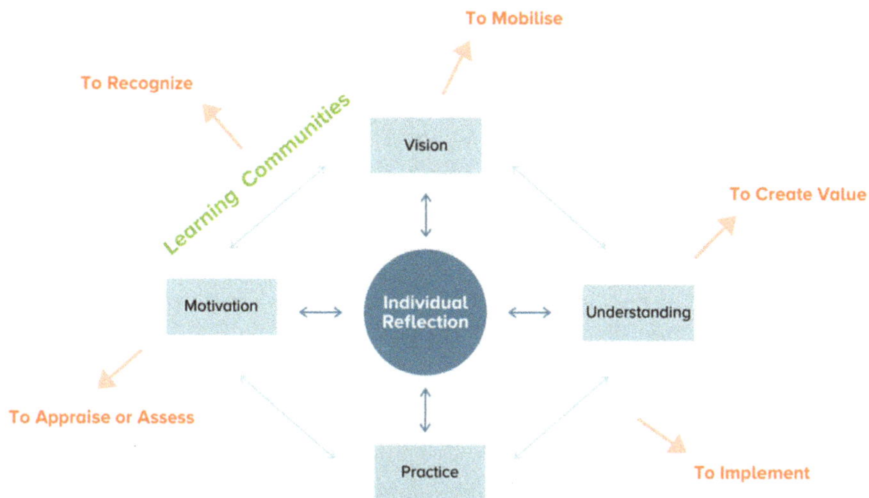

Figure 3. The framework for data collection and analysis.

To transform EE, the implementation of EntreComp in European policymaking relies upon the five goals of EntreComp: mobilize, create value, implement, appraise or assess, and recognize. However, to initiate these goals, individuals and communities also need motivation, vision, understanding of the need for reform, and how to implement these goals in practice, which are the starting points for our survey design, implementation, and data analysis.

3.3. Data Analysis Methods

The quantitative data showing a comparison of countries were analyzed by analysis of variance (Kruskall–Wallis rank sum test). The qualitative data gathered through open-ended survey questions were analyzed using content analysis following the logic of the theoretical propositions guiding the overall design of the survey based on Figure 3, which also created the basis for organizing the data to draw conclusions [52]. The data were grouped into parallel types by searching for similarities in the data. It is based on theme categorization and grouping, and is a valuable method for illustrating research problems with practical examples (e.g., [27]). By using content analysis, we aimed to find answers to the set research questions by concentrating on respondents' views on entrepreneurship education and the EntreComp framework. The content analysis was realized as follows:

1. The data collected from different countries and with different languages were translated into English.
2. Then, the survey data were read several times to try to construct an overall picture of the responses, including the elements of how entrepreneurship education was described.
3. The data were read more reflectively and analytically, organizing the data through the questions answered by the respondents.

4. The answers were mirrored against our literature review and concept definitions, involving, for example, different aspects of reflection in learning communities (e.g., vision, motivation, practice, and understanding). Similar types of answers were grouped.
5. The above data analysis was integrated, allowing analysis of the respondents' reflections in the context of entrepreneurship education and EntreComp.

4. Results

4.1. Background Variables

A country comparison was realized between the U.K., Finland, Spain, Germany, Italy, and Iceland. First, we present the respondents' job descriptions. One respondent may be involved in several fields, so the comparison of different job fields is not possible.

In Figure 4 it can be seen that the education sector is strongly represented among the respondents. Many of them (n = 155) work in the field of higher education, adult education or adult inclusion (n = 114), vocational education and training (n = 112), non-formal youth education (n = 105), and start-up or business growth support (n =104). Only 15 respondents worked in the field of human resources. There were 35 respondents who work in unspecified fields. Next, we explain respondents' understandings of EntreComp while highlighting country-specific differences—what they know about EntreComp in general, and how it relates to their work.

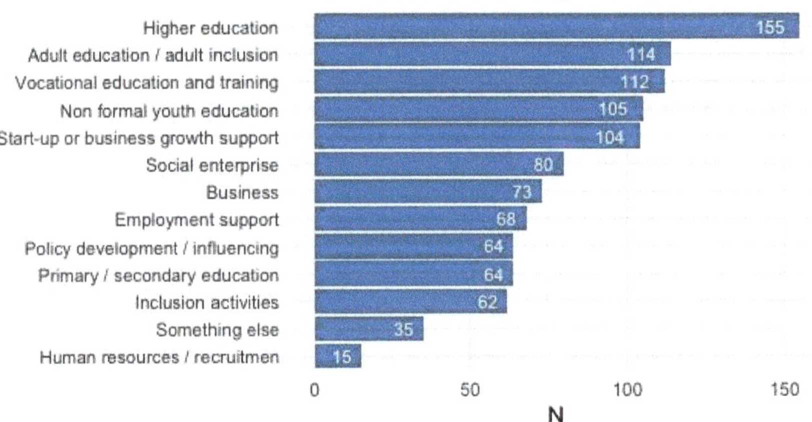

Figure 4. The respondents' professional backgrounds.

4.2. What Is the Understanding of EntreComp?

Almost half of all the respondents, 49.1 % (n = 171), were familiar with the EntreComp framework, 21.3% (n = 74) had heard something about it, and 29.3% (n = 102) of the respondents had not heard about EntreComp at all. One respondent did not answer. In the comparison of the six countries related to the question, 'Are you familiar with EntreComp?', the differences between countries were statistically significant (p-value = 2.187×10^{-11}). The significance of the comparisons was obtained using the Kruskal–Wallis rank sum test. Table 1 shows that in the U.K, (67.3%), Spain (63.9%), and Italy (76.7%), most of the respondents were familiar with EntreComp. On the other hand, in Germany (65.7%) and Iceland (70.4%), most of the respondents had not heard about EntreComp.

Table 1. Are you familiar with EntreComp?

	Finland	Germany	Iceland	Italy	Spain	U.K.
Yes	16 (43.2%)	8 (22.9%)	2 (7.4%)	23 (76.7%)	23 (63.9%)	35 (67.3%)
Heard something about it	11 (29.7%)	4 (11.4%)	6 (22.2%)	3 (10.0%)	6 (16.7%)	11 (21.2%)
No	10 (27.0%)	23 (65.7%)	19 (70.4%)	4 (13.3%)	7 (19.4%)	6 (11.5%)
					Sig. = 2.187×10^{-11}	

In the quantitative question 'EntreComp is about promoting entrepreneurial learning. How much does the concept of entrepreneurial learning link to your work?': 12.6% (n = 44) of all the respondents argued that 'This is the main theme of my work'; 32.8% (n = 114) 'Plays a big part in my work'; 35.1% (n = 122) 'Relevant and included in my work to some extent'; 12.4% (n = 43) 'Relevant but not (yet) included in my work'; and 2.3% (n = 8) 'Not relevant to my work'. 17 did not answer.

When comparing the six countries, the comparison is statistically significant with p-value = 4.32×10^{-6} (Table 2). In Finland, Germany, and Spain, most of the respondents answered that the quantitative question, 'How much does the concept of entrepreneurial learning link to your work?', is relevant and included in their work to some extent. In the U.K. and Italy most of the respondents answered that it plays a big part in their work. In Iceland most of the respondents answered that it is relevant and included in their work to some extent. It can be concluded that entrepreneurial learning does link to many respondents' work. In the U.K. almost one third thought that it is the main theme of their work.

Table 2. How much does the concept of entrepreneurial learning link to your work?

	Finland	Germany	Iceland	Italy	Spain	U.K.
This is the main theme of my work	2 5.7%	1 3.0%	2 7.7%	3 10.0%	6 17.1%	14 29.2%
Plays a big part in my work	6 17.1%	11 33.3%	6 23.1%	14 46.7%	11 31.4%	22 45.8%
Relevant and included in my work to some extent	19 54.3%	13 39.4%	9 34.6%	9 30.0%	13 37.1%	10 20.8%
Relevant but not (yet) included in my work	6 17.1%	6 18.2%	7 26.9%	4 13.3%	5 14.3%	2 4.2%
Not relevant to my work	2 5.7%	2 6.1%	2 7.7%	0 0.0%	0 0.0%	0 0.0%
					Sig. = 4.32×10^{-6}	

A majority of respondents, 61.5% (n = 214), answered 'Yes' to the quantitative question 'Does your work link to some or all of the competences highlighted in EntreComp', 29.9% (n = 104); 'Partially—in some ways', 2.3% (n = 8); 'Not yet, but interested', 0.1% (n = 3); 'No' and 5.5% did not answer (n = 19).

The results of the comparison between countries were parallel to the previous ones. Most of the respondents in Germany and Iceland answered that their work links partially to some or all of the competences highlighted in EntreComp (Table 3). In Finland, Italy, Spain, and the U.K. the most common answer was that their work links to some or all of the competences highlighted in EntreComp. Altogether, it can be seen that almost all of the respondents' work links to the competences highlighted in EntreComp in some ways.

Table 3. Does your work link to some or all of the competences highlighted in EntreComp?

	Finland	Germany	Iceland	Italy	Spain	U.K.
Yes	23 (67.6%)	15 (46.9%)	12 (46.2%)	18 (60.0%)	22 (62.9%)	42 (87.5%)
Partially—in some ways	9 (26.5%)	17 (53.1%)	13 (50.0%)	10 (33.3%)	11 (31.4%)	6 (12.5%)
Not yet, but interested	0 (0.0%)	0 (0.0%)	1 (3.8%)	2 (6.7%)	2 (5.7%)	0 (0.0%)
No	2 (5.9%)	0 (0.0%)	0 (0.0%)	0 (0.0%)	0 (0.0%)	0 (0.0%)
					Sig. = 0.002182	

In the last question (quantitative), 'If entrepreneurial learning is not in the focus of your activities, are there other learning activities related to helping people such as those which allow young people or adults to transform ideas and opportunities, by mobilizing resources, into action?' 48.3% (n = 168) of all the respondents answered 'Yes'; 26.4% (n = 92) 'Partially—in some ways'; 8.0% (n = 28) 'Not yet, but interested'; 5.7% (n = 20) 'No'; and 11.5% (n = 40) did not answer. The differences in comparison of the six countries was not statistically significant ($p = 0.0842$) (Table 4.).

Table 4. If entrepreneurial learning is not in the focus of your activities, are there other learning activities related to helping people such as those which allow young people or adults to transform ideas and opportunities, by mobilizing resources, into action?

	Finland	Germany	Iceland	Italy	Spain	U.K.
Yes	23 (67.6%)	18 (58.1%)	9 (39.1%)	16 (59.3%)	11 (35.5%)	23 (54.8%)
Partially—in some ways	9 (26.5%)	10 (32.3%)	11 (47.8%)	9 (33.3%)	13 (41.9%)	11 (26.2%)
Not yet, but interested	1 (2.9%)	1 (3.2%)	3 (13.0%)	1 (3.7%)	4 (12.9%)	4 (9.5%)
No	1 (2.9%)	2 (6.4%)	0 (0.0%)	1 (3.7%)	3 (9.7%)	4 (9.5%)
					Sig. = 0.0842	

Next, we will present the results of the research based on the theorical frame (vision, motivation, practice, and understanding).

4.3. What Is the Motivation for Using EntreComp?

In the quantitative question, 'Do you already use EntreComp as a model or as inspiration for your work on developing entrepreneurial competences?',' one could answer 'Yes', 'No, but I would like to include it in my work', or 'No'. Some respondents omitted this question. The frequencies are described as a bar plot in Figure 5. From the figure we see, for example, that slightly less than half (ca. 42%) of the respondents are developing entrepreneurial competences by using the EntreComp framework.

The comparison of the six countries shows (Table 5) that most respondents in Iceland (57.7%) did not use EntreComp as a model/inspiration for their work on developing entrepreneurial competences, but they would like to. In Finland, Italy, Spain, and the U.K. the vast majority of respondents already use EntreComp as a model/inspiration for their work on developing entrepreneurial competences. In Germany, most respondents (42.4%) answered no. The comparison is statistically significant ($p = 2.016 \times 10^{-7}$).

Table 5. Do you already use EntreComp as a model or as inspiration for your work on developing entrepreneurial competencies?

	Finland	Germany	Iceland	Italy	Spain	U.K.
Yes	15 (42.9%)	6 (18.2%)	1 (3.8%)	18 (60.0%)	16 (45.7%)	29 (60.4%)
No—but I would like to include it in my work	12 (34.3%)	13 (39.4%)	15 (57.7%)	11 (36.7%)	13 (37.1%)	14 (29.2%)
No	8 (22.9%)	14 (42.4%)	10 (38.5%)	1 (3.3%)	6 (17.1%)	5 (10.4%)
					Sig. = 2.016×10^{-7}	

In the quantitative question, 'What would encourage you to become part of our EntreComp community?', one respondent may provide several answers. The choices were 'Sharing case studies or practices linked to my work' (n = 180); 'Contribute to research on ways to develop this work further at national/EU level—inform the next steps development of EntreComp' (n = 165); 'Putting me and/or my organization on the global map of EntreComp users' (n = 163); 'Access to professional development via online training or MOOCs' (n = 136); 'Becoming recognized as an EntreComp ambassador' (n = 115); 'Profiling and sharing my work through articles, blog posts or webinars' (n = 113); and 'Other (please specify)' (n = 26). The frequencies of each answer are in parentheses after the answer alternative.

Answers to the question 'What are the best ways to connect you to an online EntreComp community?' are plotted as bars in Figure 6. The question includes quantitative responses and a qualitative response in the option 'Other'. Thus, we may conclude that LinkedIn and Facebook are the best platforms to build online connections with future community partners.

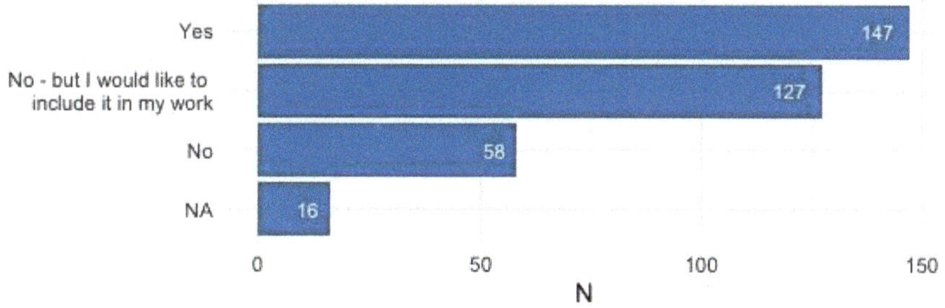

Figure 5. The utilization of EntreComp.

4.4. *How Is EntreComp Implemented in Practice?*

Because our study specifically sought to identify the factors that may make it challenging to implement EntreComp in order to create, e.g., future practices of EntreComp, we focused our questions primarily on these challenges rather than opportunities or other aspects. Therefore, our results are highlighted with an emphasis on the critical entry angle.

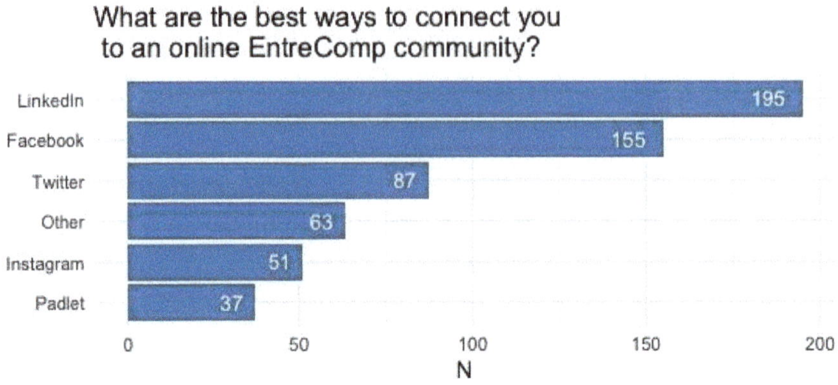

Figure 6. Social media in the community development of EntreComp.

Implementation of the EntreComp framework was reported to include some challenges. Respondents find it time-consuming, difficult to understand, and the lack of language selection is problematic. Respondents already using EntreComp use it in all five areas (to mobilize, to create value, to implement, to assess, and to recognize), but mostly 'TO MOBILIZE—raise awareness and understanding of these competences' (n = 63) and 'TO MOBILIZE—inspire or engage new people or organizations into entrepreneurial learning' (n = 60). Again, it was possible to have several answers to the same (quantitative) question.

In an open-ended response question, 40 respondents had found, or were aware of, barriers that might prevent them or others from using EntreComp. For six respondents, language hindered use of the EntreComp framework. Two of the six respondents found the language options too limited, whereas, one respondent thought that the nuances in meaning of different languages makes translation challenging. Two respondents found time consumption to be a barrier and two respondents found the framework difficult to understand. Furthermore, four respondents faced unspecified barriers and three respondents found the framework to be difficult to use: e.g., one respondent described the appendix consisting of the list of skills to be difficult to use.

In another open-ended response question, 40 respondents expressed concerns about using EntreComp, with four respondents finding it too vast, and two respondents finding that they did not fully understand it. It was suggested to try to simplify it and to have a 'use directly tool', as the framework was seen to be rather theoretical. In addition, measuring the impact of entrepreneurial learning was considered to be difficult. There were also concerns about translating the framework into non-Western cultural contexts.

4.5. How the Vision for the Future Is Developed by Using EntreComp?

Of the 70 responses to the open-ended question, 'What is your vision for using EntreComp?', 10 respondents reported their visions relating to young people recognizing their own entrepreneurial skills, with 22 respondents highlighting application of the framework to different areas of life. Seven respondents' visions emphasized the possibility of sharing good practices, consulting, and networking.

In the quantitative question, 'How can we help you?', presenting practical examples of how to support respondents to include EntreComp in their area of work, respondents may select more than one response. The choices were: 'Resources that explain the value of entrepreneurial competences', (n = 150); 'Help to mobilize others on the relevance of EntreComp in my area of work', (n = 75); 'Practical examples of how to include EntreComp in my area of work', (n = 203); 'Online training on EntreComp in my area of work', (n = 141); 'Demonstrating the value of entrepreneurial learning', (n = 136); 'Being part of a community of people interested in developing and using EntreComp or entrepreneurial competences', (n = 169); 'High profile recognition for using EntreComp at individual or organizational level', (n = 97); 'Partner searching/matching to develop new tools or projects based on EntreComp/entrepreneurial competences', (n = 155); 'Understanding how to evaluate learning or working performance related to EntreComp competences', (n = 152); 'Having guidance on how to assess progress in the EntreComp competences', (n = 142); and 'Other', (n = 20).

There were a total of 112 responses to the open-ended question: 'This online community is about making EntreComp more easy-to-understand and easy to use. Do you have any suggestions?'. 15 respondents suggested using practical cases and examples to realistically implement the concept behind the model. Five respondents recommended that the model should be available in languages other than English, also reflecting non-European cultures. Four people said that instructional videos would be valuable. Twelve respondents suggested an online forum or Skype/Teams meetings to promote better understanding and use of EntreComp. Two respondents suggested face-to-face meetings and two suggested that the terminology should be tailored to target specific groups.

In an open-ended question concerning good practices, 49 respondents had found or were aware of successes or good practices that would encourage others to use EntreComp: 15 responded 'No'; 284 did not respond at all; and 14 respondents answered 'Yes' without giving examples. Nine respondents reported using EntreComp to teach different skills in different practices. For example, they had had project-based approaches, workshops, and 'normal' classes online. The target groups mentioned ranged from young children to young apprentices, university students, and the elderly.

5. Discussion

The aims of the recent EU policy documents (e.g., [2]) to foster entrepreneurial skills can be clearly seen within many EU-funded project initiatives, such as the development of the EntreComp framework. Even though fostering entrepreneurial skills is recognized to be globally important (e.g., UN SDGs), the effective implementation, and translation into practice, of the education policy documents remains complex. Therefore, this paper focused on analyzing how a cross section of actors, using EntreComp as a European framework for entrepreneurial competences, see that entrepreneurial learning has been realized and could be further supported in transnational education contexts. The results from the online survey generated new knowledge on how the EntreComp framework can guide the collaborative creation of a future vision for education; how well it is recognized, implemented, and understood in practice; and how it motivates and supports realization of educational change, which emphasizes the value of entrepreneurial competencies. The aim was to reveal how EU policies can drive educational change in Europe illustrated by the case of EntreComp.

Our results reveal that overall awareness and understanding of entrepreneurial education are 'improving' compared with earlier works, in which similar kinds of research settings and data collection were used [15,26,50]. According to those previous studies, there is a need for the development of all aspects in reflection: vision, motivation, practice, and understanding. However, these were not the results in our case study. It can be stated that the EntreComp framework driving entrepreneurial learning is relevant to most of the respondents' work. Moreover, the majority of respondents are motivated to integrate EntreComp in their work. Therefore, we may conclude that EntreComp can work in practice to strengthen the entrepreneurial capacity of European citizens and organizations (e.g., [3]). This is also supported by Dinning's [6] and Gerbutt et al.'s [7]) previous studies on EntreComp and entrepreneurial competencies. However, there might be some sampling error, thus raising the concern of bias in the chosen data collection method. As an example, in some situations, responses came by EntreComp projects. In some cases, answers were collected randomly from the education sector, without any connection to EntreComp. Furthermore, the number of responses is still low to conclude, e.g., the European situation of implementing EntreComp in practice.

However, our study also stressed that the implementation of entrepreneurial education initiatives driven by international policy goals is challenging. For example, the respondents estimated that adapting the EntreComp framework is time-consuming and it is difficult to understand the framework conceptually. On a more practical level, the lack of widely available translations of EntreComp is problematic. Furthermore, the need for training and guidance is highlighted to promote effective implementation. The respondents that already use EntreComp in their work, use it in all five areas (to mobilize, to create value, to implement, to assess, and to recognize) but mostly 'TO MOBILIZE—raise awareness and understanding of these competences', and 'TO MOBILIZE—inspire or engage new people or organizations into entrepreneurial learning'. These results imply that currently the implementation of EntreComp is mostly related to awareness-raising on entrepreneurial education and inspiring students, curricula developers and educators, and may indicate that EntreComp is at an early stage in the implementation journey.

These same aspects were highlighted in the answers in which respondents explain their future visions related to using the EntreComp framework. However, the descriptions of the respondents' visions are relatively modest, e.g., none of them described any future changes (societal, economic, or environmental) that they would like to see resulting from entrepreneurial education. This challenge related to the lack of understanding of EE in general and is also recognized in many previous studies (e.g., [15,26]). This finding is supported by the results of the survey, which indicate that almost one third of the respondents would require more support to be able to better demonstrate the value added by EE. Therefore, the potential impact and benefits of entrepreneurship education still require more explicit articulation and promotion [21,26,50,51], as well as more solid theoretical underpinning [13]. This need was also reflected by respondents who highlighted the importance of evaluating the outcomes and impacts of entrepreneurship education.

According to the results, they key factors motivating respondents to use the EntreComp framework include the following: being psychologically and socially part of the EntreComp learning community in which recognition and support of individual and community entrepreneurial initiatives and activities are highlighted as valuable to promotion of EntreComp. It is thus important to note that the learning processes of the developers of entrepreneurship education play an important role in making progress in educational reforms in practice (e.g., [12,16]). This downplays the role and effectiveness of the traditional education policymakers, such as the state, as primary drivers of educational progress (e.g., [31]). Indeed, effective learning requires participation, e.g., feedback and encouragement (e.g., [53]), which is also relevant for the future development of such competence frameworks. Thus, further studies are still needed on how both individual and community learning processes can be integrated into the designing of policy tools, e.g., frames and roadmaps, to support a more effective implementation of the EE-related reforms. However, we may say that our study contributes the theoretical discussion in the field of entrepreneurship education by integrating 'the reflection process of learners', borrowing mainly the significant outcomes and frameworks from curriculum and curriculum reform studies (e.g., [14,15,26]) to widen and deepen the understanding of factors influencing on the successful education change.

The respondents also suggested that ideas and best practices, related to implementation of the EntreComp framework, should be shared with other developers, e.g., through case study presentations. This draws attention to effective communication mechanisms within and between different learning communities. As an example, by highlighting individual success stories, significant psychological support and encouragement can be provided, which, according to the survey results, is one of the key areas for future development of the EntreComp framework. This learning proceeds most effectively if it is accompanied by metacognitive awareness and analysis of one's learning processes and is supported by membership in a learning community (e.g., [17–20,54]).

Adapting the community learning approach might be beneficial for further enhancing students' learning processes as well as to the learning processes of, e.g., educators, other practitioners, parents, policymakers, and stakeholders [14,23]. Respondents highlight the importance of 'belonging and recognition' throughout the survey. This implies that a reciprocal relationship between community-based learning practices and national and local public policies is needed for seamless enforcement of educational aims [49]. From this perspective, the development of the learning communities in a transnational context should also be given more attention. The learning community can often only be perceived too narrowly, only relating to the physically close, e.g., regional, community. However, the learning communities have become global and they are not solely tied to certain locations and can be digital, particularly in the COVID era when interaction is increasingly online. As a result, these learning communities can be global in reach and involve actors working at all levels, e.g., educators, stakeholders, policymakers, students, and private sector partners. Therefore, further studies are needed on how to promote the psychosocial learning processes of international communities, e.g., through adapting new technologies?

This becomes apparent, especially when looking at country differences as revealed by the results. This raises questions of whether global entrepreneurial learning communities could be developed more effectively, despite the country divergence, across lifelong learning policy and practice? Based on the survey results, we were able to identify three progress scenarios related to the implementation of the EntreComp framework: (a) in the U.K. and Italy, most of the respondents answered that EntreComp already plays a big part in their work; (b) in Finland, Germany, and Spain, most of the respondents feel that EntreComp is relevant and included in their work to some extent; and (c) in Iceland the respondents think their work links partially to some or all of the competences highlighted in EntreComp. However, EntreComp is not greatly utilized in their practical work. Thinking about these three scenarios, it would be valuable to examine how different cultural paths and educational policy factors guide the implementation of similar frameworks on a national level. In this study, we were able to indicate country-specific differences, but not the reasons behind them. Indeed, studying educational reform is not a straightforward process: it involves developing understanding of different cultures, policies, rationality, and emotional dimensions, which play a high role when adopting new approaches to education (e.g., [25]). Our findings might provide a new starting point for further investigation of future design and implementation of education policies; as a next step, we propose collecting a broader data set from a more comprehensive, global sample of respondents, which would further develop the understanding of how entrepreneurial education initiatives can be efficiently and effectively implemented in different regions, countries, and cultures.

Since, e.g., The United Nations' Sustainable Development Goals and The European Union promote entrepreneurial competencies as a means of supporting young people to innovate, start businesses and create jobs, and creating welfare and economic sustainability, we focused our study on this area. Therefore, we conclude that our findings support that our educational initiatives are on 'the right track'. However, more research and practical implications are needed to promote a 'real' change in education. In this context, we highlight, e.g., communities' reflection and learning processes, thus supporting the development of a concrete vision for a 'better sustainable world' and pedagogical and practical ideas to take into use.

6. Conclusions

This paper extensively focused on the development of entrepreneurship education at the European level by examining how the EntreComp framework can act as an engine for transnational policy implementation driving entrepreneurial competences. This is the first time an entrepreneurship education study has respondents from so many different countries (46). Such major international studies have previously been, e.g., global reviews of studies conducted on entrepreneurship education in teacher education (e.g., [55]). Furthermore, previous country comparisons were limited to fewer countries (e.g., [15]). In that regard, our findings provide new insights into the overall progress of entrepreneurship education in the European context. However, its key contribution is linked to previous entrepreneurship education research by integrating the Shulman and Shulman [14], Seikkula-Leino [26], and Seikkula-Leino et al. [21,50,51], theoretical framework to other relevant studies [26] and the conceptual EntreComp framework's goals: mobilize, create value, implement, appraise or assess, and recognize [3] as summarized in Figure 7.

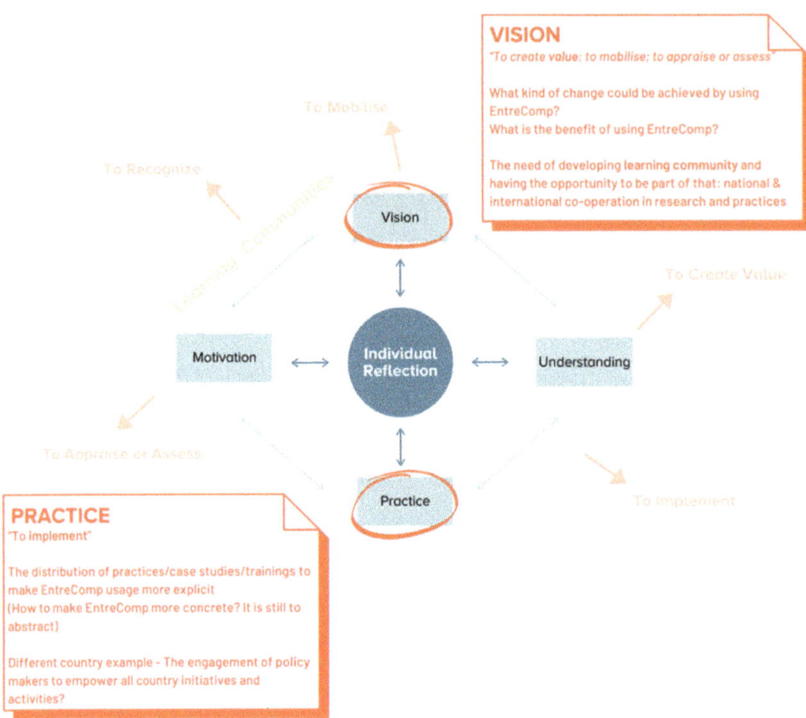

Figure 7. EntreComp as an engine for transnational policy implementation driving entrepreneurial competencies.

The results highlight that the European competence framework EntreComp can increase motivation and understanding of entrepreneurship education in different transnational contexts. Moreover, implementation has used all five goals. However, recognition of and support for learners' (e.g., educators, other practitioners, policymakers, and stakeholders) roles within various learning communities, require further study. Educational practices (e.g., training, tools, and concepts) need to be developed to support the reflection and learning of learners to empower development of their vision and practices related to entrepreneurship education. Some practical examples to promote these are presented in Figure 7 (e.g., case studies, other explicit models, country-specific examples, vision development through questions, and emphasis on learning communities). Further studies are still needed to identify the gaps in terms of required leverage towards coherent education change that promote entrepreneurial competences and the overall understanding of the EE within Europe. This would also provide new avenues to investigate how sustainable development could be promoted at the global, country, and local levels of education. This would also generate new knowledge on how to promote sustainable development within the private sector, for example in staff training.

Our case study has also broadened the general understanding of the way in which European strategies are guiding the development of entrepreneurial education. Overall, this kind of case study provides a suitable platform for investigating how these global goals can be detected in individual members' attitudes and beliefs in different country contexts. As Cohen et al. [27] argues, the generalizability of such single experiments (e.g., case and pilot studies) can be further extended through replication or multiple experiment strategies, which allows single case studies to contribute to the development of a growing pool of data for eventually achieving a wider generalizability of the key findings. Therefore, we suggest similar types of studies to be conducted to identify how policy goals can be successfully translated into frameworks, and what are the best practices for their successful implementation. Furthermore, a series of large-scale international studies could be useful in detecting how entrepreneurial education can be driven through policy framework, but also in demonstrating the added value of entrepreneurship education, as an example, by evaluation and case studies that would emphasize the learning and reflection processes within and across learning communities. This would also enable the design of more practical concepts and tools to support community learning processes to further strengthen entrepreneurial competences.

However, our research has certain limitations that need to be considered. For example, the number of respondents differed to some extent in different countries. Although the measure utilized previous bases, it could be further developed and validated based on the theoretical basis of Shulman and Shulman [14], and Seikkula-Leino [26], and Seikkula-Leino et al. [21,50,51]. In addition, we could further develop the metrics to guide the target group to respond more precisely to issues related to sustainable development.

Undoubtedly, our research has significant value in finding out how, in practice, the European framework promotes practical change in teaching and learning. We highlight concrete proposals that could be considered in the future. In addition, we have opened up the theoretical discussion of entrepreneurship education in the direction of education science by utilizing the results obtained in this field and theoretical entry angles by stressing learning communities and their reflection. With this research, we contribute to the development of entrepreneurship education in many ways, both in theory and in practice, and globally, thus providing a sustainable ground for developing entrepreneurial society by education.

Author Contributions: Conceptualization, J.S.-L. and E.M.; methodology, J.S.-L.; formal analysis, J.S.-L., M.S., S.R.J., E.M., and H.I.;—original draft preparation, J.S.-L., M.S., S.R.J., E.M., and H.I.; writing—review and editing, J.S.-L., M.S., S.R.J., E.M., and H.I.; visualization, E.M. All authors have read and agreed to the published version of the manuscript.

Funding: The investigative process and data findings supporting this research paper were funded through the EntreComp360 project co-financed through the Erasmus+ programme of the European Union.

Institutional Review Board Statement: Not applicable.

Informed Consent Statement: Not applicable.

Data Availability Statement: The dataset generated for this study will not be made publicly available because of the sensitive nature of the questions. All study participants were assured that the data will remain confidential and will not be shared. All requests concerning the access to the dataset should be directed to the corresponding author.

Conflicts of Interest: The authors declare no conflict of interest.

References

1. United Nations. Transforming Our World: The 2030 Agenda for Sustainable Development. 2015. Available online: https://sdgs.un.org/sites/default/files/publications/21252030%20Agenda%20for%20Sustainable%20Development%20web.pdf (accessed on 10 February 2021).
2. European Commission. European Skills Agenda for Sustainable Competitiveness, Social Fairness and Resilience. 2020. Available online: https://ec.europa.eu/social/BlobServlet?docId=22832&langId=en (accessed on 5 February 2021).
3. Bacigalupo, M.; Kampylis, P.; McCallum, E.; Punie, Y. Promoting the Entrepreneurship Competence of Young Adults in Europe: Towards a Self-Assessment Tool. In Proceedings of the 9th International Conference of Education, Research and Innovation, Seville, Spain, 14–16 November 2016; pp. 611–621. [CrossRef]

4. European Commission. Green Paper "Entrepreneurship in Europe". 2003. Available online: https://eur-lex.europa.eu/legal-content/EN/TXT/PDF/?uri=CELEX:52003DC0027&from=EN (accessed on 12 December 2020).
5. Borbás, L. A critical analysis of the 'Small Business Act' for Europe. In Proceedings of the 7th International Conference on Management, Enterprise and Benchmarking, Budapest, Hungary, 5–6 June 2009.
6. Dinning, T. Articulating entrepreneurial competencies in the undergraduate curricular. *ET* **2019**, *61*, 432–444. [CrossRef]
7. Garbutt, J.; Antes, A.; Mozersky, J.; Pearson, J.; Grailer, J.; Toker, E.; DuBois, J. Validating curricular competencies in innovation and entrepreneurship for biomedical research trainees: A modified Delphi approach. *J. Clin. Transl. Sci.* **2019**, *3*, 165–183. [CrossRef] [PubMed]
8. Joint Research Centre. *EntreComp: The Entrepreneurship Competence Framework*; Publications Office of the European Union: Luxembourg, 2016; Available online: https://data.europa.eu/doi/10.2791/593884 (accessed on 25 May 2021).
9. Weicht, R.; Ivanova, I.; Gikopoulou, O. The CRADLE teaching methodology: Developing foreign language and entrepreneurial skills in primary school pupils. *Entrep. Educ.* **2020**, *3*, 265–285. [CrossRef]
10. Jónsdóttir, S.R. Entrepreneurship Education at the Upper-Secondary Level in ICELAND in 2020. Available online: https://entrecomp360.eu/entrepreneurship-education-upper-secondary-level-iceland-2020/ (accessed on 12 June 2021).
11. Teneva, I. Entrepreneurial Competences: Comparison of Proacademy and EntreComp. Tampereen Ammattikorkeakoulu, 2018. Available online: https://www.theseus.fi/handle/10024/158476?show=full (accessed on 7 October 2020).
12. Education, Audiovisual and Culture Executive Agency. *Entrepreneurship Education at School in Europe: Eurydice Report*; Publications Office of the European Union: Luxembourg, 2016; Available online: https://data.europa.eu/doi/10.2797/731298 (accessed on 25 May 2021).
13. Ruskovaara, E.; Pihkala, T.; Seikkula-Leino, J.; Järvinen, M.R. Broadening the resource base for entrepreneurship education through teachers' networking activities. *Teach. Teach. Educ.* **2015**, *47*, 62–70. [CrossRef]
14. Shulman, L.S.; Shulman, J.H. How and what teachers learn: A shifting perspective. *J. Curric. Stud.* **2004**, *36*, 257–271. [CrossRef]
15. Seikkula-Leino, J. Developing entrepreneurship education in Europe: Teachers' commitment to entrepreneurship education in the UK, Finland and Spain. In *The Role and Impact of Entrepreneurship Education*; Edward Elgar Publishing: Cheltenham, UK, 2019; pp. 130–145. [CrossRef]
16. Kelchtermans, G. Teachers' emotions in educational reforms: Self-understanding, vulnerable commitment and micropolitical literacy. *Teach. Teach. Educ.* **2005**, *21*, 995–1006. [CrossRef]
17. Cooper, S.; Bottomley, C.; Gordon, J. Stepping Out of the Classroom and up the Ladder of Learning: An Experiential Learning Approach to Entrepreneurship Education. *Ind. High. Educ.* **2004**, *18*, 11–22. [CrossRef]
18. Fayolle, A.; Gailly, B. From craft to science: Teaching models and learning processes in entrepreneurship education. *J. Eur. Ind. Train.* **2008**, *32*, 569–593. [CrossRef]
19. Henderson, R.; Robertson, M. Who wants to be an entrepreneur? Young adult attitudes to entrepreneurship as a career. *Educ. Train.* **1999**, *41*, 236–245. [CrossRef]
20. Pittaway, L.; Cope, J. Simulating Entrepreneurial Learning: Integrating Experiential and Collaborative Approaches to Learning. *Manag. Learn.* **2007**, *38*, 211–233. [CrossRef]
21. Seikkula-Leino, J.; Ruskovaara, E.; Hannula, H.; Saarivirta, T. Facing the Changing Demands of Europe: Integrating Entrepreneurship Education in Finnish Teacher Training Curricula. *Eur. Educ. Res. J.* **2012**, *11*, 382–399. [CrossRef]
22. Pittaway, L.; Hannon, P. Institutional strategies for developing enterprise education: A review of some concepts and models. *J. Small Bus Ente. Dev.* **2008**, *15*, 202–226. [CrossRef]
23. Jónsdóttir, S.R.; Macdonald, M.A. The feasibility of innovation and entrepreneurial education in middle schools. *JSBED* **2019**, *26*, 255–272. [CrossRef]
24. Hannay, L.M.; Earl, L. School district triggers for reconstructing professional knowledge. *J. Educ. Change* **2012**, *13*, 311–326. [CrossRef]
25. Perrotta, C. Beyond rational choice: How teacher engagement with technology is mediated by culture and emotions. *Educ. Inf. Technol.* **2017**, *22*, 789–804. [CrossRef]
26. Seikkula-Leino, J. The implementation of entrepreneurship education through curriculum reform in Finnish comprehensive schools. *J. Curric. Stud.* **2011**, *43*, 69–85. [CrossRef]
27. Cohen, L.; Manion, L.; Morrison, K. *Research Methods in Education*, 8th ed.; Routledge: New York, NY, USA; London, UK, 2018.
28. EntreComp360 Project Online. Available online: https://entrecomp360.eu/ (accessed on 6 October 2020).
29. Mccallum, E.; Weicht, R.; Mcmullan, L.; Price, A. *EntreComp into Action—Get Inspired, Make it Happen: A User Guide to the European Entrepreneurship Competence Framework*; Bacigalupo, M., O'keeffe, W., Eds.; Publications Office of the European Union: Luxembourg, 2018; Available online: https://data.europa.eu/doi/10.2760/574864 (accessed on 25 May 2021).
30. Šlaus, I.; Jacobs, G. Human Capital and Sustainability. *Sustainability* **2011**, *3*, 97–154. [CrossRef]
31. Torfi Jónasson, J. Policy and reality in educational development: An analysis based on examples from Iceland. *J. Educ. Policy* **2002**, *17*, 659–671. [CrossRef]
32. Jónsdóttir, S.R.; Gunnarsdóttir, R. The Road to Independence: Emancipatory Pedagogy. 2017. Available online: https://doi.org/10.1007/978-94-6300-800-6 (accessed on 25 May 2021).

33. Jónsdóttir, S.R.; MacDonald, A. Settings and Pedagogy in Innovation Education. In *The Routledge International Handbook of Innovation Education*; Shavinina, L.V., Ed.; Routledge: Abingdon, UK, 2013; pp. 273–287. Available online: http://site.ebrary.com/id/10691776 (accessed on 25 May 2021).
34. European Commission. Effects and Impact of Entrepreneurship Programmes in Higher Education. 2012. Available online: https://ec.europa.eu/docsroom/documents/375/attachments/1/translations/en/renditions/native (accessed on 20 January 2021).
35. Smyth, J. Schooling and enterprise culture: Pause for a critical policy analysis. *J. Educ. Policy* **1999**, *14*, 435–444. [CrossRef]
36. Sukarieh, M.; Tannock, S. Putting school commercialism in context: A global history of Junior Achievement Worldwide. *J. Educ. Policy* **2009**, *24*, 769–786. [CrossRef]
37. Mccallum, E.; Mcmullan, L.; Weicht, R.; Kluzer, S. *EntreComp at Work*; Bacigalupo, M., Ed.; Publications Office of the European Union: Luxembourg, 2020; Available online: https://data.europa.eu/doi/10.2760/673856 (accessed on 25 May 2021).
38. Patton, K.; Parker, M. Teacher education communities of practice: More than a culture of collaboration. *Teach. Teach. Educ.* **2017**, *67*, 351–360. [CrossRef]
39. Wenger-Trayner, E.; Fenton-O'Creevy, M.; Hutchinson, S.; Kubiak, C.; Wenger-Trayner, B. *Learning in Landscapes of Practice: Boundaries, Identity, and Knowledgeability in Practice-Based Learning*; Routledge: London, UK; New York, NY, USA, 2015.
40. Lave, J.; Wenger, E. *Situated Learning: Legitimate Peripheral Participation*, 1st ed.; Cambridge University Press: Cambridge, UK, 1991. [CrossRef]
41. Chaurasia, S.S.; Kaul, N.; Yadav, B.; Shukla, D. Open innovation for sustainability through creating shared value-role of knowledge management system, openness and organizational structure. *JKM* **2020**, *24*, 2491–2511. [CrossRef]
42. Ibrahim, M. The use of community based learning in educating college students in Midwestern USA. *Procedia Soc. Behav. Sci.* **2010**, *2*, 392–396. [CrossRef]
43. Chung, E.Y. Facilitating learning of community-based rehabilitation through problem-based learning in higher education. *BMC Med. Educ.* **2019**, *19*, 433. [CrossRef]
44. Lundy, B.L. Service Learning in Life-Span Developmental Psychology: Higher Exam Scores and Increased Empathy. *Teach. Psychol.* **2007**, *34*, 23–27. [CrossRef]
45. Ngai, S.S.-y. Service-learning, personal development, and social commitment: A case study of university students in Hong Kong. *Adolescence* **2006**, *41*, 165–176. [PubMed]
46. Phan, M.-H.; Ngo, H.Q.T. A Multidisciplinary Mechatronics Program: From Project-Based Learning to a Community-Based Approach on an Open Platform. *Electronics* **2020**, *9*, 954. [CrossRef]
47. Sather, P.; Weitz, B.; Carlson, P. Engaging Students in Macro Issues through Community-Based Learning: The Policy, Practice, and Research Sequence. *J. Teach. Soc. Work* **2007**, *27*, 61–79. [CrossRef]
48. Joseph, C.; Said, R. Community-Based Education: A Participatory Approach to Achieve the Sustainable Development Goal. In *Quality Education*; Leal Filho, W., Azul, A.M., Brandli, L., Özuyar, P.G., Wall, T., Eds.; Springer International Publishing: Cham, Switzerland, 2020; pp. 101–111. [CrossRef]
49. Community-based learning for sustainable development. *UNESCO Institute for Lifelong Learning*. 2017. Available online: https://unesdoc.unesco.org/ark:/48223/pf0000247569 (accessed on 8 August 2020).
50. Seikkula-Leino, J.; Ruskovaara, E.; Ikavalko, M.; Mattila, J.; Rytkola, T. Promoting entrepreneurship education: The role of the teacher? *Educ. Train.* **2010**, *52*, 117–127. [CrossRef]
51. Seikkula-Leino, J.; Ruskovaara, E.; Ikävalko, M.; Kolhinen, J.; Rytkölä, T. Teachers' Reflections on Entrepreneurship Education: Their Understanding and Practices. In *Conceptual Richness and Methodological Diversity in Entrepreneurship Research*; Fayolle, A., Kyrö, P., Mets, T., Venesaar, U., Eds.; Edward Elgar: Cheltenham, UK; Northampton, MA, USA, 2013; pp. 146–171.
52. Yin, R.K. *Case Study Research: Design and Methods*, 5th ed.; SAGE: Los Angeles, CA, USA, 2014.
53. Ruskovaara, E.; Pihkala, T.; Seikkula-Leino, J.; Rytkölä, T. Creating a measurement tool for entrepreneurship education: A participatory development approach. In *Developing, Shaping and Growing Entrepreneurship*; Edward Elgar Publishing: Cheltenham, UK, 2015; pp. 40–59. [CrossRef]
54. Hynes, B.; Richardson, I. Entrepreneurship education: A mechanism for engaging and exchanging with the small business sector. *Educ. Train.* **2007**, *49*, 732–744. [CrossRef]
55. Deveci, İ.; Seikkula-Leino, J. A Review of Entrepreneurship Education in Teacher Education. *MJLI* **2018**, *15*, 105–148. [CrossRef]

MDPI

Article

The New Progression Model of Entrepreneurial Education—Guideline for the Development of an Entrepreneurial University with a Sustainability Approach

Agnieszka Klucznik-Törő [1,2,3,4]

[1] Department of Public Policy and Administration, Jesuit University Ignatianum, 31-501 Krakow, Poland; agnieszka.klucznik@ignatianum.edu.pl
[2] Horizon Europe, European Innovation Council and SMEs Executive Agency (EISMEA), 1049 Brussel, Belgium
[3] Green Deal Pilot, Horizon 2020, European Innovation Council and SMEs Executive Agency (EISMEA), 1049 Brussel, Belgium
[4] International Center for Entrepreneurship, 56-321 Bukowice, Poland

Abstract: Entrepreneurship is becoming understood as a set of competencies needed for many professions and, as a result, requires to be integrated into higher education even in such seemingly distant areas as, e.g., public administration, sport, agriculture, tourism, etc. Therefore, there is a need for research-based guidance on how to introduce and develop entrepreneurial education as an enabling approach to the transition in higher education that could serve as an integral part of a paradigm shift towards an entrepreneurial university. This paper aims to support that transition and to address related challenges by the presentation of a new progression model, which provides guidelines for the development of courses at the tertiary level with an entrepreneurial university approach. The construction of the new applicable model is central to the purpose of this study and based on a systematized literature review. Additionally, the input–process–output–outcome framework, originally constructed for the evaluation of educational programs, was adapted to the incorporation of an overall framework into the new model. In the results, the paper redefines some of the relevant core terms, such as "entrepreneurial education" and its "progression model". The research outcomes offer broad practical and theoretical applicability to a range of stakeholders—educators, students/learners, industry/business, policy makers, and researchers.

Keywords: entrepreneurial education; entrepreneurial education model; entrepreneurial university; entrepreneurship; graduate entrepreneurship; progression model; sustainability in entrepreneurial education modeling

Citation: Klucznik-Törő, A. The New Progression Model of Entrepreneurial Education—Guideline for the Development of an Entrepreneurial University with a Sustainability Approach. *Sustainability* **2021**, *13*, 11243. https://doi.org/10.3390/su132011243

Academic Editors: Jaana Seikkula-Leino, Mats Westerberg, Priti Verma and Maria Salomaa

Received: 13 August 2021
Accepted: 4 October 2021
Published: 12 October 2021

Publisher's Note: MDPI stays neutral with regard to jurisdictional claims in published maps and institutional affiliations.

1. Introduction

1.1. Context

The paradigm of higher education is shifting from the traditional towards an entrepreneurial university, giving different meanings of this term [1–4]. As the role of academic institutions in stimulating, therefore contributing to, the development of the modern knowledge-based economy has become widely accepted [5], the entrepreneurial university concept in practice has been facilitated by the collaboration between universities, government entities, and industrial partners, as outlined by a triple helix model [6]. So far, the main research has been focused on industry-university relations with respect to technology transfer [7] or university–business cooperation [8]. This approach to the entrepreneurial university is sometimes expanded by introducing novel elements into its conceptual framework such as the creation of new spin-off firms [9].

However, the author is convinced that the entrepreneurial university should be much more than just selling knowledge to the industry, gaining commercial orders from external stakeholders, or even more than new venture creation. Instead, the entrepreneurial university approach should be further expanded to drive societal change responding to sustainability requirements. The expansion of the concept would also allow for the application of the latest scientific interpretative frameworks such as the evolving nature of innovation models from the triple towards quadruple and quintuple helix models [10] in response to various challenges, not only for affected stakeholders, but society at large and its natural environment.

Moreover, entrepreneurship is increasingly understood as a set of competencies needed for many professions to be integrated into education and training, even in such seemingly distant areas as, e.g., public administration [11], sport [12], agriculture [13], or tourism [14] (p.186). This tendency towards an understanding of entrepreneurship in a broader way and, at the same time, a growing need for inclusion of sustainability issues in education, should consequently lead to the expansion of the concept of the entrepreneurial university.

So far, sustainability challenges faced by academic and economic stakeholders of higher education institutions (HEIs), society, and the natural environment have been recognized by the innovation models such as the quintuple helix model. The HEIs response has been, for example, an inclusion of knowledge transfer in their processes and a focus on creation of value for others in teaching/learning topics. It may take place as a part of projects, internships, assignments, etc. and lead to the development of valuable skills, such as the ability to recognize opportunities and find inspiration for new initiatives and innovation where most others face only problems and difficulties (such as climate change, pandemic, etc.). Simultaneously, these problems require new methods of teaching/learning (e.g., action learning vs. theoretical learning) and new theories development (e.g., experiential learning theory). In a response, the author draws on several concepts from experiential learning and action learning to provide descriptions of different learning activities undertaken in the course of the transformation from traditional education into entrepreneurship education.

Consequently, the main focus of the paper is on conceptual approach and includes explorative analysis with policy implications.

1.2. *The Relevance and Currency of the Research Questions to Be Examined*

The result of the higher education transition towards an entrepreneurial university—hardly explored so far—is the inclusion of entrepreneurial education into many other possible academic studies and courses far beyond the range of business studies and economics. That process has been initiated in higher education at the national level by some institutions including the Danish Foundation for Entrepreneurship [15]. So far, however, it has remained prevalent that educators are often unprepared to support the development of entrepreneurial competencies and skills in the course of teaching their subjects. The existing publications in favor of this approach and concept do not provide guiding principles for practitioners on the applicable methods [16], which is especially valid for higher education [17].

Additionally, entrepreneurial education models are rare [18], and the existing ones have not been intended as means of supporting the implementation of an entrepreneurial university concept with sustainability approach inclusion. Therefore, research-based guidance is needed on the possible ways for the introduction and development of entrepreneurial education as an enabling approach to the transition in higher education that could support and reinforce a desirable paradigm shift towards the practice of sustainable-based Entrepreneurial Universities.

This paper aims to support that transition and address related challenges by the presentation of a new progression model, which provides guidelines for the development of courses at the tertiary level with that new approach.

1.3. Aims and the Objective

The related core research questions have been framed in the following way: 1. How can a model of entrepreneurial and sustainable education be formulated? 2. What should define the integral parts of that model in order to permit its broad application?

In response to these questions, the paper intends to argue for a progressive model for entrepreneurial higher education (also called 'the Model' or the MEHE hereinafter) with the inclusion of sustainability issues and presents the relevant background research findings. The integrative scope of the Model is manifested in the combination of a selected range of elements—educational inputs, processes, outputs, and socio-economic outcomes—into its frame of analysis and evaluation as well as in the very construct of it, which is not country-specific, but instead, it offers a wide range of applicability.

Taking the research questions into consideration, the thesis of the paper is formulated as follows: sustainability is under-represented in the educational models and in the progression models of entrepreneurial education. As a result, there is a need for a new progression model with a sustainability approach.

The approach inherent in the above statement does not simply call for a new pedagogical method, but rather promotes a re-conceptualization of how educators and students could become co-constructors of the learning experience in order to develop entrepreneurial competencies while gaining course-specific knowledge at the same time.

Consequently, the objective of the paper is to critically reflect on how theories, concepts, methods, and findings from other bodies of inquiry can be applied to improve an understanding of entrepreneurial education while filling the gap in the entrepreneurial education modelling with respect to the inclusion of the sustainability-related aspects, and how it can be implemented in a new context of an entrepreneurial university framework.

1.4. Main Conclusions of the Research

The paper redefines some of the relevant core terms related to the subject matter, such as "entrepreneurial education" and its "progression model". The construction of the new applicable model is central to the purpose of this study and based on an original methodology, combining both a systematized literature review and a detailed analysis of the progressive models of entrepreneurial education with a focus on the inclusion of sustainability elements.

The research result offers broad practical and theoretical applicability to a range of stakeholders—educators, students/learners, industry/business, policy makers, and researchers.

2. Literature Review

The author has identified three strands of scientific literature incorporated into research on entrepreneurial education. The first aspect is connected to the conceptualization of the terms entrepreneurship and entrepreneur and their typologies [3,19,20]. The second one examines the development of a framework that highlights the role of educational inputs to achieve more entrepreneurial outputs, analyzed mainly with respect to entrepreneurial intentions [21,22]. A third avenue is also brought into the scope of study to examine the predominant influence of demand for educated graduates with more entrepreneurial perspectives prepared by the higher education system [3,23,24], stimulating the development of the concept of an entrepreneurial university and the share of experiences coming from its implementation [3,4].

2.1. Educational Inputs

The conducted and accomplished literature review proves that the definition of an entrepreneur has expanded over the last two decades from its originally narrow understanding of "a company owner or a self-employed person" [25], to a broader perspective by denoting a person who looks for possibilities to launch a new undertaking, able to spot market opportunities and exploit those by means of their own firm or as a part of their profit-oriented/business-motivated private activities [26,27].

A number of literature reviews on the concepts of an entrepreneur and entrepreneurship connect them with entrepreneurial competencies [3,28]. Those competencies are affected by certain personal characteristics, which can be developed and strengthened within the system of education. Among those characteristics are, e.g., creativity [29,30], innovation [31], imagination [29], problem-solving skills [28,32], degree of risk aversion [31,33–35], alertness [29], motivation [29], and willingness to take calculated risks [30].

Furthermore, the orientation of entrepreneurs towards action and engagement plays a significant role in the perspective of educational inputs, which requires the inclusion of some elements of experience and discovery in the entrepreneurial learning process [28,36] often associated with experiential learning [13]. It underlines the necessity to incorporate "effectuation development" in the educational process, especially at the higher education level [28]. Certain observers, such as Mansoori and Lackéus [37], have stressed that entrepreneurs learn through value creation and action, which has been confirmed by other authors [13].

Some important correlation between successful entrepreneurship and efficient entrepreneurial education can be identified in the value and relevance of teamwork for both aspects [37,38], as demonstrated by the fact that the majority of successful firms have been started by teams rather than solo entrepreneurs. Likewise, the experiential entrepreneurial learning process (mentioned above) is significantly more efficient when accompanied by a cohort-based system [38,39]. The reason is that "social learning", as it is sometimes called [13], facilitates the development of social capital. This arguably enables access to resources, including knowledge and skills. The ongoing multiple perspectives exposure can also significantly facilitate an easier absorption and application of knowledge into the learning process.

With respect to examples of entrepreneurial topics to be incorporated into a broad range of courses at the higher education level, there is a proven need for learning and understanding the importance of opportunity recognition, work–life balance, emotion management, learning from failures, and entrepreneurial mindset [28] in addition to other issues, such as knowledge transfer, directly associated with the implementation of an entrepreneurial university concept.

Interestingly, an entrepreneurial education that includes in-built inspirational part(s) producing positive attitudes and intentions [40] could be expected to prove especially effective in increasing the chances that students eventually set up their own businesses and launch enterprises.

It has also become clear that entrepreneurial self-efficacy can be strengthened by appropriate training and is fundamentally important in the activation of entrepreneurship [22,41].

2.2. *Desired Outputs of Entrepreneurial Higher Education*

The existing research on the outputs of entrepreneurial education has been almost entirely limited to the narrow understanding of entrepreneurship [28], and specifically, to the launch of new business ventures, self-employment, faculty-led start-up ventures, spin-off firms, and start-ups launched by entrepreneurial students and graduates of higher education systems [3].

The development of the MEHE, which corresponds to the wider sense of entrepreneurship and an entrepreneur, aligns this paper with the understanding that the most difficult and important question that entrepreneurial higher education needs to solve is how to make students more entrepreneurial during and after their higher education advancement.

A growing number of research results indicate that higher education in itself might develop entrepreneurial competencies, which increase the competitiveness of the firms [21,28,42] most efficiently taking benefit of those learned skills and perspectives. In particular, the entrepreneurial skills acquired during advanced education have been proven to highly correlate with entrepreneurial ability to enter the market and face competition [43]. This is so because higher education develops a certain "personal theory of practice" in university students (understood as the ability to apply theory in a practical context): it provides them with tools to minimize risks and to develop systematic decision-making skills. In essence, advanced education enables graduates to face uncertainty with greater confidence through the skillful allocation of resources and the exploitation of market opportunities—competencies that are conventionally attributed to business entrepreneurs [21,44]. Research indeed confirms that longer and more intensive studying, which is associated with a higher education level, increases the chances to develop entrepreneurial skills, e.g., critical thinking, opportunity recognition, evaluation and exploitation [28], teamwork, communication, etc. [14,43]. In addition, educational achievements are indicative of high levels of ambition, motivation, and endurance [43], which contribute to a positive entrepreneurial attitude and mindset. Accordingly, it can be inferred that advanced education fosters the development of business acumen, even if the knowledge and skills gained in formal education are not directly related to entrepreneurship.

The existing analyses confirm a positive association of entrepreneurial education with human capital assets considered essential for entrepreneurship such as knowledge, skills, positive perceptions of entrepreneurship, and entrepreneurial intentions [14,45]. Several sources in the relevant literature have also started to identify the effectuation as the desired output of entrepreneurial education [28,37].

The results of EE also involve innovation and venture creation, which may include not only forming a new organization but also a new activity such as the launch of a new project [14,46].

In addition to the provision of teaching, higher education also offers access to important social networks such as alumni networks or student organization networks [30]. In the context of business formation or any other entrepreneurial activity, this translates into access to scarce resources, potential key suppliers, and clients, which are paramount for entrepreneurial success.

Interestingly, the most advanced educational levels—such as Ph.D. or postdoc positions—tend to discourage graduates from launching their own companies. The most probable explanation may be found in the qualifications and cognitive advantage many students and graduates can acquire in the form of skills and marketable professional knowledge during their BA and MA studies. Their endowment of "educational capital" may easily open avenues to explore and pursue valuable opportunities as employees of others without the risk of their own business venture, especially in an economic environment of high labor demands. This applies particularly to the fast-growing and innovative activities driven by venture capital investment in enterprises—start-ups or more established companies—that recruit a workforce with the seeds and strands of entrepreneurial qualifications and skills [43,47]. As a consequence, graduates of higher education are likely to launch companies in knowledge-based industries (technologies, finance, real estate, insurance, etc.) and innovative businesses [32], while graduates of even more advanced educational levels are more likely to become employed by those companies. Both types of graduates, nonetheless, can be entrepreneurs in the wider sense.

2.3. Desired Entrepreneurial Education Outcomes

The aggregate outcomes of entrepreneurial education are socio-economic developments and human welfare [46,48]. Both educators and policymakers recognize entrepreneurial education as a means of macroeconomic growth and job creation [49]. It also supports an expansion of a knowledge-based society and a promotion of entrepreneurial economy and innovation culture [14]. Furthermore, the broad benefits comprise the creation of added value to societies while promoting social awareness and engagement from all actors involved, in addition to strengthening entrepreneurial behavior and entry [32].

2.4. Entrepreneurial Education Models

To address the aforementioned challenges and expectations towards entrepreneurial education in terms of its inputs, associated processes, desired outputs, and broad outcomes, both the linear and progression models of EE have been reviewed and critically analyzed for the purposes of this paper. The author recognizes that these models are rare, especially their progression representations intended to solve the problem of differing definitions of entrepreneurship, intended learning outcomes, and pedagogical approaches [18].

A common understanding of the progression models of entrepreneurial education relies on a renewed organizational perspective, and "progress" is understood in terms of the incorporation of various aspects and issues of entrepreneurship into the successive stages of formal education, usually starting, in some form, in primary school and finishing at college or university level.

3. Materials and Methods

3.1. Data Collection

The data for the literature review were collected from articles published in the following scientific journals: *Journal of Business Venturing* (JBV), *Entrepreneurship Theory and Practice* (ET&P), *Small Business Economics* (SBM), *Entrepreneurship and Regional Development* (E&RD), and *Higher Education* (HE).

3.2. The Assessment Frame and Research Questions

The qualitative data gathered through literature review were analyzed using content analysis following the logic of the IPO process [50] and guiding the overall design of the new progression model, which also created the basis for organizing the data to draw conclusions.

3.3. Data Analysis Methods

The data were grouped on the basis of the content analysis, which was realized as follows:

- Firstly, an in-depth systematized literature review [51] was accomplished to gain insights into the conceptual framework, inputs, processes, outputs, and outcomes of entrepreneurial education (EE).
- The criteria for the selection of articles to be reviewed for the study were based on certain words included in titles, abstracts, or keywords, such as "entrepreneurship", "entrepreneurialism", "entrepreneurial education", "entrepreneurship education", "entrepreneurial university", or "entrepreneurial universities".
- In the end, forty-one articles that satisfied the criteria were rigorously studied for the qualitative content analysis in view of the purposes of this paper, and the resulting relevant findings constitute the ground for the observations in the first part of the paper.
- The underlying information was drawn from a database along with the subject matters and aspects of IPO that are the desired entrepreneurial education inputs, processes, outcomes, and conceptual framework. The convenience sampling technique was applied.
- Secondly, the discussion delves more deeply into the subject of entrepreneurial education modelling, presenting the existing approaches with a focus on the inclusion of sustainability issues in the progression models of entrepreneurial education.
- Thirdly, a new progression model of entrepreneurial education is constructed, and the research findings are presented in a comprehensive and systematic way, starting from educational inputs, processes, and outputs and concluding with societal as well as economic outcomes.

The study is built on three theories incorporated in the foundations of the new model: the theory of entrepreneurial learning processes [52,53], the theory of students' entrepreneurial competency development [28,42,54], and the theory of entrepreneurial intentions [41,55].

The first theory centers on three crucial aspects of the learning process in entrepreneurial education encompassing skills development, entrepreneurial mindset preparation, and theories development.

The second theory builds on and connects to the previous one but focuses on the development of entrepreneurial competencies understood as the combination of skills, attitudes, and knowledge.

The third theory connects entrepreneurial education with the necessity to strengthen entrepreneurial intention, understood as a combination of entrepreneurial knowledge, perceived desirability towards entrepreneurship (attitudes and social norms), and perceived self-efficacy.

The main concepts and models used from the literature include the learning loop process and the progression models of education. Among constituted milestones of the learning loop process are theorization, experience, action, and reflection [56]. Additionally, the progression models in the context of education are understood as the successive stages of formal education, starting from primary and finishing at college or university level.

Since collaboration between universities, governmental entities, and business remains at the foundation of the entrepreneurial university, knowledge transfer becomes integrated into the new paradigm of higher education. As a result, the applied triple helix model [6] becomes superior to the model's other processes and elements by placing special emphasis on the collaborative interactions between university stakeholders.

The article proceeds along the following line of analysis: first, the results of systematized literature review in relation to prescriptive work in entrepreneurial education are elaborated; then, the comparative analysis of progression EE models is presented. The result, together with findings from the literature review, leads to the construction of the new model. A number of issues arising from this model are discussed, followed by an articulation of implications for the theory, practice, education, and policy.

4. Results

Taking the presented results of the literature review into consideration, the author defines entrepreneurial education as an essential pedagogy-driven dimension of a transition of HEIs towards a practice of a more entrepreneurial university model by incorporating the development of entrepreneurial competencies. In the author's assessment, it requires the inclusion of entrepreneurial skills, knowledge, and attitudes into a broad range of academic curricula, not just entrepreneurship or business studies.

Entrepreneurial education at the university level, understood and implemented on the ground of the above defining qualities, can be expected to create "entrepreneurial graduates" who are prepared by the higher education system to demonstrate entrepreneurial competencies in their professional activities. The model presented below is constructed and presented with the purpose to exemplify how to implement that concept.

The learning process lends support to the construction of the Model as it encompasses three types of provision: training aimed at skills development, entrepreneurial mindset preparation, and theory development [28,52].

Contrary to a current understanding of the progression models of EE, the author offers an understanding by which the progressive character of the proposed Model is associated with a different pedagogical approach and purpose of entrepreneurial education rather than with formal educational levels and, as a consequence, progress, in the form of enabling transformation, can be manifested within a single educational stage—in the case of this paper—namely the last one: higher education. Consequently, a "progression" model of EE is defined by the author with that new focus as a pedagogy-driven approach involving learning through the successive stages of a learning loop process comprising theorization, experience, action, and reflection. The pace of the process and its starting point is subject-specific and dependent on its main stakeholders—academics and students.

4.1. Comparison between the Models

Evaluation of whether sustainability is represented in the educational models required the performance of detailed comparison. The progression models of entrepreneurial education were considered for this purpose as they represent the most advanced form of EE modelling. The analysis of the four models was developed in ten steps and included an in-depth comparison between three existed models and the newly constructed one—the MEHE (see Table 1).

Table 1. Sustainability in the progression models of entrepreneurial education.

	British-Based Model [17]	Danish-Based Model [15]	Integrated Model [18]	MEHE
INPUTS				
Educational inputs:	Detailed for every educational level.	Include action, creativity, attitude, environment.	Assignments resulting in teamwork and value creation.	Inputs in a form of problems to be solved come from external stakeholders and lead to value creation for the environment, society, etc.
Entrepreneurial competencies:	Detailed for every educational level.	Importance of innovative and entrepreneurial competencies.	Focus on value creation.	Combination of 1. skills, i.a., critical thinking, problem-solving, creating values for others, teamwork; 2. attitudes, i.a, creativity, imagination, innovation, empathising with stakeholders needs; 3. subject-specific knowledge.
Teaching topics:	not specified.	not specified.	not specified.	Includes, i.a., environmental knowledge and interaction with the outside world, work–life balance.
Pedagogical approaches:	Includes, i.a., work and society related model.	Includes, i.a., value creation for others, creativity development, environmental connectivity.	Includes, i.a., taking action by addressing societal challenges, sustainable venture creation.	Includes, i.a., experiential learning, social learning, value creation for external stakeholders, creativity development.
Stakeholders:	A general approach to stakeholders.	A general approach to stakeholders.	A general approach to stakeholders.	Both internal and external stakeholders, i.a., environment.
Highlighted features:	Building the networking capacity.	Action, creativity, attitude, environment.	Team-based approach, value creation, outside-world connectivity.	Environmental knowledge and interactions, transfer of knowledge (ToK), value creation for external stakeholders.
PROCESSES				
Process(es) included in the model:	Start-up process simulation.	Value-creating entrepreneurial processes, entrepreneurial and innovative processes.	The educational process starts from *"Education through entrepreneurship"*, proceeds with *"Education about and through entrepreneurship"*, and ends with *"Education about, for and through entrepreneurship"*.	Application of: 1. IPO process, 2. Entrepreneurial learning process—the *learning loop*, 3. Process of knowledge transfer.
Transfer of knowledge (ToK):	ToK only mentioned.	ToK is indirectly mentioned in the pedagogical approaches by recommendation to connect to the environment outside the school.	ToK mentioned in relation to value creation.	ToK is applied.
OUTPUTS				
Approach to entrepreneurship:	Wide	Wide	From narrow to wide	Wide
OUTCOMES				
Main outcomes:	Not specified	Not specified	Jobs creation and economic growth.	Socio-economic development, human welfare, expansion of knowledge-based society, strengthened entrepreneurial behavior and entry, promotion of entrepreneurial economy, innovation culture, social awareness, and engagement.

Source: Authors' elaboration.

The comparison between the models leads to the conclusion that sustainability is under-represented in the existed EE models, especially with respect to

- Educational inputs,
- Recognition of main stakeholders of EE,
- Entrepreneurial competencies to be developed,
- Teaching topics propositions, which could touch upon sustainability issues, and
- Desired main outcomes of EE.

4.2. Presentation of the Model

The construction of the model starts from the input–process–output–outcome framework, in which the identified elements critical for EE are incorporated together with the relationships between them, whenever it was found supportive of better communication of the model's logic to its stakeholders.

5. Discussion

The postulated expansion of an entrepreneurial university approach creates a new response to some of the concept's criticism about the perceived growing dependency of higher education institutions from the industrial and business partners or sponsors. That new answer comes from the shift of the main focus from HEIs' external stakeholders towards internal stakeholders—educators and students. It is important to emphasize scientific advances and conceptual improvements in the understanding and explanation of the evolving nature of innovation models—triple, quadruple, and quintuple helix models—in response to challenges facing academic and economic stakeholders, society, and the natural environment, respectively, which are included in the Model by its emphasis on knowledge transfer through the creation of value for others. It may take place as a part of projects, internships, assignments, etc., leading to the development of valuable skills such as the ability to recognize opportunities and find inspiration for new initiatives and innovation where most others face only problems and difficulties (such as climate change or pandemic). With respect to value creation, the Model encourages the main stakeholders—educators and students— to consider not only financial rewards, but instead, recognize the possibility for the creation of broadly understood economic, social, and ecological values.

Since entrepreneurship is understood as a set of competencies needed for many professions, higher education requires the integration of entrepreneurial approaches into the educational methods of possibly all suitable courses in BA and MA program. Consequently, understanding the benefits and importance of teaching, sharing, and transferring certain entrepreneurship skills and competencies in broader terms would lead to the expansion of an entrepreneurial university framework by enriching its concept with pedagogical elements, means, and methods oriented towards that purpose.

In order to avoid the usual pitfall of entrepreneurial education research by neglecting the theory-based methodology [28], the current study was built on three theories of entrepreneurial education incorporated in the foundations of a new model. These theories include competencies theory, theory of learning processes, and entrepreneurial intentions theory. In the Model, the author has identified conceptual connections between all three of them. That integration was possible as the competencies theory distinguishes knowledge, skills, and abilities as its constituents, the theory of learning processes details how to develop competencies for outward-facing, task-oriented, and socially beneficial activities, and the entrepreneurial intentions theory aims at the student's internal entrepreneurial development. Additionally, according to the resulting new Model, the conceptual and methodical foundations for the definition, development, and transfer of knowledge relevant for entrepreneurial initiatives and accomplishments represent the theoretical aspect of entrepreneurship education, whereas skills and attitude development are recognized as practical aspects (see Table 2).

Table 2. Entrepreneurial competencies development.

Competencies	Skills	Attitudes	Knowledge
	practice-oriented aspects of entrepreneurship ↕		theory-oriented aspects of entrepreneurship ↕
Learning process provisions:	training aimed at skills development ↕	entrepreneurial mindset preparation ↕	theories for the development and transfer of knowledge ↕
Entrepreneurial intentions/ motivations:	perceived feasibility understood as self-efficacy	perceived desirability (attitudes, social norms)	acquisition of entrepreneurial knowledge

Source: Authors' elaboration based on Fayolle [28], and Linan and Chen [55].

Direct linkages between elements of the theory of the learning processes and those of the entrepreneurial intentions theory offer the opportunity to increase the results of entrepreneurial education by purposefully leveraging the three types of learning provisions to strengthen the entrepreneurial intentions/motivations. The extent to which opportunities for synergies are identified and harnessed depends not only on the entrepreneurial competencies of educators but also on the principles that guide the educational process in the design of the applied methods and choice of teaching approaches in addition to the teaching interactions with students.

Moreover, as outlined below in Figure 1, the resulting model manifests the recognition that the four stages of the "learning loop process" are recurrent and that the learning loop may start at different stages, not just at the stage of theorization.

Educators and students constitute the internal stakeholders of the Model. The skills of educators may prove decisive as the implementation of entrepreneurial education to a curriculum could greatly depend on their specialized and entrepreneurial knowledge, skills, motivation, competencies, but also on the choice of teaching approaches and teaching interactions with students, together with the educators' sense of ownership in that process. At the same time, the students are active participants in entrepreneurial learning by adding to the process their motivation, earlier experiences, and competencies. Both stakeholders are bound in the pedagogical process conceived to instill appropriate competencies, which include entrepreneurial skills and attitudes, besides specialized knowledge. Most importantly, the university support for educators and students would be required to smoothly adjust to new educational challenges by organizing appropriate entrepreneurial training.

At the curriculum level, the entrepreneurial process correlates with the educational design and simultaneously, entrepreneurial perspectives in learning/teaching are central to a sound transfer of knowledge.

The adaptability of the model to any subject is facilitated by the inclusion of teaching topics that are subject-specific but framed from an entrepreneurial perspective.

The intended educational outputs should be defined as the development of adequate entrepreneurial competencies and the proper understanding of social capital. That can further translate into a direct connection to important social networks, which can be crucial for entrepreneurial success, as it could greatly facilitate access to key resources, suppliers, potential clients, and valuable market information.

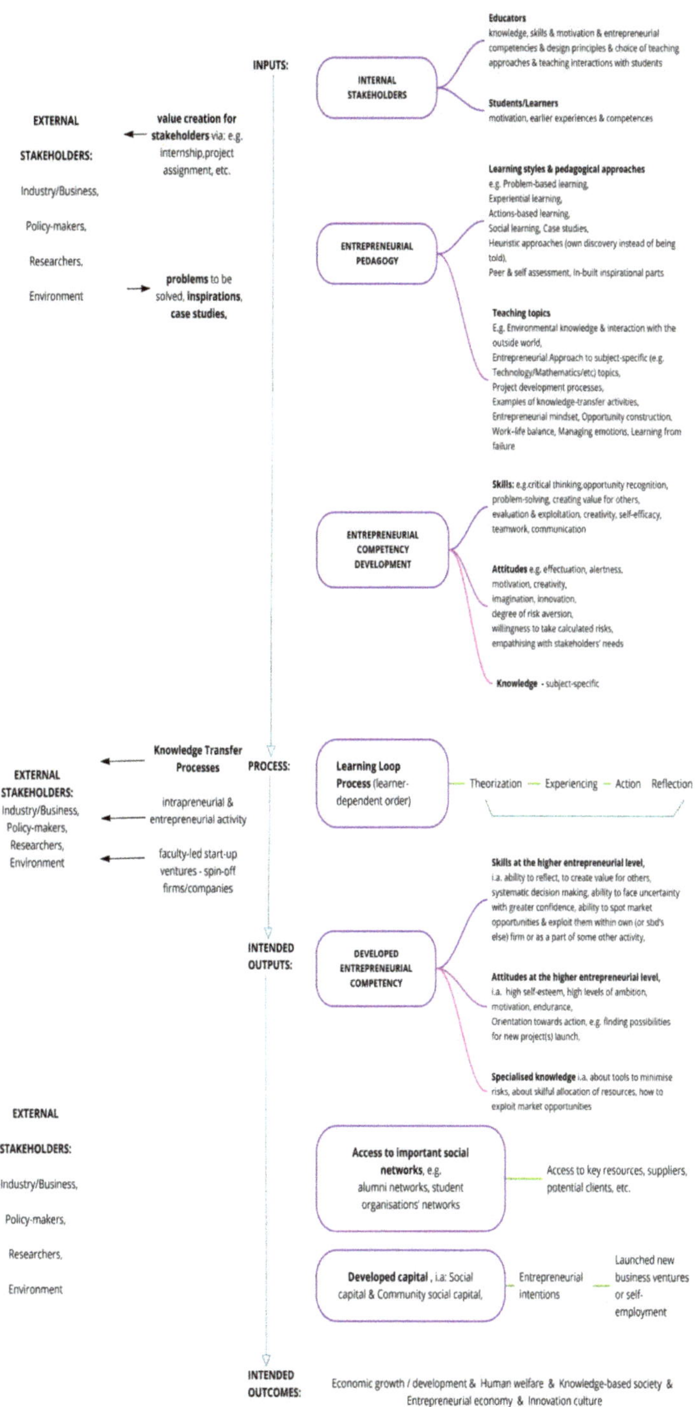

Figure 1. Progression Model for Entrepreneurial Higher Education.

The author draws on several concepts from experiential learning and action learning to provide descriptions of different learning activities undertaken in the course of the transformation of traditional education into entrepreneurship education.

The transformation of traditional education into entrepreneurship education can be connected to the experiential learning process for both the education system and its stakeholders. Experimentation encourages students to step outside assumptions taken for granted and requires them to enter an uncertain and ambiguous context. Uncertainty, as a feature of an educational program, replicates the circumstances in which an entrepreneur launches a business because starting a venture is an uncertain endeavor.

The iteration of the stages of an experiential learning process and their repetition as indicated by the theories of learning loop increases the educational efficiency of the whole process.

In the Model, the intended outcomes of entrepreneurial education have more general nature than the identified desired outputs and combine larger social, economic, and cultural aspects.

When it comes to future research directions, they might include:

- Research on the formation of an entrepreneurial mindset—motivation towards various types of entrepreneurship, e.g., social, green, etc.;
- Analysis of teaching topics, other than those mentioned in the Model, e.g., mission formulation with the social impact approach, social value creation, sustainable business model development, etc.;
- If any other methodical/pedagogical approaches increase the efficiency of entrepreneurial education and, consequently, whether they should be included in the pedagogical repertoire of educators;
- The impact of entrepreneur networks in strengthening social entrepreneurship; and
- Research on the development of a new educational model for social entrepreneurship supporting sustainable social change.

6. Conclusions

The author has summarized, organized, and adapted the research findings on entrepreneurial education modelling. The results are applied to the construction of the new progression model for entrepreneurial higher education (MEHE), with the intention to propose a conceptual and applicable tool for higher education policy making, management, and teaching for the development and introduction of entrepreneurial education as an operational perspective for the application of knowledge acquired in various higher education courses. The selection of the Model's key components was determined by methodology-based, targeted, and systematized literature review.

The answer to the question of how a model of an entrepreneurial education can be formulated to support a paradigm shift towards a more entrepreneurial university in order to prepare entrepreneurial graduates with appropriate competencies can be found in the construction of a new model representing a progression approach. Simultaneously, the progression model of EE was reassessed and defined from a renewed perspective, which stresses the relevance of the applied entrepreneurial education methods rather than formal educational levels.

The answer to the other research question of what would define the integral parts of MEHE was supported by the combination of the relevant entrepreneurial theories, system-based framework, and the analytical findings, which constitute the desired educational inputs, processes, outputs, and outcomes of the Model.

So far, the research on entrepreneurial education has separately examined the subject of entrepreneurship and the development of entrepreneurial competencies by higher education in general. Therefore, its EE modelling has been mainly linear. The Model outlined in this study combines the relevant research insights and maps out the relationships between the individual elements of EE in the new context of HE paradigm shifts towards a concept of an entrepreneurial university by introducing an innovative approach into the research on EE modelling.

In the study, the need for the application of a broader novel perspective to the concept of the entrepreneurial university has been emphasized rather than the traditional one defined by the commercialization of research, knowledge transfer, or university–industry relations. The perspective provided by the triple helix model stressing the role of the stakeholders' system in the creation of innovation was found supportive for the inclusion of students and educators in their relational science-driven efforts in strengthening the university paradigm shift towards a more entrepreneurial university.

Consequently, the views offered by the quadruple and quintuple helix models emphasizing the role of society and the environment in the innovation systems enabled the inclusion of value creation into the Model.

The systematized literature review and comparative analysis undertaken in the paper has positioned research observations and conclusions in the context of entrepreneurial education modelling as an emerging field in the academic scholarship on entrepreneurship. Instead of applying entrepreneurship in the narrow sense e.g., by only a knowledge transfer between university and its external stakeholders, the Model could be used as a vehicle for broadening its scope by involving teaching activities in supporting students to become more entrepreneurial regardless of their specialization or subject of study. This could be a pragmatic way to make the entrepreneurial university concept more applicable, inclusive, and as a result, more beneficial for a broad range of stakeholders. The implementation of such an approach would indispensably call for collaboration between HE managers, educators, learning coaches, and entrepreneurship trainers.

6.1. Implications for Education

The presented model responds to the question of how to formulate educational programs to support the application of the entrepreneurial university concept in the broad sense, that is, by investing graduates with entrepreneurial competencies. The Model also establishes several important linkages that enable the development and introduction of entrepreneurial education as a mode of teaching that can be applied to most of the subjects of higher education.

The Model includes methods of learning that have received little attention or have been neglected in entrepreneurial education at higher levels, such as reflection [57] or inspiration. Implementation of these methods might be facilitated or bolstered by learning coaches or business mentors and include, for example, immerse learning journeys to varied locations and settings [58] to strengthen an educational effort in deep learning.

6.2. Implications for Theory

A key theoretical implication of this study is that scholars can use the key identified findings of the Model and of the comparative analysis of EE models to include (or improve) aspects of entrepreneurial approaches into the teaching of their subjects that require theoretical and practical development.

Advancing beyond the current state of entrepreneurial education modelling in the context of an entrepreneurial university concept could include the adoption of a more integrated approach by creating a comprehensive meta-method(s) supported by theory and examined empirically in a broad variety of contexts and situations.

On the theoretical level, the strengths of the constructed Model could be used to develop other educational models. Findings from this article can thus hopefully aid entrepreneurship scholars to improve their perspectives and can create new avenues for developing entrepreneurial education model(ing).

6.3. Implications for Practice

The Model offers a theoretical framework suggestion, the validity of which must be tested in practice. However, the practical implications of the Model are its capability to support higher education institutions to adapt their educational offers across many subjects to the paradigm shift towards an entrepreneurial university. Consequently, scholars and managers of HEIs could reflect critically on the possibilities suggested in the paper and consider the suitability of the Model for their purposes.

Simultaneously, an advantage of the proposed model comes from its applicability as a set of guidelines for the development of new entrepreneurial education programs within academic courses. That supports the aim to build an entrepreneurial university in an extended form not only with more emphasis on knowledge transfer but also by incorporating appropriate entrepreneurial methodological approaches.

Another possible area of the practical application of this paper may be derived from the evaluation method offered by the Model for academic courses, as its general framework is adapted from the IPO approach, which was originally created for training evaluation purposes.

6.4. Implications for Policy

Policy makers could address the shortcomings identified in the comparative analysis of entrepreneurial education models in terms of educational inputs, processes, intended outputs, and outcomes demanding more practical relevance in teaching and scholarly activity in the course of the transition in higher education. Policy makers could encourage stakeholders (e.g., researchers, educators, managers of incubator and accelerator programs, learning coaches, university officials, entrepreneurship consultants) to raise the expectations about rigorous conceptual developments and at the same time, for contextual relevance and applicability.

6.5. Research Limitations

The research was supported by a systematized literature review covering forty-one articles. The range of those sources aimed to ensure their relevance and currency for the elaboration of the Model outlined for a more entrepreneurial university concept and practice. The reviewed articles were selected from various scientific journals incorporated simultaneously in the *Social Science Citation Index* and the *Entrepreneurship Journals Rankings* for that purpose and not intended to provide the basis for a comprehensive literature review.

The resulting Model represents the author's intention to formulate an applicable interpretive and analytical framework despite the inherent limitations of practice at the early phase of desirable transformation and adaptation of universities to the needs of more entrepreneurial higher education. The Model can certainly be adjusted and refined at a later stage depending on the availability of more data, observations from practice, and patterns of adaptive evolution of higher education institutions. In the light of more data and information, the future improvement of the Model may include the extension of some of its composite elements in the "input" and "output" dimensions. At the current phase of construction, the Model deliberately incorporated only those aspects that could be reliably based on available data and related analysis.

Funding: The APC was funded by a grant for the maintenance of research potential by the Jesuit University Ignatianum in Krakow granted by the Polish Minister of Education and Science in 2021.

Institutional Review Board Statement: Not applicable.

Informed Consent Statement: Not applicable.

Data Availability Statement: Data supporting reported results can be found via the ResearchGate portal www.researchgate.net (accessed on 30 September 2021).

Conflicts of Interest: The author declares no conflict of interest.

References

1. Clark, B.R. The entrepreneurial university: Demand and response. *Tert. Educ. Manag.* **1998**, *4*, 5–16. [CrossRef]
2. Etzkowitz, H.; Webster, A.; Gebhardt, C.; Terra, B.R.C. The future of the university and the university of the future: Evolution of ivory tower to entrepreneurial paradigm. *Res. Policy* **2000**, *29*, 313–330. [CrossRef]
3. Mars, M.M.; Rios-Aguilar, C. Academic entrepreneurship (re)defined: Significance and implications for the scholarship of higher education. *High. Educ.* **2010**, *59*, 441–460. [CrossRef]
4. Sam, C.; Van Der Sijde, P. Understanding the concept of the entrepreneurial university from the perspective of higher education models. *High. Educ.* **2014**, *68*, 891–908. [CrossRef]
5. Chankseliani, M.; Qoraboyev, I.; Gimranova, D. Higher education contributing to local, national, and global development: New empirical and conceptual insights. *High. Educ.* **2021**, *81*, 109–127. [CrossRef]
6. Etzkowitz, H.; Leydesdorff, L. The dynamics of innovation: From National Systems and "Mode 2" to a Triple Helix of university–industry–government relations. *Res. Policy* **2000**, *29*, 109–123. [CrossRef]
7. Hackett, S.M.; Dilts, D.M. A Systematic Review of Business Incubation Research. *J. Technol. Transf.* **2004**, *29*, 55–82. [CrossRef]
8. Adamsone-Fiskovica, A.; Kristapsons, J.; Tjunina, E.; Ulnicane, I. Moving beyond teaching and research: Economic and social tasks of universities in Latvia. *Sci. Public Policy* **2009**, *36*, 133–137. [CrossRef]
9. Rothaermel, F.T.; Agung, S.D.; Jiang, L. University entrepreneurship: A taxonomy of the literature. *Ind. Corp. Chang.* **2007**, *16*, 691–791. [CrossRef]
10. Carayannis, E.G.; Barth, T.D.; Campbell, D.F. The Quintuple Helix innovation model: Global warming as a challenge and driver for innovation. *J. Innov. Entrep.* **2012**, *1*, 2. [CrossRef]
11. Mazzucato, M. The Entrepreneurial State: Debunking Public vs Private Sector Myths (an excerpt). *J. Econ. Sociol.* **2021**, *22*. [CrossRef]
12. Ratten, V.; Thukral, E. Sport Entrepreneurship Education. In *Sport Entrepreneurship*; Emerald Publishing Limited: Bentley, UK, 2020; pp. 151–160.
13. Zamani, N.; Mohammadi, M. Entrepreneurial learning as experienced by agricultural graduate entrepreneurs. *High. Educ.* **2018**, *76*, 301–316. [CrossRef]
14. Hu, R.; Wang, Y.; Bin, P.; Ye, Y. Marta Peris-Ortiz, Jaime Alonso Gómez, José, M. Merigó-Lindahl, Carlos Rueda-Armengot (eds.): Entrepreneurial universities: Exploring the academic and innovative dimensions of entrepreneurship in higher education. *High. Educ.* **2018**, *76*, 183–186. [CrossRef]
15. Rasmussen, A.; Nybye, N. *Entrepreneurship Education: Progression Model*; The Danish Foundation for Entrepreneurship—Young Enterprise: Odense, Denmark, 2013; Available online: https://eng.ffe-ye.dk/media/785762/progression-model-en.pdf (accessed on 16 November 2020).
16. Pittaway, L.; Cope, J. Entrepreneurship Education. *Int. Small Bus. J. Res. Entrep.* **2007**, *25*, 479–510. [CrossRef]
17. Gibb, A. Entrepreneurship and Enterprise Education in Schools and Colleges: Insights from UK Practice. *Int. J. Entrep. Edu.* **2008**, *6*, 101–144.
18. Lackéus, M. *Entrepreneurship in Education: What, Why, When, How*; OECD: Paris, France, 2015.
19. Katz, J.A. Endowed Positions: Entrepreneurship and Related Fields. *Entrep. Theory Pr.* **1991**, *15*, 53–68. [CrossRef]
20. Katz, J.A. The Institution and Infrastructure of Entrepreneurship. *Entrep. Theory Pr.* **1991**, *15*, 85–102. [CrossRef]
21. Ferrante, F. Revealing Entrepreneurial Talent. *Small Bus. Econ.* **2005**, *25*, 159–174. [CrossRef]
22. Wilson, F.; Kickul, J.; Marlino, D. Gender, Entrepreneurial Self–Efficacy, and Entrepreneurial Career Intentions: Implications for Entrepreneurship Education. *Entrep. Theory Pr.* **2007**, *31*, 387–406. [CrossRef]
23. Jessop, B. Varieties of academic capitalism and entrepreneurial universities. *High. Educ.* **2017**, *73*, 853–870. [CrossRef]
24. Tassone, V.C.; O'Mahony, C.; McKenna, E.; Eppink, H.J.; Wals, A. (Re-)designing higher education curricula in times of systemic dysfunction: A responsible research and innovation perspective. *High. Educ.* **2017**, *76*, 337–352. [CrossRef]
25. Gartner, W.B. What are we talking about when we talk about entrepreneurship? *J. Bus. Ventur.* **1990**, *5*, 15–28. [CrossRef]
26. Mueller, S.L.; Thomas, A.S. Culture and entrepreneurial potential: A nine country study of locus of control and innovativeness. *J. Bus. Ventur.* **2001**, *16*, 51–75. [CrossRef]
27. Van Wennekers, A.; Thurik, R.; Reynolds, P. Nascent Entrepreneurship and the Level of Economic Development. *Small Bus. Econ.* **2005**, *24*, 293–309. [CrossRef]
28. Fayolle, A. Personal views on the future of entrepreneurship education. In *A Research Agenda for Entrepreneurship Education*; Edward Elgar Publishing: Cheltenham, UK, 2013; pp. 127–138. [CrossRef]
29. Burke, A.; Fitzroy, F.R.; Nolan, M.A. Self-employment Wealth and Job Creation: The Roles of Gender, Non-pecuniary Motivation and Entrepreneurial Ability. *Small Bus. Econ.* **2002**, *19*, 255–270. [CrossRef]
30. Kuratko, D.F. The Emergence of Entrepreneurship Education: Development, Trends, and Challenges. *Entrep. Theory Pract.* **2005**, *29*, 577–597. [CrossRef]
31. Verheul, I.; Uhlaner, L.; Thurik, R. Business accomplishments, gender and entrepreneurial self-image. *J. Bus. Ventur.* **2005**, *20*, 483–518. [CrossRef]
32. Backes-Gellner, U.; Werner, A. Entrepreneurial Signaling via Education: A Success Factor in Innovative Start-Ups. *Small Bus. Econ.* **2006**, *29*, 173–190. [CrossRef]

33. Begley, T.M. Using founder status, age of firm, and company growth rate as the basis for distinguishing entrepreneurs from managers of smaller businesses. *J. Bus. Ventur.* **1995**, *10*, 249–263. [CrossRef]
34. Sexton, D.L.; Bowman, N. The entrepreneur: A capable executive and more. *J. Bus. Ventur.* **1985**, *1*, 129–140. [CrossRef]
35. Stewart, W.H., Jr.; Watson, W.E.; Carland, J.C.; Carland, W. A proclivity for entrepreneurship: A comparison of entrepreneurs, small business owners, and corporate managers. *J.Bus. Ventur.* **1999**, *14*, 189–214. [CrossRef]
36. McKelvie, A.; Haynie, J.M.; Gustavsson, V. Unpacking the uncertainty construct: Implications for entrepreneurial action. *J. Bus. Ventur.* **2011**, *26*, 273–292. [CrossRef]
37. Mansoori, Y.; Lackéus, M. Comparing effectuation to discovery-driven planning, prescriptive entrepreneurship, business planning, lean startup, and design thinking. *Small Bus. Econ.* **2020**, *54*, 791–818. [CrossRef]
38. Arenius, P.; De Clercq, D. A Network-based Approach on Opportunity Recognition. *Small Bus. Econ.* **2005**, *24*, 249–265. [CrossRef]
39. Ni, H.; Xu, X. Sherry Hoskinson and Donald, F. Kuratko (eds): Innovative pathways for university entrepreneurship in the 21st century. *High. Educ.* **2016**, *74*, 1117–1120. [CrossRef]
40. Souitaris, V.; Zerbinati, S.; Al-Laham, A. Do entrepreneurship programmes raise entrepreneurial intention of science and engineering students? The effect of learning, inspiration and resources. *J. Bus. Ventur.* **2007**, *22*, 566–591. [CrossRef]
41. Wang, J.-H.; Chang, C.-C.; Yao, S.-N.; Liang, C. The contribution of self-efficacy to the relationship between personality traits and entrepreneurial intention. *High. Educ.* **2016**, *72*, 209–224. [CrossRef]
42. Man, T.W.; Lau, T.; Chan, K. The competitiveness of small and medium enterprises: A conceptualization with focus on entrepreneurial competencies. *J. Bus. Ventur.* **2002**, *17*, 123–142. [CrossRef]
43. Kim, P.H.; Aldrich, H.E.; Keister, L.A. Access (Not) Denied: The Impact of Financial, Human, and Cultural Capital on Entrepreneurial Entryin the United States. *Small Bus. Econ.* **2006**, *27*, 5–22. [CrossRef]
44. Shepherd, D.A.; DeTienne, D.R. Prior Knowledge, Potential Financial Reward, and Opportunity Identification. *Entrep. Theory Pr.* **2005**, *29*, 91–112. [CrossRef]
45. Martin, B.; McNally, J.J.; Kay, M.J. Examining the formation of human capital in entrepreneurship: A meta-analysis of entrepreneurship education outcomes. *J. Bus. Ventur.* **2013**, *28*, 211–224. [CrossRef]
46. Carlsson, B.; Braunerhjelm, P.; McKelvey, M.; Olofsson, C.; Persson, L.; Ylinenpää, H. The evolving domain of entrepreneurship research. *Small Bus. Econ.* **2013**, *41*, 913–930. [CrossRef]
47. Grilo, I.; Irigoyen, J.-M. Entrepreneurship in the EU: To Wish and not to be. *Small Bus. Econ.* **2006**, *26*, 305–318. [CrossRef]
48. Baumol, W.J. Formal entrepreneurship theory in economics: Existence and bounds. *J. Bus. Ventur.* **1993**, *8*, 197–210. [CrossRef]
49. Wong, P.K.; Ho, Y.P.; Autio, E. Entrepreneurship, Innovation and Economic Growth: Evidence from GEM data. *Small Bus. Econ.* **2005**, *24*, 335–350. [CrossRef]
50. Bushnell, D.S. Input, Process, Output: A Model for Evaluating Training. *Train. Dev. J.* **1990**, *44*, 41–43.
51. Clarke, A.; Crane, A. Cross-Sector Partnerships for Systemic Change: Systematized Literature Review and Agenda for Further Research. *J. Bus. Ethics* **2018**, *150*, 303–313. [CrossRef]
52. Politis, D. The Process of Entrepreneurial Learning: A Conceptual Framework. *Entrep. Theory Pr.* **2005**, *29*, 399–424. [CrossRef]
53. Fayolle, A. Three types of learning processes in entrepreneurship education. *Int. J. Bus. Glob.* **2008**, *2*, 198. [CrossRef]
54. Fiet, J.O. The pedagogical side of entrepreneurship theory. *J. Bus. Ventur.* **2001**, *16*, 101–117. [CrossRef]
55. Liñán, F.; Chen, Y. Development and Cross–Cultural Application of a Specific Instrument to Measure Entrepreneurial Intentions. *Entrep. Theory Pr.* **2009**, *33*, 593–617. [CrossRef]
56. Pittaway, L.; Thorpe, R. A framework for entrepreneurial learning: A tribute to Jason Cope. *Entrep. Reg. Dev.* **2012**, *24*, 837–859. [CrossRef]
57. Higgins, D.; Refai, D.; Keita, D. Focus point: The need for alternative insight into the entrepreneurial education paradigm. *J. Small Bus. Entrep.* **2019**, *31*, 225–242. [CrossRef]
58. Hancox, I.; Heron, K.; Klucznik-Törő, A. Coaching & facilitating young entrepreneurs from the perspectives of a coach, an education manager and a jury member—The case study of the Climate-KIC Journey programme. In Proceedings of 8th International Conference of Education, Research and Innovation, Seville, Spain, 18–20 November 2015; pp. 2430–2442.

 *sustainability*

Article

Leveraging a Community of Practice to Build Faculty Resilience and Support Innovations in Teaching during a Time of Crisis

Taryn Mead [1,*], Carlie Pietsch [2], Victoria Matthew [3], Surbhi Lipkin-Moore [3], Ellen Metzger [2], Ilya V. Avdeev [4] and Nancy J. Ruzycki [5]

1 School of Business, Western Colorado University, Gunnison, CO 81230, USA
2 Geology Department, San José State University, San Jose, CA 95152, USA; carlie.pietsch@sjsu.edu (C.P.); ellen.metzger@sjsu.edu (E.M.)
3 VentureWell, Hadley, MA 01035, USA; vmatthew@venturewell.org (V.M.); slipkinmoore@venturewell.org (S.L.-M.)
4 College of Engineering & Applied Science, University of Wisconsin-Milwaukee, Milwaukee, WI 53211, USA; avdeev@uwm.edu
5 Department of Materials Science and Engineering, University of Florida, Gainesville, FL 32611, USA; nruzycki@mse.ufl.edu
* Correspondence: tmead@western.edu

Citation: Mead, T.; Pietsch, C.; Matthew, V.; Lipkin-Moore, S.; Metzger, E.; Avdeev, I.V.; Ruzycki, N.J. Leveraging a Community of Practice to Build Faculty Resilience and Support Innovations in Teaching during a Time of Crisis. *Sustainability* **2021**, *13*, 10172. https://doi.org/10.3390/su131810172

Academic Editors: Jaana Seikkula-Leino, Mats Westerberg, Priti Verma and Maria Salomaa

Received: 22 July 2021
Accepted: 5 September 2021
Published: 11 September 2021

Abstract: Amidst the COVID-19 upheaval to higher education, a grantor-led community of practice (CoP) supported faculty members to deliver an innovative, sustainability-oriented entrepreneurship curriculum and maintain resiliency as teaching professionals. This paper discusses how through engagement in the CoP, this group of faculty from across engineering, material science, business, and geosciences demonstrated resilience, adaptability, and pivoted to create curriculum for students in real time, as the events of the COVID-19 pandemic unfolded throughout 2020 and impacted face-to-face learning. The role the community of practice played in sustaining and supporting the faculty will be discussed. Case studies from faculty members will demonstrate how sustainable design and social responsibility can be integrated into entrepreneurially focused classes and student experiences across disciplines. The primary contribution of this research is the important role that an emergent learning framework can play in informing how best to optimize the CoP format and approach in a way that leverages and addresses individual member strengths, challenges, and experiences, and supports the needs of CoP members during a time of significant change and crisis.

Keywords: faculty community of practice; resilience; COVID-19; sustainability-oriented entrepreneurship education; teaching cases

1. Introduction

While dissemination of best practices at conferences and through papers remains one of the leading approaches to fostering the adoption of innovative teaching practices, there is limited evidence that such practices do little more than raise awareness of said innovations [1]. This is true even when innovative teaching practices have been proven to be effective [2]. Rogers [3] posits that this hesitancy in adoption is due to a sense of uncertainty about the innovation, and this uncertainty can be overcome by connecting with others that have adopted said innovation. In a campus context, the adoption of innovative teaching practices can thus be positively impacted by peers, whether it be colleagues on campus that have adopted similar approaches [4] or members of a community of practice (CoP), which can include colleagues from other institutions [5].

When VentureWell—a not-for-profit organization with a mission to cultivate a pipeline of inventors, innovators, and entrepreneurs driven to solve the world's biggest challenges—set out to promote the integration of sustainable design on university campuses, these challenges to adoption were considered. An approach was therefore designed that combined faculty grant seed funds with an experiential in-person workshop and a one-year long CoP.

In addition to the anticipated complexities of incorporating new and experimental sustainable design strategies into coursework, the faculty CoP also experienced the unanticipated challenges of delivering coursework in a global pandemic approximately halfway through the first year. This disruption led to rapidly evolving curriculum changes, shifting delivery modalities, content modifications, and topical changes to reflect a shifting social, economic, and ecological context. The stressors of teaching in a global pandemic demanded a different type of engagement than was initially intended for this CoP, as its participants navigated this call to greater resilience, creativity, and pedagogical innovation in their course delivery.

This paper examines the way the CoP, designed by VentureWell, helped foster the integration of sustainable design and provided the faculty with the support needed to persist and innovate in the face of a global pandemic. We employed multiple methods, including an analysis of an emergent learning approach and autobiographical case studies, to examine the ways in which engagement in a CoP supported the implementation of sustainable-design educational innovations.

1.1. Communities of Practice in a Higher Education Context

Teacher collaboration and communities of practice within professional development of teachers is found broadly within research literature [6–13] and take various forms [10] (p. 69) as "co-teaching, mentoring, reflecting on lessons, group discussions of student work, a book club, a teacher network, or a study group." These methods are applicable to faculty professional development and can be used to improve student experiences in a post-secondary setting. Using collaborative learning in professional development of faculty is supported by social constructivism precepts and the work of Vygotsky [14] that learning is a deeply social process with extended Piagetian framings of an individual's cognitive processes by introduction of the zone of proximal development, where learning is a shared and social experience. The social nature of learning is not reserved for young learners; teachers as learners can similarly benefit from access to others [15]. Engaging activities and active learning occur when faculty are involved in their learning, rather than passively sitting through lectures or demonstrations [10]. Additionally, faculty benefit from introduction to concepts of life-long learning, an essential aspect in producing self-directed high-quality faculty for program sustainability [16]. These practices can be built through communities of practice who meet on a regular basis [17] to develop and implement engaging instruction for students.

By participating in communities of practice, faculty benefit from connecting with and feeling supported by individuals who "share a concern or passion for something they do" [17] (p. 1). The value experienced includes the exchange of innovative practices, and how best to contextualize those practices, such that these practices can be adopted and adapted on a variety of campuses [18]. As described by Schreurs et al., the face of faculty development is shifting from formal learning models to a mix of formal and informal learning models, with CoPs contributing to that informal learning aspect [19]. Benefits of participating in a CoP as a form of faculty development include, "a sense of community, self-awareness, motivation and validation of current practices and beliefs" [20] (p. 1).

As noted in the previous section and other research, the value of CoPs is well-documented in times of 'business as usual' [21]. However, there is a growing consensus that CoPs can be effective structures for developing resilience in practitioners in a number of fields, including medicine [22,23], disaster management [24,25], and particularly amongst educators [26]. Most universities abruptly transitioned to online learning in the spring of 2020 as a result of the COVID-19 pandemic and CoPs have been highlighted as an effective strategy to support faculty amidst COVID-19-related disruptions [27–31].

In highly dynamic situations and in times of crisis, decisive, rapid, and agile actions of faculty are needed to ensure the continuity of the educational experience [32]. For many faculty, the disruption caused by COVID-19 included a transition to online learning in just a matter of days [33,34]. In 'normal' times, faculty generally resist the transition to online learning and are told to "plan ahead" as a means to minimize disruption to the learning experience [35]. However, in these circumstances, the majority of faculty were forced to modify at least one aspect of their courses during the transition to remote instruction, including modalities of delivery of content, class communications, and strategies for assessment [34]. A large study of 897 faculty and administrators and at 672 U.S. institutions also discussed other modifications described as "emergency teaching and learning approaches", which included making changes to course expectations, assignments, and exams [34] (p. 6). Given these sudden changes, "faculty and administrators identified a need for assistance related to student support, greater access to online digital materials, and guidance for working from home" [34] (p. 6).

In the time since the disruptions caused by COVID-19, the stamina and resilience of university faculty have been repeatedly tested by ever-changing demands of the teaching environment, the breakdown in student mental health [36,37], and the larger social, political, and economic uncertainties at a global scale [38]. "Experiences during the early months of the pandemic were described [by faculty] as being overwhelming and exhausting, and participants described as being stuck in a cycle of never-ending repetitiveness, sadness and loss, or managing life, teaching, and other professional responsibilities with little sense of direction" [39] (p. 1306). This raises questions about the factors that contribute to resilience at the individual level and in society, more broadly. "Resilience definitions address issues of being prepared for unexpected events, recovering after them, picking up early warning and weak signals, learning from past events, addressing conflicts and information sharing between actors, working on institutional weaknesses, educating managers and leveraging social networks, all while serving citizens whose routines, emotional and physical stability and livelihoods are interrupted in minor and major ways. Resilience as an adaptive quality of the people ... " [40] (p. 118). At the individual level, resilience is improved by sufficient social support [41] and at scale, social capital leads to greater community resilience [42]. Teo et al. [43] (p. 136) define organizational resilience "as the process of developing relational networks that allow the organization to adapt and restore function". These definitions and circumstances related to resilience provide a useful lens through which to view the role of the VentureWell CoP in supporting faculty grantees throughout the COVID-related disruptions.

1.2. *Design and Evolution of the Sustainability Curriculum Community of Practice*

VentureWell (VW) aims to support faculty and students in developing innovations to make positive social and environmental impacts. To that end, VentureWell provides grants to faculty ("Faculty Grants program") to support the development of educational innovations. In spring 2019, VentureWell piloted the explicit inclusion of sustainable design into the request for proposals (RFP). The RFP sought applications from faculty seeking to develop new courses or programs, or strengthen existing courses or programs that focus on the incorporation of principles and frameworks of sustainability, with the end goal of novel sustainable designs and/or sustainable technologies. Following an external review process, eight faculty grantee teams were selected from the applicant pool from institutions from across the country and across disciplines including design, engineering, and geology (Figure 1).

As a part of the application process, grantees agreed to set aside time and grant funds to participate in a two-day Green Launchpad Educators Workshop and follow-on monthly CoP meetings. The workshop and the CoP meetings were integrated into the grant cycle in an effort to maximize the adoption of sustainable design teaching practices. The overall framing for the design of the workshop and CoPs was guided by Wenger's definition of CoPs as being, "groups of people who share a concern or a passion for something they do and learn how to do it better as they interact regularly" [17] (p. 1).

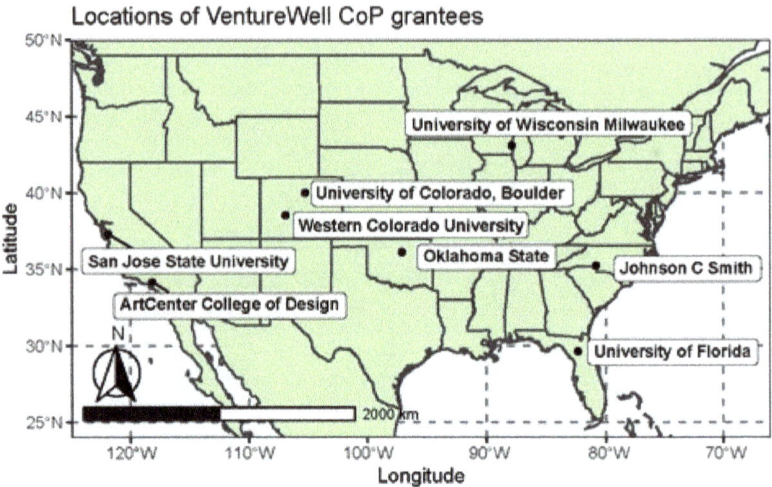

Figure 1. Locations of VentureWell CoP grantees.

The two-day in-person workshop was centered around key sustainability concepts, tools, and methodologies participants might adopt in their classroom. Participants shared their plans for integrating sustainability at the outset, shared helpful resources midway, and provided project updates at the close of the workshop that integrated concepts learned during the workshop. Additionally, plentiful unprogrammed breaks were integrated to provide participants with an opportunity to connect personally. The workshop approach was designed to enable participants to experience the value of connecting with peers working on a shared area of concern or interest, as well as the value of engaging with workshop content, and the project plans and resources they shared with each other.

Before the close of the workshop, two CoPs were formed comprising participants from four different institutions, with one to two team members from each institution. The specific participant makeup of each CoP was based on preferences expressed by participants at the end of day 1 regarding which individuals and teams they felt they would most benefit from working with. Monthly CoP meetings were hosted and facilitated via Zoom by the two facilitators. Facilitators were also responsible for scheduling the meetings, assembling the agenda, distributing meeting notes, and capturing key learnings about the CoP.

As suggested by previous literature, the monthly meetings were used by the facilitators to explore the value of the CoP from the immediate social-emotional support of faculty participants to the long-term realized value of systems-wide integration of sustainability curriculum into university programming [44]. The facilitators were also able to use emergent learning practices to adapt the framework of the CoP in response to the shifting educational landscape during the COVID-19 pandemic.

2. Theory and Methods: Understanding, Evolving, and Assessing the Value of the CoP

Communities of practice (CoP) are referenced with some regularity in faculty development literature [45,46]. However, when looking across different implementations, approaches range from large-scale online listservs to top-down knowledge management efforts, to homegrown small group convenings. One can argue that such variations are necessary, given the distinct goals and contexts of each CoP. However, the variations can leave CoP practitioners uncertain about how best to design and continue to evolve a CoP. An emergent learning framework was therefore adopted to consistently monitor and optimize the efficacy of the CoP. The role of the CoP was to support individual faculty in the shared goal of implementing sustainable design into their curricula and creating lasting system-level changes in university programming. A value creation framework was additionally used to evaluate how the CoP provided value to faculty engaged in redesigning and instructing courses that integrate sustainable design.

The research contained two phases: (1) Individual interviews with faculty grantees and (2) Autobiographical case studies from faculty grantees. Participants were chosen based on their receipt of faculty grants from VentureWell's grants program and their participation in the community of practice throughout the year of the grant cycle. All participants were invited to participate in the two-part study. All participated in the first interview phase and only a subset participated in the second case study phase.

2.1. Emergent Learning Framework

Emergent learning is an iterative framework where independently operating individuals develop and test strategies toward accomplishing a shared goal and through dialogue and interaction, they realize solutions to seemingly insurmountable challenges [47]. An emergent learning approach was utilized for facilitators to explore the question: "What will it take for faculty to learn from each other through a CoP-and for them to apply that learning to foster systems-level change at their university?"

An emergent learning table (a four-quadrant process tool that supports a team to come together around a specific question and share their stories (data) and their interpretation of these data) was developed by the two VentureWell meeting facilitators (Figure 2) and the VentureWell program evaluator. Learning logs (a document where the VentureWell team tracks their assumptions, learning, and insights around the framing question as the hypotheses are tested) were used to reflect on and track patterns in monthly CoP meetings. In an initial meeting, the learning log was populated with learnings grounded in the facilitators' past experiences with CoPs, and sub-hypotheses were developed to test as part of the emergent learning strategy. Data including the structure, format, and content of each CoP meeting was collected via tracking notes. Iterative analysis of these tracking notes was conducted on a quarterly basis using a collective learning log and discussions. Facilitators met with a VentureWell program evaluator on a regular cadence, providing time to reflect upon their observations. This process enabled the facilitators to optimize the CoP meetings in real time, providing for both a flexible and resilient approach to supporting the CoP participants and their work. Findings were documented in the learning log to track the ways in which strategy yielded hypothesized outcomes.

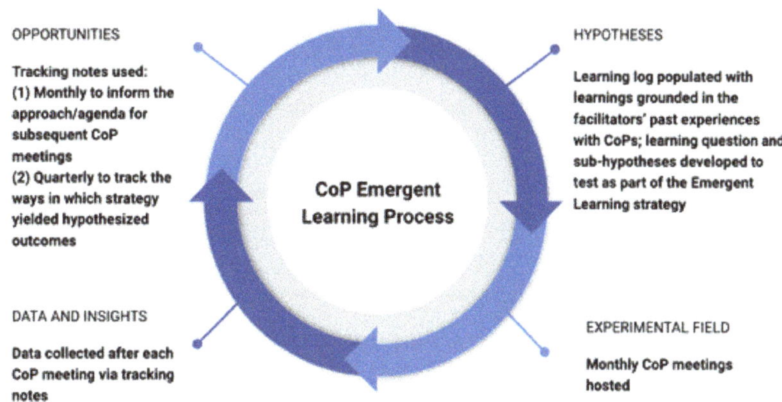

Figure 2. The emergent learning process employed for the VentureWell community of practice.

2.2. Value Creation Framework

Wenger-Trayner's cycles of value creation framework captures how value may evolve over time as a part of a CoP [48]. Leveraging this framework, the current study explored if engagement in a CoP added value to the faculty practice of developing a sustainable design course or program. The research sought to explore: the individual level gains faculty received from their participation in the CoP; if and how they applied their learnings to changes within their course, program, or institution; and what enabling or orienting factors supported the functioning of the CoP. Additionally, the study explored to what extent faculty processes aligned with the Wenger-Trayner's cycles of value creation framework.

Faculty members who met the criteria of currently working at the institution that was funded as part of the faculty grants program ($n = 8$) were contacted to participate in a short interview to understand if and how CoP members applied learnings from the CoP into their curricular change efforts and what that change looked like. Among the eight possible participants, seven ($n = 7$) members completed a 20-min interview with S. Lipkin-Moore, a program analyst at VentureWell. The semi-structured interview protocol was grounded in the following exploratory questions:

1. To what extent did faculty report changes in their knowledge, motivation, and sense of connection through their participation in the CoP?
2. In what ways did faculty apply their learnings from the CoP to modify their course or program?
3. In what ways did faculty apply their learnings from the CoP to catalyze institutional level changes?
4. In what ways do faculty experiences align with the Wenger-Trayner cycles of value framework?

Interviews were recorded on Zoom between December 2020 and January 2021. Interviews were transcribed by a transcription service and checked by S. Lipkin-Moore. Transcripts were imported in a qualitative data analysis software package, Dedoose, for coding. A start list of codes was generated based on an evaluative codebook developed for the faculty grants program, guided by a program level theory of change, as well as codes generated on an initial review of the CoP learning logs and interview transcripts. An initial cycle of coding was conducted on two interviews using this coding framework, where additional codes were generated, and a finalized codebook was developed. Using the final coding scheme, S. Lipkin-Moore reviewed the data and assigned codes to each unit (e.g., words, sentences) within Dedoose. Prevalent unique codes and co-occurring codes were identified, and were grouped into themes, following the methodological procedures established by Braun and Clarke [48]. Trustworthiness of the analysis [49] was assessed in multiple ways: (a) responses between participants were cross-checked (*triangulation*), (b) interviewees collaborated on the reporting of the analysis (*member checks*), and (c) staff members reviewed the interpretation of the data to ensure accuracy (*peer review*). The themes generated were validated through the analysis of the learning logs and subsequent case studies. Lastly, we used an exploratory autobiographical case study approach [50] to explore the ways in which individual faculty member experience aligned with the theory underlying the value creation framework [44]. Five case studies were used to illustrate how engagement in CoP supported curricular change.

3. Results

Overall, the results suggest—through engagement with the CoP—all faculty engaged in a process of learning that resulted in impacts at the student or course level. Across all interviews, the pathway through which (at least one) meaningful change occurred was: (1) building strong relationships, (2) sharing stories of successes and challenges in course development, (3) shifting individual ways of thinking or approaching a course or program, and (4) adapting and applying learning or knowledge to their course or program. As illustrated below, unique interpersonal and structural supportive factors enabled faculty member resilience through participation in the CoP through the difficult early months of the COVID-19 pandemic.

The following sections explain: How the meetings of the CoP were structured (Section 3.1); the experiences of the participants as analyzed through a third-party interview process (Section 3.2); and the experiences of the participants through autobiographical case studies (Section 3.3). Each of these sections contributes to a more complete picture of the CoP and the experiences of the participants and its facilitators.

3.1. *Emergent Learning: CoP Optimization and Adaptation to Support Faculty during the COVID-19 Pandemic*

For the first six monthly meetings of the CoP, 15 min were allocated to each institution to provide a brief project update along with a request to the group for assistance with a challenge associated with project implementation. This design was informed by a desire to create both a sense of forward momentum and accountability, as well as to firmly situate the CoP conversations in the practice of each participant.

After six meetings, the two smaller CoPs were merged. Facilitators had observed this shift to a single group might be useful due to the overlapping topics of conversation across the groups, and the desire expressed by participants to hear the perspectives of colleagues from other disciplines. However, the need for this shift was galvanized by the emergence of COVID-19 and the shared challenge experienced by participants of adapting their teaching, in real time, to modalities that continued to shift between blended, in person, and online. Participants were eager to learn how to adapt to these challenges with a broader group of colleagues with different campus contexts and disciplinary areas of expertise. Given the increased workload and sense of uncertainty faculty experienced due to COVID-19, pre-work was limited for participants by eliminating reports on project implementations. Social-emotional support was instead emphasized by including time for participants to share and empathize on current stressors and challenges. Participants also co-created the agenda by adding their own desired topics or engaged in a process of nominating and upvoting topics such that a meeting could have a single topical focus. This approach continued for the remainder of the two-year grant period, because it provided participants with social-emotional support needed, and the flexibility to co-create an agenda that helped address the immediate COVID-related challenges at hand, i.e., adapting teaching practices to different and fluctuating classroom modalities, including online, hybrid learning, and learning in classrooms, with the added complexity of mask wearing and social distancing.

3.2. *Analysis of Value Creation through a CoP*

Data in this section was collected by a program analyst from Venturewell who was not directly involved in the CoP throughout. The participants were asked to recount their experience for the purposes of program evaluation and improvement. Their interview responses were then compiled into this synopsis by the program analyst.

3.2.1. Shared Purpose and Connection as a Precursor to Meaningful Sharing (Immediate Value)

Across all interviews, relationships with other members were named as central to the success of the group's CoP. Several interviewees described that the opportunity to meet face to face (during an intensive 2-day *Green Launchpad* training) provided an opportunity for members to develop personal connections, build trust, and develop a shared purpose:

> *"The Green Launchpad was an amazing launchpad to kick it off, and the fact that we met in person, I think it's what sustained the Community of Practice, it's much harder to do it when you've never met in person."*

Another faculty member reported:

> *"I think that is intrinsic to the nature of a Community of Practices that you have this sense of shared purpose. And so, because there's a sense of shared purpose, you have a sense that you're there to help your peers also achieve that sense of purpose."*

The personal connection and "sense of solidarity" among the group, enabled a deep and more authentic level of sharing:

> *Faculty: In terms of emotional support, it's [the CoP] been great. In terms of real details, everyone's really open, just shares everything. And so, that's been fantastic.*
>
> *Interviewer: So, there's that level of comfort ... ?*

> *Faculty: Oh yeah. They're really generous, would just give away anything that they had made . . . . Like a syllabus, an idea. And then, there's been great connections . . . [Now] I have these ideas and now I have some actual resources behind them.*

The group had a shared understanding of what they were working toward and developed the personal connections necessary for the group members to have trust and be vulnerable with their work.

3.2.2. Variability in Individual Strengths, Challenges, and Experiences to Enhance Group Resilience (Potential Value)

Four interviewees named that the variability of members' backgrounds, expertise, and experiences made it such that each member could provide and learn about new and different ideas. Importantly, learning from others about what's worked and what's not worked in their specific contexts enabled faculty members to apply those learnings to their own context. For example, one faculty member described how they thought about student group work (and the need for scaffolding within the group):

> *I didn't know, probably everything I should have known to help prepare E-Teams and especially the younger students, right? So how do I plan that into not only my curriculum, but [also] provide the external supports that those students need, right?*

> *The Community of Practice is really nice because you realize that different participants that you're working with have different strengths and weaknesses. And there's always great ideas that you can steal from other people. So I think they're really good for that because even if it's just one part of it, there's different clever pieces that people have that make it really useful.*

Moreover, because the CoP outcomes were emergent, it allowed for members to be responsive to the most immediate needs, which enabled faculty to lend support where others experienced challenges:

> *For me the greatest benefit is . . . I don't know what I don't know. And if I'm just talking to people on my campus, that gets to be a bit of an echo chamber. So it's nice when I'm talking to people spread across the US to get an idea of what they're doing and the tools that they're using and resources they have. The people that they contact on their campus when they have questions or issues. And it's those kinds of things where I wouldn't know the questions to ask, but just being part of that conversation. I'm like, "Oh, there's a little golden nugget that fell out. I can do that here or maybe I cannot do that here but modify that for something else.*

The variability in disciplines, stages of the work, strengths, and weaknesses emerged as an important aspect of what made the group "work":

> *The student examples have been helpful to me . . . [because] I never knew what a student project could really look like, or should look like, what level is appropriate . . . .all of the faculty who are involved really do have different expertise themselves . . . and they have all different experiences and how they've built student teams, the kind of projects they worked on, the way they've structured their courses . . . .So just hearing all these different strategies to motivate students, help them emotionally work through this, in their teams to build successful teams with different personality types. To make everyone feel really welcome in the team. It's just been really helpful to just see everyone's accumulated experience and distill that into like, okay, what do I think is going to work for us?*

Consistent with these experiences, knowledge sharing between CoP members is reflected in the presence of key information shared between members, captured in the learning logs. Synthesis of the learning log suggests that all CoP meetings included sharing between members, and involved "advice", "sharing materials and tools", and "ideas" in four areas: (1) navigating campus policies, (2) expanding the container (e.g., activities outside of the classroom, drawing in other faculty, connecting with other courses), (3) content (lesson plans, videos, readings), and course design (pedagogy, flipped classroom).

3.2.3. Consistent Structure and Cadence Foster Accountability and Sustained Momentum (Applied and Realized Value)

Relationships were integral to fostering a space whereby faculty felt comfortable to share. However, an essential element that moved faculty members from sharing stories to applying their learnings relates to the consistency, expectations, and structure (facilitation) of the CoP meetings themselves. The CoP met once a month, and enabled faculty to get into a rhythm needed to progress in their practice:

"The community of practice is amazing, because suddenly, there is accountability and pacing."

Other faculty members explained:

I think the other thing too was making people [CoP members] think about it [the topics] in advance ... then also the expectation that you're going to show up to help other people. It was very clear that we were going to first talk about what we needed and then everyone else was going to help us with what we needed and there was an opportunity and almost a responsibility for everyone to be able to do that on every call with varying levels of depth, depending on what we all had to contribute. I think that was a big part of it.

It's a little bit of ... implicit peer pressure ... [you're] ... always thinking like oh okay, I might not do much this month because I'm busy with other things, but on the other hand, I want to be able to share something with the group, so it's like, might as well do something and not push it for next month. So that's in a sense, accountability and because the meetings were regular, that creates the pacing because it's predictable ...

Moreover, two interviewees named the importance of the group facilitators in creating and supporting a structure that allows for continuity. As one interviewee said:

I will tell you that sustained interaction is extremely important .... And I think it's probably more meaningful than even participants understand because it provides the accountability measure, it provides the sort of guide on the side ... [And] you have two great facilitators that have a lot of really good ideas, [and] are super enthusiastic.

The results indicate that the structure and cadence of monthly meetings was important to achieving meaningful change. This was largely because faculty were able to share and discuss challenges in real time, and were able to learn from each other's experiences as they tested new approaches. Further, the implicit pressure and accountability encouraged members to engage in their work in between meetings, which enabled faculty to have more content and experiences to share with the CoP.

As such, across interviewees, the most common realized value faculty described was a course-related impact. Faculty identified specific strategies, approaches, and tools that they learned from other members of the CoP that strengthened their course. As one faculty member described, the group supported each other in testing ways to enhance collaboration among their students in a virtual environment:

Interviewer: You've mentioned having ways to collaborate was one of the things you learned in the community of practice. [Did you mean] having ways to collaborate in a virtual platform?

Faculty: Yeah .... .We've been learning in our community of practice about various tools you can use for [collaboration] ... We brainstorm with students in person and they just put post-it notes on a whiteboard and made notes. And so we've learned some ways to translate that into the online environment.

Further, four faculty added the ways in which the addition of sustainability-focused class projects impacted students. Faculty were also able to share with the CoP COVID-19-related adaptations for engaging students' design, fabrication, and team work across multiple modalities. Specifically, faculty noted that students gained from solving for real problems:

"So the students really liked it and I think there were a couple of factors that went into that. One, I really pushed them to figure it out. They felt like they were doing

something towards a real world solution and knowing that we had the incubator, they were competing for the incubator. They all really hustled to try to get there. So I think that was a big part of it."

Ultimately, the appreciation for the group and perspectives was summarized in a learning log entry from one of the concluding meetings of the group: *"most of the conversation [in the meeting] focused on next steps for the group and gratefulness for the CoP and all that they gained from it in terms of learning and connections. [They] also talked about taking the facilitation techniques [of the CoP] and approaches into other contexts."* Beyond the content received from the group, the structure and facilitation of the CoP was an additional learning faculty could apply to their own work.

3.3. *Case Studies*

This section contains a series of reflective autobiographical case studies written by the participants in the CoP to more specifically address how their personal resilience was enabled through their experiences with the CoP as they implemented their grant activities and courses amidst the first round of COVID-19 disruptions.

3.3.1. Case Study: Western Colorado University

The VentureWell GreenLaunchpad grant supported the development of a new course entitled "Waste = Food: Science-Driven Innovation for Rural Circular Economies" at Western Colorado University in Gunnison, CO in the fall of 2020. The intent of the class was for student teams to develop new product concepts from locally available waste materials that could be upcycled to create novel forms of value using circular sourcing and supply chain strategies. Throughout the semester, the teams developed their product concepts using various circular economy tools and compiling their ideas using the Business Model Canvas [51]. The course was cross-listed to attract students from environment and sustainability, business, biology, and honors departments to create an interdisciplinary team experience for one semester at the ICELab, Gunnison County's economic development center (pictured in Figure 3).

The majority of course planning occurred in the application process for the grant submission, so much of the conceptual work was completed early on in the course development cycle. However, the unknown modalities of delivery and the uncertainty of the semester schedule for much of the summer of 2020 before the course made planning tenuous, at best. Additionally, the ICELab's plans to deliver their Incubator program were also uncertain with COVID-19-related disruptions. Ultimately, the in-class content remained similar to the original plans and the class was able to take field trips to the local landfill and recycling center. Unfortunately, an overnight trip to companies working in the circular economy was not possible and the class instead took a local overnight camping trip for an immersive biomimicry workshop. Additionally, the intended live pitch event was replaced with team video submissions to an online voting platform and distributed to the campus community and their personal social networks for a competitive voting process. The successful team recently completed the incubator program, developing their concept to make messenger bags and wallets from discarded whitewater raft material.

One valuable cultural component of the course experience was the establishment of class values, which served as a periodic touchpoint to align the students and provide a framework for shared values in a time of great social and cultural uncertainty. This concept came from the sharing of teaching resources with other members of the CoP and the class values that this other faculty member provided served as an example. Class values were:

- Repeated failure evolves our thinking and progresses our ideas.
- Creativity, ingenuity, and innovativeness are learned skills, not inborn gifts.
- Do not expect anyone to know the answer to your questions.
- Be solutions-oriented but ensure that you fully understand the problem first.
- Maintain integrity despite uncertainty.
- Be ready to adapt, pivot, and become more resilient.

- Trust each other and our intentions.
- Let play and curiosity guide our actions.

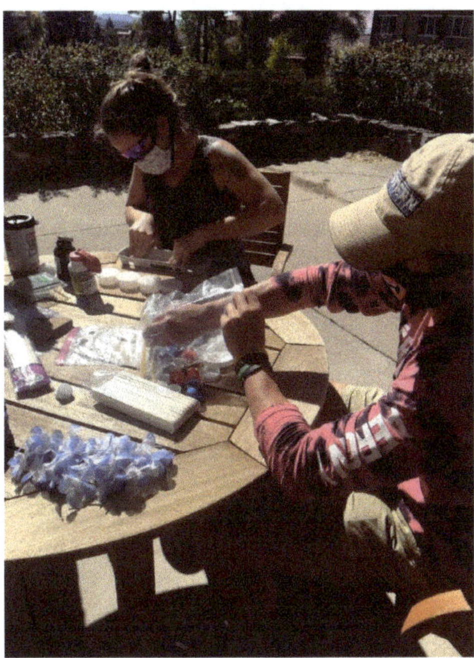

Figure 3. Students prototyping circular design concepts in a makeshift outdoor classroom.

The course has since developed into two other programmatic activities related to innovation in the circular economy. The first will be a student design competition in partnership with the new School of Engineering on the Western Colorado University campus to be delivered in the 2021–2022 academic year. The second evolution of the grant is a proposed new interdisciplinary master's degree in circular design and manufacturing, a first of its kind program.

Overall, the development of this course had a powerful impact on the students, faculty, and affiliated organizations to experiment with circular economy mechanisms in a rural mountain community. For the faculty and students involved, it demonstrated the enriching experiences that can emerge in unpredictable and co-creative learning settings. It also enabled the faculty member to demonstrate how these concepts can be effectively used to advance economic development, potentially providing opportunities for the larger community, and prototyping how the university can support student entrepreneurs in sustainability ventures. Finally, it set the stage for the implementation of a longer-term design competition and master's-level degree program, which will attract new kinds of students to Western from diverse fields of design, engineering, and business for sustainability.

3.3.2. Case Study: University of Florida

The course EMA3000L Sophomore Materials Laboratory at the University of Florida, Herbert Wertheim College of Engineering, Department of Materials Science and Engineering was designed to address gaps students had for application of materials knowledge to real world problems and to help these students have a hands-on experience early in their academic pathway. Funding from Venturewell Green Launchpad allowed the class to explore biomaterials design and application for additive manufacturing and resources from the Venturewell "Green Toolkit" (https://venturewell.org/inventing-green-toolkits/, accessed on 11 June 2019) were used to support the class in design thinking.

Having a CoP as part of the support from Venturewell allowed for the sharing of practices and ideas across disciplines and helped to bring in diverse voices and practices, including from arts and design courses, geology courses, sustainability courses, and business courses. Having instructors from varying disciplines all applying sustainable practices led to a rich sharing of ideas, and proved particularly useful when COVID shut down face-to-face classes and remote learning environments dominated the landscape. Through the CoP, best practices for remote learning were shared and refined, including those for student groupings, breakout room activities, remote design teams, prototyping, and pitch presentations. Having CoP meetings allowed for a constant iteration of the class and provided support for what directions seemed to be supporting student engagement more than other pathways.

These practices became important in spring 2020, and were applied strategically to the design of the spring 2021 remote laboratory course, including development of specific team building experiences early in the semester to promote remote teamwork practices and development of strong frameworks and rubrics for use throughout the course. The Venturewell Green Toolkit also supported remote learning and student work on the early design pieces and LCA analysis for the design process.

Students were able to work remotely in groups to create a business canvas for their work redesigning more sustainable light sources (flashlights, headlamps), and the remote nature of the design work led to more group accountability for students to assure that project deliverables were met. Student teams utilized agile principles for the design process, which allowed for more resilience in the teams for design changes and remote communication issues.

3.3.3. Case Study: San José State University

The VentureWell grant supported further development of a new team-taught Earth system science course at San José State University in San Jose, CA, USA. Earth system science instructors and students have a unique appreciation of the long reach of time and the interconnected nature of the Earth's spheres: hydrosphere, geosphere, biosphere, and atmosphere [52]. With the knowledge of the processes and the rates by which Earth has changed in the past, geoscientists realize human impacts on all Earth spheres are likely unprecedented and demand immediate intervention [53–55], but students enrolled in traditional geoscience programs receive little training on approaches to sustainability. Support from the CoP allowed SJSU instructors to initially develop and implement a sustainability-focused class project and to reconfigure the project to adapt to the online learning environment. Following experiences in the CoP during the fall 2019 Green Launchpad meeting, the final project for the ongoing Earth system science course was immediately reconstructed to focus on a sustainable product redesign. Students developed a systems map of a product they were interested in re-developing for sustainability (Figure 4A), conducted an abbreviated life-cycle analysis, and decided on business model goals in order to complete an evaluation of their sustainable project solutions. In spring 2021 when the class was taught a second time, the course was reframed to center the sustainability product redesign by frontloading concepts and goals for global sustainability and by including more in-class work time and intermediate project deadlines. This new project scaffolding emerged through a year of personal reflections and conversations with the CoP. The CoP was also a crucial source of tools and resources for evaluating sustainability, including systems mapping, life cycle analysis, and design thinking principles as an iterative approach toward sustainable product redesign. The 2021 course was adapted for a completely online remote experience. Online resources and exercises developed by VentureWell provided springboards for lesson plans and student projects. Conversations with the CoP helped to troubleshoot solutions for design projects taught remotely.

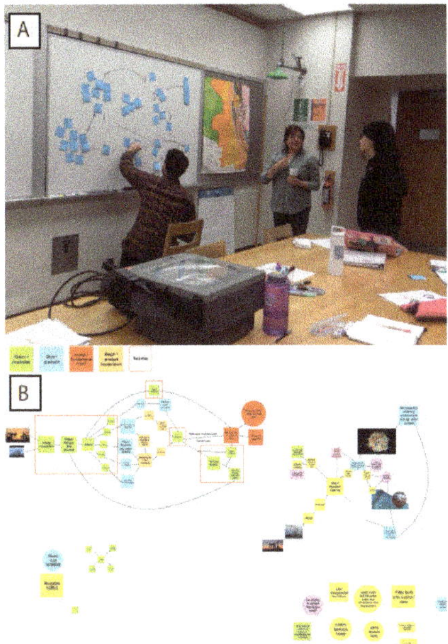

Figure 4. (**A**) Students in the fall 2019 course working on a systems map. The physical space does not always support equal participation of all group members; (**B**) A systems map constructed remotely and iteratively throughout the semester allowing for simultaneous participation by all team members.

Discussions throughout the year with the CoP supported building effective student teams including surveys on personal assets and goals for growth, methods for managing teams through team contracts and the RACI model [56] and tools for collaboration (Mural) and creative thinking including six thinking hats and the creative problem solving process [57]. Online collaborative documents including Mural, Google Drive, and Canvas discussion boards proved to be more amenable to teamwork than analog counterparts (post-its, whiteboards, notebooks) by allowing students to revisit and iterate on ideas, collaborate remotely on their own schedules, and democratized participation, whereby there was "space" and "tools" for everyone at the drawing board (Figure 4A,B).

Conversations with the CoP also provided invaluable suggestions for how to support our geoscience students through the unfamiliar process of sustainable product redesign by connecting to engineering faculty, the on-campus sustainability board, and the broader SJSU entrepreneurial ecosystem, which includes an innovation club and annual business and design competition. In the 2021 course, a self-reflection survey was assigned to students to allow the instructors to build resilient student teams that balanced personality, working style, and topic interest [1,58–60]. The CoP continues to inspire the instructors to incorporate more creativity-driven exercises and to provide more design experiences using on-campus maker spaces at the reestablishment of face-to-face instruction. Students did reflect on a continuing commitment to sustainable living and a new quantitative appreciation for the broad scale of systemic changes needed for Earth system sustainability.

A considerable broadening of perspectives has been an overarching outcome of faculty participation in the VentureWell CoP. Finding solutions to problems arising from unsustainable ways of living demands not only scientific understanding of the Earth system, but consideration of the socio-economic factors that drive individual and collective decision-making and actions. The geosciences are inherently interdisciplinary, but Earth scientists do not traditionally engage in investigation of complexly interacting planetary and human systems. Engagement with the CoP has introduced new colleagues, conferences, programs, and literature that interweave scientific concepts with ideas and practices drawn from other disciplines, including sustainable design and social responsibility. This more holistic view of the Earth system not only shaped the development of a new Earth system science course, but also helped to catalyze a new undergraduate degree in Earth system science and helped to guide the selection of appropriate electives from beyond the College of Science to include courses from engineering, business, and environmental studies.

3.3.4. Case Study: University of Wisconsin-Milwaukee

In this project at the University of Wisconsin-Milwaukee (UWM), the multidisciplinary teams of faculty were set to investigate two questions: (1) "how might we design a Community of Practice for 200+ UWM instructors teaching sustainability-related classes" and (2) "what would happen if we started experimenting with the learning experience 'containers', i.e., courses, workshops, etc.?"

To explore the first question of designing a faculty CoP, the UWM team proposed to facilitate a series of faculty workshops around a theme of "Hacking for Sustainability". The "hook" for the faculty was: how might we help the students who are passionate about sustainability topics to form collaborative teams across disciplines and to channel their passions into project-based explorations? The first two-hour workshop (12 June 2019) focused on empathy, i.e., understanding the aspirations and needs of a diverse group of instructors across colleges and disciplines around topics of sustainability. Through peer conversations and inferences, a group of 25 participants offered a first look at the opportunity space in the form of generative questions that looked like this:

- "How might we help Portia, professor of film, to disseminate knowledge and sustain research projects?
- "How might we help Michael, freshwater sciences, to encourage his collaborators to take the risk to embrace new big ideas?"
- "How might we help Bob, urban planning, to engage and empower community members to advocate for their own safety in their neighborhoods?"

Grouping questions together revealed a few themes outlining faculty/instructor needs and aspirations that emerged from the workshop (in the form of *needs*):

- To identify collaborators/establish and maintain relationships.
- To create a network/connect disciplines/establish collaborative environment.
- To drum up interest/disseminate knowledge/reach more students.
- To transform industry/engage community members.
- To find time/secure funding.

In parallel, the UWM team started investigating the second question: what shape, i.e., "a container", might the new teaching and learning experiences take for the faculty and students? Instead of designing an entire new course (there are 400+ courses on sustainability topics in the UWM course catalog), the initial plan was to develop a not-for-credit summer workshop, "Hacking for Sustainability", designed and delivered by a group of self-selected faculty across disciplines. Faculty would invite their students who would form small teams, start ideas exploration, get connected to the UWM entrepreneurial ecosystem, and develop through various launch pads (VentureWell E-Teams, NSF I-Corps Site, UWM Startup Challenge, etc.). Unfortunately, this concept was not possible with pandemic-related disruptions.

Fortunately, with the help of this CoP, the UWM team pivoted towards rapid low-resolution experimenting in the form of an in-class mini hackathon around these six topics of food scarcity and insecurity. The team picked an existing multidisciplinary course, *ME-405: Product Realization*, and the hackathon partners that had vested interest in this topic area—*UWM Office of Sustainability* and the *Milwaukee RiverWest Food Pantry*. The team tested three key hypotheses:

Hypothesis 1 (H1). *Students will be able to self-organize and form rapid teams around six topics tied to the hackathon theme based on their internal motivations and passions.*

Hypothesis 2 (H2). *Students will be able to explore the problem space on-line with the help of area experts and identify a problem to work on and hack for solutions in 2.5 h.*

Hypothesis 3 (H3). *We can assemble a teaching team for this experiment, instead of a solo instructor.*

The experiment was conducted in the fall 2020 using a 100% online teaching and learning environment. Instructors used two virtual collaboration environments for the hackathon: Zoom and Mural. Prior to the hackathon, students had experienced two design sprints and were exposed to design thinking methods and mindsets. The hackathon included three blocks (see Figure 5): (1) a 60-min problem exploration that included interviewing experts and doing some background research on the problem; (2) an 80-min design sprint (hacking) around the problem statement; and (3) a 20-min showcase of their low-resolution virtual prototypes. Here is what we learned.

Figure 5. Timeline of the Hackathon event.

Students were able to identify the topic area of their interest and self-organize into six teams. They then conducted interviews (example in Figure 6) around shopper experience in the neighborhood food pantry.

During the hackathon, instructors intentionally blurred the lines between ideation and prototyping (those are usually separated as distinct steps of the design process for novice learners) and encouraged hacking. Figure 7 provides an example of the solution idea prototyped graphically in Mural addressing a problem of promoting food pantry in the community beyond just food.

The teaching team hypothesis was validated and we were able to assemble a team of experts for interviews from the food pantry and the office of sustainability. The class had already been taught by a team of two instructors and the results suggest that this method could be used to meaningfully connect instructors from various courses.

PROBLEM SPACE:

PROBLEM STATEMENT:

Questions

What did the community base for the pantry look like prior to COVID and after COVID?

What is the most common item people buy?

Which items last the longest and which last the shortest?

What neighborhoods are using the pantry the most?

How do people interact with the volunteers? What is the atmosphere?

How do you encourage people to choose healthier food options?

What kind of people will be shopping at the food pantry

how many people are going through every day?

What is the community view of the pantry and its users?

What kind of home resources does the average shopper have?

Should we be focused on community health?

how much face to face contact happen since COVID

From Discussion

we have less people now compared to pre COVID. the past few months is approaching the pre covid mark

Do not have food drives from Catholic parishes. Virtual food drives. Other suppliers are based on headcount and are steady

food drives gave canned goods or hygiene products. dont have a shelf life

Wants to educate about experience and culture of food

Help find job opportunities and rent resources

get food from own farm and local produce. focus on healthy choices for their customers

Ideas

Recipe cards/ideas for shoppers

Classes for how to properly use fruits/vegetables

Drive thru menu that features available items by category

Speaker projected for different groups at different times

Figure 6. Hackathon process outcomes.

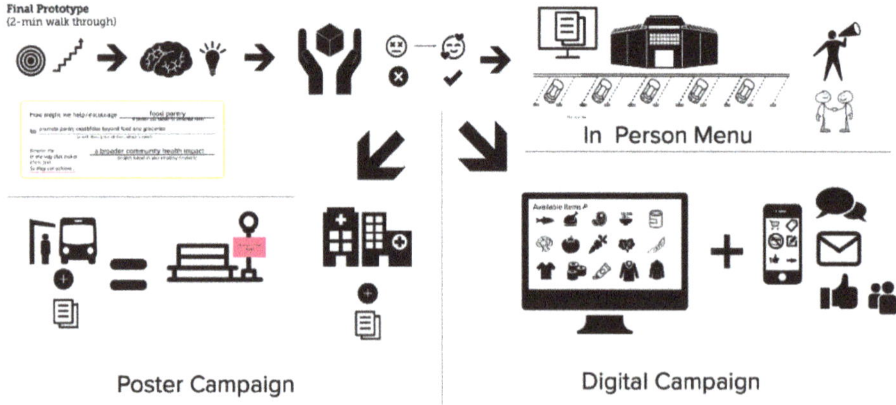

Figure 7. Sample prototype from Hackathon.

4. Discussion

At the outset, the CoP was designed to support a group of faculty grantees as they worked to integrate sustainable design and entrepreneurship into the curriculum. The CoP supported this effort, and due to the emergent learning approach, was also able to flexibly adapt to meet the exigent needs of the group. The CoP thus effectively supported faculty through the disruptions related to the COVID-19 pandemic. Achievements in teaching and mentoring—despite the pandemic—include faculty establishing, revising, and implementing student support systems inspired by the CoP, including explicit class values, highly scaffolded team construction, and frequent student feedback and subsequent course adjustments. Achievements in classroom innovation include multiple student-centered design projects and class competitions that incorporate "real-world" scenarios including food bank accessibility, upcycling outdoor recreation gear, and responding to natural hazards. Accomplishments in university-wide program development includes the initiation of an interdisciplinary University of Wisconsin-Milwaukee sustainability CoP, partnerships between the School of Business and School of Engineering as well as a new MS degree in circular design and manufacturing at Western Colorado University, and a new BS in Earth systems science at San José State University.

The faculty interview quotes and case studies, while diverse, support existing literature which suggests "engaging with a multidisciplinary community of practice can ... provide more than online educational skills; they foster a sense of togetherness and a safe environment to share concerns and challenges on both a professional and personal level" [29]. The immediate value experienced, including personal connections, trust, and shared sense of purpose, developed throughout the CoP paved the way for faculty to generously share their promising practices (potential value) and co-mentor each other in a way that enabled the resilient adoption and adaptation of promising practices on their own campus, in the face of unpredictable social and economic disruption of the COVID-19 pandemic (applied value and realized value). We contend that a CoP, organized around a shared concern, that leverages an emergent learning framework can allow (1) CoP facilitators to adapt the CoP format and approach to meet evolving needs, which in turn (2) enables CoP participants to equip themselves with the social-emotional support and social learning needed to be resilient in the face of disruption (Figure 8).

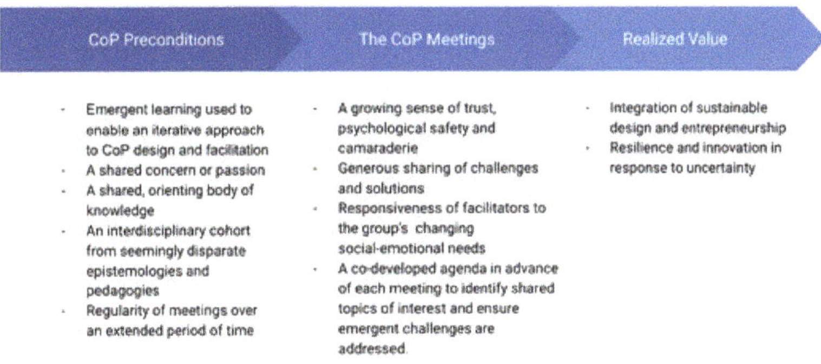

Figure 8. Summary of what this study has identified as the effective CoP preconditions, e.g., approaches to CoP design, group configuration, features of the CoP meetings including how meetings were run by facilitators and experienced by participants, and ultimately the realized value experienced by participants.

According to Teo et al. [43] (p. 136), "Resilience may be framed as the capacity to bounce back to a state of normality, or as an emergent property, when an organization learns to adjust to adversity and in the process, strengthens its capability to overcome future challenges." One barrier often cited with regard to the introduction of curricular innovations is competing priorities and time [1]. The added disruption of the COVID-19 pandemic produced an interference phenomenon for VentureWell CoP participants, wherein innovative class delivery formats, new structures for student support, teambuilding, project scaffolding, and novel scholarly programs emerged as a result, as suggested is necessary in previous studies, e.g., [32]. Together the faculty and facilitators of the CoP, while learning to adjust to the complexities associated with the COVID-19 pandemic, overhauled their courses, and made vast improvements to the student experience. Given that resilience has been described as "an adaptive quality of the people" [40] (p. 118), it is clear that the CoP contributed to the ability of the participants to adapt to the rapidly changing conditions of the work environment, social engagement, and tools of learning and teaching. These faculty and their courses will not be reverting to "normal" in future semesters but instead have used the disruption to build better courses that integrate new skills and approaches, and are addressing the needs of their local environment and community stakeholders. New interdisciplinary frameworks established throughout the duration of the CoP provide infrastructure support for continued innovation and future integration of sustainability into university systems. Establishing and facilitating faculty-centered CoPs are an effective approach to support faculty when navigating disruption and may have the added benefit of catalyzing organizational transformation. As the interviews and case studies exemplify, the resilience of the individuals involved was enhanced by the social support of the group, in alignment with existing literature (i.e., [41]).

The types of transitional states the CoP participants experienced have also been described as periods of liminality in other research related to organizational resilience (e.g., [43]). Liminality is a period when routines are disrupted, and new relational connections are "made to allow members to adjust psychologically, emotionally, and socially, to activate resilience. Within the liminal period, leaders influence(d) the formation of new connections through mutual and swift trust and utilize(d) these networks to enable collective meaning-making and sensemaking. If resilience is a necessary component and hoped for outcome of group dynamics, it is critical that educators, leaders, and facilitators become attuned to emerging moments of liminality, and integrate that awareness into the process of group facilitation. In times of crisis, it is necessary for leaders to recognize that a moment of liminality is emerging and to "communicate mindfully via these networks to promote positive emotional connections among members" [43] (p. 136). The CoP and its facilitators clearly supported the period of liminality of the 2020–2021 academic year.

Figure 8 below captures the characteristics and features and outcomes of the supportive approach used.

5. Conclusions

This analysis adds to the growing body of literature that suggests that CoPs can help professionals, particularly educators, to adopt innovative teaching practices, even in times of significant uncertainty. The autobiographical quotes and cases provide detailed examples of how resilience was demonstrated by a cohort of eight grantees participating in a VentureWell CoP. While each case was unique, they all demonstrate how the addition of this more informal social learning modality for faculty development [19] had the added benefit of fostering support, camaraderie, and open sharing of challenges and ideas around the practice of teaching and learning, which in turn enables faculty to successfully pivot and execute on their grant-funded projects, despite regular and consistent disruptions. One might argue that the CoP not only fostered resilience, but it also ensured faculty benefited from the challenges of the pandemic by supporting them to experiment with their teaching approach and build interdisciplinary connections, in much the same way that participatory managers support their employees to take risks as they seek out new innovative approaches [61]. Indeed, several CoP participants maintained that the teaching and learning innovations that resulted would endure beyond the life of the pandemic. Additionally, the student experience may have benefited overall from the uncertainty and subsequent need to pivot, since students engaged in new ways of thinking and working, and were able to observe the agile way in which their faculty adjusted their approach in real time in response to a constantly changing landscape.

This study highlights the important role that the emergent learning framework plays in ensuring the CoP itself is poised to flex and demonstrate resilience to changing circumstances in much the same way the faculty members are supported to be adaptive and resilient amidst turbulent circumstances. The lessons learned from this study may support faculty development practitioners in advocating for and integrating this more informal and social, and yet highly effective, approach to faculty development. It also provides a potential approach for other leaders and organizations seeking ways to productively support and sustain the work and creativity of others during times of change.

This analysis was limited by the number of faculty that were able to participate in the post-hoc analysis of the CoP through the writing and analysis process. Additionally, it only includes data from a single CoP and would benefit from a comparative analysis of various CoPs in higher education that operated during a similar time frame to develop a more holistic perspective of the impact on personal resilience in times of disruption.

Recommendations for further research include the integration of a comparison group of non-CoP participating faculty, to help better determine the degree to which participation in the CoP versus participation in informal networks of colleagues fostered resilience. Additionally, while this paper demonstrates the positive impact of a CoP on participants' ability to resiliently adopt and adapt innovative teaching practices during a time of significant change, it is also important to research and better understand the degree to which a shared period of liminality in turn impacts participants' sense of connectedness and willingness to share promising practices with other CoP participants. Given the important role that an emergent learning approach played in the responsive design of the CoP, a methods paper that captures that process and tools developed would also serve as an important and useful addition to the field. Finally, universities may consider professional development policies that explicitly reward participation in CoPs as they have been demonstrated as being overall beneficial for faculty members' overall resilience, life-long learning, and engagement in advance curriculum development.

Author Contributions: Methodology, T.M., V.M. and S.L.-M.; Writing—original draft, T.M., C.P., V.M., S.L.-M., E.M., I.V.A. and N.J.R.; Writing—review and editing, E.M. All authors have read and agreed to the published version of the manuscript.

Funding: This research received no external funding beyond the grant-funded community of practice under study which was funded by VentureWell. Publication was not a requirement of the grant for all grantees.

Institutional Review Board Statement: Ethical review and approval were waived for this study, due to the nature of the data collected being from faculty grantees.

Informed Consent Statement: Informed consent was obtained from all subjects involved in the study.

Data Availability Statement: Interview data can be made available upon request of the corresponding author.

Acknowledgments: Numerous stakeholders from across all of the funded universities and VentureWell, including students, faculty, and staff, contributed to the success of this effort and research.

Conflicts of Interest: The authors declare no conflict of interest.

References

1. Taylor, C.; Spacco, J.; Bunde, D.P.; Zeume, T.; Butler, Z.; Barnas, M.; Bort, H.; Maiorana, F.; Hovey, C.L. Promoting the Adoption of Educational Innovations. In *Proceedings of the 23rd Annual ACM Conference on Innovation and Technology in Computer Science Education (ITiCSE 2018)*; Association for Computing Machinery: New York, NY, USA, 2018; p. 368. [CrossRef]
2. Seymour, E. Tracking the Processes of Change in US Undergraduate Education in Science, Mathematics, Engineering, and Technology. *Sci. Educ.* **2002**, *86*, 79–105. [CrossRef]
3. Rogers, E.M. *Diffusion of Innovations*, 5th ed.; The Free Press: New York, NY, USA, 2003.
4. Roberts, F.D.; Kelley, C.L.; Medlin, B.D. Factors Influencing Accounting Faculty Members' Decision to Adopt Technology in the Classroom. *Coll. Stud. J.* **2007**, *41*, 423–436.
5. Herman, G.L.; Mena, I.B.; West, M.; Mestre, J.; Tomkin, J.H. Creating Institution-Level Change in Instructional Practices through Faculty Communities of Practice. In Proceedings of the 2015 ASEE Annual Conference & Exposition, Seattle, DC, USA, 14–17 June 2015.
6. Brown, J.S.; Collins, A.; Duguid, P. Situated Cognition and the Culture of Learning. *Educ. Res.* **1989**, *18*, 32–42. [CrossRef]
7. Bruckman, A. Community Support for Constructionist Learning. *Comput. Support. Coop. Work.* **1998**, *7*, 47–86. [CrossRef]

8. Bruckman, A. Learning in Online Communities. In *The Cambridge Handbook of the Learning Sciences*; Sawyer, K., Ed.; Cambridge University Press: New York, NY, USA, 2006; pp. 461–472.
9. Collins, A. How Society Can Foster Self-Directed Learning. *Hum. Dev.* **2006**, *49*, 225–228. [CrossRef]
10. Desimone, L.M. A Primer on Effective Professional Development. *Phi Delta Kappan* **2011**, *92*, 68–71. [CrossRef]
11. Lave, J.; Wenger, E. *Situated Learning: Legitimate Peripheral Participation*; Cambridge University Press: Cambridge, UK, 1991.
12. Main, K.; Pendergast, D. Core Features of Effective Continuing Professional Development for the Middle Years: A Tool for Reflection. *RMLE Online* **2015**, *38*, 1–18. [CrossRef]
13. Rogoff, B. Developing Understanding of the Idea of Communities of Learners. *Mind Cult. Act.* **1994**, *1*, 209–229.
14. Vygotsky, L.S. Socio-Cultural Theory. *Mind Soc.* **1978**, *6*, 52–58.
15. Fishman, B.J.; Davis, E.A. *Teacher Learning Research and the Learning Sciences*; Cambridge University Press: Cambridge, UK, 2006.
16. Candy, P.C. *Self-Direction for Lifelong Learning. A Comprehensive Guide to Theory and Practice*; Jossey-Bass: San Francisco, CA, USA, 1991.
17. Wenger, E. *Communities of Practice: A Brief Introduction STEP Leadership Workshop*; University of Oregon: Eugene, OR, USA, 2011.
18. Gehrke, S.; Kezar, A. The Roles of STEM Faculty Communities of Practice in Institutional and Departmental Reform in Higher Education. *Am. Educ. Res. J.* **2017**, *54*, 803–833. [CrossRef]
19. Schreurs, M.-L.; Huveneers, W.; Dolmans, D. Communities of Teaching Practice in the Workplace: Evaluation of a Faculty Development Programme. *Med. Teach.* **2016**, *38*, 808–814. [CrossRef]
20. Steinert, Y. Viewpoints-Faculty Development: Yesterday, Today, and Tomorrow. in McLean, M., Cilliers, F. and Van Wyk, J.M. Faculty development: Yesterday, today and tomorrow: AMEE Guide 33. *Med. Teach.* **2008**, *30*, 555–584.
21. Bolisani, E.; Fedeli, M.; Bierema, L.; De Marchi, V. United We Adapt: Communities of Practice to Face the CoronaVirus Crisis in Higher Education. *Knowl. Manag. Res. Pract.* **2020**, 1–5. [CrossRef]
22. Delgado, J.; de Groot, J.; McCaffrey, G.; Dimitropoulos, G.; Sitter, K.C.; Austin, W. Communities of Practice: Acknowledging Vulnerability to Improve Resilience in Healthcare Teams. *J. Med. Ethics* **2021**, *47*, 488. [CrossRef] [PubMed]
23. Delgado, J.; Siow, S.; de Groot, J.; McLane, B.; Hedlin, M. Towards Collective Moral Resilience: The Potential of Communities of Practice during the COVID-19 Pandemic and Beyond. *J. Med. Ethics* **2021**, *47*, 374–382. [CrossRef] [PubMed]
24. Amaratunga, C.A. Building Community Disaster Resilience through a Virtual Community of Practice (VCOP). *Int. J. Disaster Resil. Built Environ.* **2014**, *5*, 66–78. [CrossRef]
25. Gimenez, R.; Hernantes, J.; Labaka, L.; Hiltz, S.R.; Turoff, M. Improving the Resilience of Disaster Management Organizations through Virtual Communities of Practice: A Delphi Study. *J. Contingencies Crisis Manag.* **2017**, *25*, 160–170. [CrossRef]
26. Davies, C.; Hart, A.; Eryigit-Madzwamuse, S.; Stubbs, C.; Aumann, K.; Aranda, K.; Heaver, B. Communities of Practice in Community-University Engagement: Supporting Co-Productive Resilience Research and Practice. In *Communities of Practice*; Springer: Berlin/Heidelberg, Germany, 2017; pp. 175–198.
27. Jettpace, L.; Miller, L.; Frank, M.A.; Clemons, M.L.; Goldfarb, N. With a Little Help from Our Friends: Teaching Collectives As Lifelines in Troublesome Times. *J. Teach. Learn. Technol.* **2021**, *10*, 127–134.
28. Van Deusen-Scholl, N. The resilience of a community of practice during the COVID-19 crisis. *Second Lang. Res. Pract.* **2020**, *1*, 144–148.
29. Sadiq, K. Communities of Practice as a Multidisciplinary Response in Times of Crisis: Adapting to Successful Online Learning Practices. *Account. Res. J.* **2020**, *34*, 134–145. [CrossRef]
30. Grunspan, D.Z.; Holt, E.A.; Keenan, S.M. Instructional Communities of Practice during COVID-19: Social Networks and Their Implications for Resilience. *J. Microbiol. Biol. Educ.* **2021**, *22*, i1–i2505. [CrossRef] [PubMed]
31. Felvegi, E.; Barascout, R.; Belinne, J.K.; Sebastijanovic, M.; Miljanic, O.; del Pino Kloques, D. Remote Communities of Practice, Resilience, and Mentorship During the COVID-19 Pandemic. In *Innovate Learning Summit 2020*; Association for the Advancement of Computing in Education (AACE): Waynesville, NC, USA, 2020; pp. 90–97.
32. Burde, D.; Kapit, A.; Wahl, R.L.; Guven, O.; Skarpeteig, M.I. Education in Emergencies: A Review of Theory and Research. *Rev. Educ. Res.* **2017**, *87*, 619–658. [CrossRef]
33. Rupnow, R.L.; LaDue, N.D.; James, N.M.; Bergan-Roller, H.E. A Perturbed System: How Tenured Faculty Responded to the COVID-19 Shift to Remote Instruction. *J. Chem. Educ.* **2020**, *97*, 2397–2407. [CrossRef]
34. Johnson, N.; Veletsianos, G.; Seaman, J. US Faculty and Administrators' Experiences and Approaches in the Early Weeks of the COVID-19 Pandemic. *Online Learn.* **2020**, *24*, 6–21. [CrossRef]
35. Cook, J.P. Online Education and the Emotional Experience of the Teacher. *New Dir. Teach. Learn.* **2018**, *153*, 67–75. [CrossRef]
36. Grubic, N.; Badovinac, S.; Johri, A.M. Student Mental Health in the Midst of the COVID-19 Pandemic: A Call for Further Research and Immediate Solutions. *Int. J. Soc. Psychiatry* **2020**, *66*, 517–518. [CrossRef]
37. Wang, X.; Hegde, S.; Son, C.; Keller, B.; Smith, A.; Sasangohar, F. Investigating Mental Health of US College Students during the COVID-19 Pandemic: Cross-Sectional Survey Study. *J. Med. Internet Res.* **2020**, *22*, e22817. [CrossRef]
38. Yu, Z.; Razzaq, A.; Rehman, A.; Shah, A.; Jameel, K.; Mor, R.S. Disruption in Global Supply Chain and Socio-Economic Shocks: A Lesson from COVID-19 for Sustainable Production and Consumption. *Oper. Manag. Res.* **2021**, 1–16. [CrossRef]
39. VanLeeuwen, C.A.; Veletsianos, G.; Johnson, N.; Belikov, O. Never-Ending Repetitiveness, Sadness, Loss, and "Juggling with a Blindfold on:" Lived Experiences of Canadian College and University Faculty Members during the COVID-19 Pandemic. *Br. J. Educ. Technol.* **2021**, *52*. [CrossRef]
40. Doerfel, M.L.; Prezelj, I. Resilience in a Complex and Unpredictable World. *J. Contingencies Crisis Manag.* **2017**, *25*, 118–122. [CrossRef]

41. Southwick, S.M.; Sippel, L.; Krystal, J.; Charney, D.; Mayes, L.; Pietrzak, R. Why Are Some Individuals More Resilient than Others: The Role of Social Support. *World Psychiatry* **2016**, *15*, 77–79. [CrossRef]
42. Aldrich, D.P.; Meyer, M.A. Social Capital and Community Resilience. *Am. Behav. Sci.* **2015**, *59*, 254–269. [CrossRef]
43. Teo, W.L.; Lee, M.; Lim, W.-S. The Relational Activation of Resilience Model: How Leadership Activates Resilience in an Organizational Crisis. *J. Contingencies Crisis Manag.* **2017**, *25*, 136–147. [CrossRef]
44. Wenger-Trayner, E.; Wenger-Trayner, B. *Learning to Make a Difference: Value Creation in Social Learning Spaces*; Cambridge University Press: Cambridge, UK, 2020.
45. Abigail, L.K.M. Do Communities of Practice Enhance Faculty Development? *Health Prof. Educ.* **2016**, *2*, 61–74. [CrossRef]
46. Stark, A.M.; Smith, G.A. Communities of Practice as Agents of Future Faculty Development. *J. Fac. Dev.* **2016**, *30*, 59–67.
47. Darling, M.; Guber, H.; Smith, J.; Stiles, J. Emergent Learning: A Framework for Whole-System Strategy, Learning, and Adaptation. *Found. Rev.* **2016**, *8*, 8. [CrossRef]
48. Braun, V.; Clarke, V. Using Thematic Analysis in Psychology. *Qual. Res. Psychol.* **2006**, *3*, 77–101. [CrossRef]
49. Lincoln, Y.S.; Guba, E.G. But Is It Rigorous? Trustworthiness and Authenticity in Naturalistic Evaluation. *New Dir. Program Eval.* **1986**, *1986*, 73–84. [CrossRef]
50. Lune, H.; Berg, B.L. *Qualitative Research Methods for the Social Sciences*, 9th ed.; Pearson: London, UK, 2017.
51. Osterwlder, A.; Pigneur, Y. *Business Model Generation: A Handbook for Visionaries, Game Changers, and Challengers*; Wiley & Sons: Hoboken, NJ, USA, 2010.
52. Bjornerud, M. *Timefulness*; Princeton University Press: Princeton, NJ, USA, 2018.
53. Crutzen, P.; Foley, J. Planetary Boundaries: Exploring the Safe Operating Space for Humanity. *Ecol. Soc.* **2009**, *14*, 32.
54. Rockstrom, J.; Steffen, W.; Noone, K.; Persson, A.; Chapin III, F.S.; Lambin, E.; Lenton, T.M.; Scheffer, M.; Folke, C.; Schellnhuber, H.J. Planetary Boundaries: Exploring the Safe Operating Space for Humanity. *Ecol. Soc.* **2009**, *14*, 1–33. [CrossRef]
55. Steffen, W.; Richardson, K.; Rockström, J.; Cornell, S.E.; Fetzer, I.; Bennett, E.M.; Biggs, R.; Carpenter, S.R.; De Vries, W.; De Wit, C.A. Planetary Boundaries: Guiding Human Development on a Changing Planet. *Science* **2015**, *347*, 6223. [CrossRef] [PubMed]
56. The RACI Model. Available online: https://racichart.org/the-raci-model/ (accessed on 22 August 2021).
57. The Creative Thinking Process. Available online: https://www.creativeeducationfoundation.org/what-is-cps/ (accessed on 22 August 2021).
58. Wolfe, J. *Team Writing: A Guide to Working in Groups*; Bedford/St. Martin's: Boston, MA, USA, 2010.
59. Medin, D.; Lee, C.D.; Bang, M. Particular Points of View. *Sci. Am.* **2014**, *311*, 44–45. [CrossRef]
60. Stoddard, E.; Wobbe, K.; Bass, R. *Project-Based Learning in the First Year: Beyond All Expectations*, 1st ed.; Stylus Publishing: Sterling, VA, USA, 2019.
61. Chen, L.; Wadei, K.A.; Bai, S.; Liu, J. Participative Leadership and Employee Creativity: A Sequential Mediation Model of Psychological Safety and Creative Process Engagement. *Leadersh. Organ. Dev. J.* **2020**, *41*, 741–759. [CrossRef]

Article

The Mediating Role of Entrepreneurial Mindset between Intolerance of Uncertainty and Career Adaptability

Ahram Lee [1] and Eunju Jung [2,*]

1 Department of Education, Sejong University, Seoul 05006, Korea; leeyalan@sejong.ac.kr
2 Graduate School of Education, Sejong University, Seoul 05006, Korea
* Correspondence: doduli@sejong.ac.kr

Abstract: Entrepreneurship education has been employed broadly in higher education, and one of the most popularly targeted outcomes is enhancing entrepreneurial mindset. However, the role of entrepreneurial mindset has not been examined in relation to career adaptability, which has been acknowledged as an important resource for adjustment, especially with the increased uncertainty caused by COVID-19. The current study investigated the relations among intolerance of uncertainty—specifically its sub-factors, prospective anxiety and inhibitory anxiety—career adaptability, and entrepreneurial mindset in 274 Korean college students facing school-to-work transition during COVID-19. The study conducted path analysis and evaluated the mediating effect of entrepreneurial mindset, using a 95% bootstrapping confidence interval, to better understand the role of entrepreneurial mindset in general career development in the context of uncertainty. The results found that inhibitory anxiety had an inverse association with career adaptability and entrepreneurial mindset, while prospective anxiety had a positive relation with career adaptability and entrepreneurial mindset. Entrepreneurial mindset and career adaptability showed a significantly positive relation, and the mediating role of entrepreneurial mindset between intolerance of uncertainty and career adaptability was also supported. Discussions on the role of entrepreneurial mindset are initiated and practical implications for entrepreneurship education are proposed along with the limitations of the study.

Keywords: entrepreneurial mindset; career adaptability; intolerance of uncertainty; prospective anxiety; inhibitory anxiety; college student

Citation: Lee, A.; Jung, E. The Mediating Role of Entrepreneurial Mindset between Intolerance of Uncertainty and Career Adaptability. *Sustainability* **2021**, *13*, 7099. https://doi.org/10.3390/su13137099

Academic Editors: Jaana Seikkula-Leino, Mats Westerberg, Priti Verma and Maria Salomaa

Received: 31 May 2021
Accepted: 23 June 2021
Published: 24 June 2021

Publisher's Note: MDPI stays neutral with regard to jurisdictional claims in published maps and institutional affiliations.

1. Introduction

1.1. Background and Purpose of the Study

Research on the education and development of entrepreneurship has been gaining increased attention [1]. In modern society, in which the business environment has become more dynamic and technology is advancing rapidly, entrepreneurship, which leads to the creation of new opportunities and businesses, can lead to a revitalization of the economy [2]. Since entrepreneurial competence can be developed and learned through education [3], entrepreneurship education has expanded rapidly within the field of higher education [4].

Entrepreneurship education does not merely encourage learners to initiate their own business but helps them to acquire the necessary skills that would enable them to identify and pursue new opportunities [5]. In other words, entrepreneurship education "leads to openness to change, willingness to adapt to new situations, and ability to work in an uncertain environment" [5] (p. 216). As such, entrepreneurship education fosters important competence necessary for individuals not just to create ventures but to adjust to an unpredictable and rapidly changing environment. Therefore, entrepreneurship education has been extended to the curricula of diverse disciplines and extracurricular activities [6].

In Korea, entrepreneurship education began to gain interest following the IMF financial crisis, during which the unemployment rate rocketed and the number of young venture start-ups plummeted [7]. The government and academia invested in promoting entrepreneurship education in higher education in order to bring economic recovery, but the initial education mainly focused on venture start-ups, which has been evaluated as not leading to the actual cultivation of entrepreneurship among students [7]. Today, there is a growing emphasis on entrepreneurship education to foster factors embedded in entrepreneurship, such as innovativeness, initiative, and risk-taking [8], but the relevant research is still limited [9].

While the majority of previous studies have focused on the impact of entrepreneurship education on factors such as entrepreneurial intent, knowledge, skills, and behavior [10], increased attention is being given to entrepreneurial mindset to be considered in entrepreneurship education [11,12]. Recent studies have found that entrepreneurship education positively influences entrepreneurial mindset [13,14]. Entrepreneurial mindset refers to the cognitive ability that allows individuals to flexibly adapt to a dynamic, uncertain environment [15]. It involves creativity, innovativeness, and risk-taking necessary for adjustment, creating new values, and utilizing new opportunities [16]. These properties of the entrepreneurial mindset are applicable not just to venture creation but to general career development, and there are studies investigating entrepreneurial mindset in relation to general career-related factors such as career awareness [17] and career decision-making self-efficacy and career maturity [18]. However, there are no studies directly investigating the relation between entrepreneurial mindset and career adaptability.

Career adaptability refers to coping resources that enable individuals to tackle complex and unfamiliar problems in the context of an uncertain and unstable labor market environment [19]. Career adaptability involves competences such as planning, decision-making, exploring, and problem-solving [19], while the entrepreneurial mindset influences managing resources, making decisions, and taking control [20]. Since both entrepreneurial mindset and career adaptability are related to navigating through and adjusting to an uncertain environment to pursue one's own career, the present study intends to investigate their relationship empirically.

The current study also examines the influence of intolerance of uncertainty. The labor market has increasingly become more unpredictable with the advancement of society and technology, but this has been amplified due to the outbreak of COVID-19. As the pandemic is prolonged, global and domestic economic damage is evident, and the unpredictable labor market conditions have affected the career trajectory of many people [21]. In particular, the level of uncertainty has increased for university students who are facing school-to-work transition amidst the pandemic. In such a precarious environment, an individual's level of tolerance to uncertainty would play a critical role. Since entrepreneurial mindset and career adaptability cannot be examined separately from the context in which they manifest, it would be important to examine the influence of intolerance of uncertainty.

In sum, the current study intends to investigate the relations among intolerance of uncertainty, career adaptability, and entrepreneurial mindset. The study focuses on college students in Korea who are facing school-to-work transition in the midst of COVID-19. By examining the relations among the variables, the current study aims to provide a better understanding of the scope of influence that entrepreneurial mindset has on general career development in the context of an uncertain environment, which may lead to practical implications for entrepreneurship education focusing on cultivating an entrepreneurial mindset.

1.2. *Previous Studies and Hypotheses*

1.2.1. Intolerance of Uncertainty and Career Adaptability

The influence of intolerance of uncertainty has been gaining attention as the COVID-19 outbreak continues for over a year. Various studies have examined intolerance of uncertainty in relation to loneliness [22], mental wellbeing [23], and positivity [24]. Intolerance of uncertainty is defined as "the predisposition to react negatively to an uncertain event or situation, independent of its probability of occurrence and of its associated consequences" [25] (p. 678). Freeston et al. [26] initially developed a 27-item scale to measure intolerance of uncertainty, based on which Carleton et al. [27] created a short version consisting of 12 items with two factors, namely prospective anxiety and inhibitory anxiety. Prospective anxiety reflects the tendency to take active measures to reduce uncertainty, while inhibitory anxiety entails withdrawing and paralyzing in uncertain situations [28]. Most of the previous studies have used the total score of the two factors to assess intolerance of uncertainty, but in a study examining intolerance of uncertainty and behavioral decision making in an uncertain situation, prospective anxiety and inhibitory anxiety showed slightly different results. For example, performance on both executive functioning and risk-gaining tasks was inversely related to both inhibitory and prospective anxiety, whereas the magnitude of the relations was greater for prospective anxiety than inhibitory anxiety [29]. Such different results were also noted in another study examining whether intolerance of uncertainty predicts a startled response while anticipating temporally uncertain aversive shock; although the direction of the relations was the same for both subscales, the strength of the relation to the anticipatory startle responses was greater for inhibitory anxiety than for prospective anxiety [30]. Since prospective anxiety and inhibitory anxiety reflect rather contrasting responses to uncertainty—that is, prospective being more approach-oriented, and inhibitory being more avoidance-oriented—the two factors may have a different magnitude of influence on the variables being investigated in this study.

Although there is a limited number of studies examining the relationship between intolerance of uncertainty and career adaptability, it has been found that intolerance of uncertainty has a negative association with career adaptability [31,32]. However, these studies have not examined prospective and inhibitory anxiety separately; hence, the following hypotheses are considered in the present study.

Hypothesis 1 (H1). *Prospective anxiety, a sub-factor of intolerance of uncertainty, will have a negative correlation with career adaptability.*

Hypothesis 2 (H2). *Inhibitory anxiety, a sub-factor of intolerance of uncertainty, will have a negative correlation with career adaptability.*

1.2.2. Intolerance of Uncertainty and Entrepreneurial Mindset

Thus far, no previous study seems to have investigated the direct relationship between intolerance of uncertainty and entrepreneurial mindset. However, there are several studies that have focused on the perceived ambiguity and uncertainty of the learners, which leads to entrepreneurial learning or the development of entrepreneurial competence. Specifically, a study was conducted using mixed methods to identify emotional events and entrepreneurial competencies that are developed within these emotion-laden situations [33]. In the study, uncertainty and confusion in the learning environment was identified as one of the emotionally intense events that led to the development of entrepreneurial competencies, such as increased uncertainty/ambiguity tolerance and self-efficacy [33]. Although not explicitly mentioned, the accounts of the participants indicated that they were positively reacting to uncertain situations. In another study, ambiguity and uncertainty were created by exposing students to a learning setting in a foreign culture, and students in the ambiguity-induced situation were able to become more entrepreneurial and develop entrepreneurial self-efficacy by coping with the novelty [34]. The study implicitly indicated that negative emotional arousal such as fear is negatively associated with self-efficacy [34], which may, in turn, affect entrepreneurial learning. Although these studies did not specifically focus on individuals' intolerance of uncertainty and entrepreneurial mindset, they indicate that ambiguity and uncertainty do lead to emotional arousal, but coping well with such emotionally laden situations provides opportunities to enhance entrepreneurial competence. Since intolerance of uncertainty entails negatively reacting to uncertain situations, the following hypotheses are considered.

Hypothesis 3 (H3). *Prospective anxiety, a sub-factor of intolerance of uncertainty, will have a negative correlation with entrepreneurial mindset.*

Hypothesis 4 (H4). *Inhibitory anxiety, a sub-factor of intolerance of uncertainty, will have a negative correlation with entrepreneurial mindset.*

1.2.3. Entrepreneurial Mindset and Career Adaptability

There is growing interest and recognition in the industry as well as academia of the importance of entrepreneurial mindset [20]. Entrepreneurial mindset can be defined as "a cognitive perspective that enables an individual to create value by recognizing and acting on opportunities, making a decision with limited information, and remaining adaptable and resilience in conditions that are often uncertain and complex" [20] (p. 6). Since entrepreneurial mindset is associated with an array of facets and characteristics, there have been various attempts to assess it using different measurements, but Jung and Lee [35] developed and validated the College Students' Entrepreneurial Mindset Scale (CS-EMS) specifically to assess the entrepreneurial mindset of college students. The scale was developed to reflect the goals and outcomes of entrepreneurship education in Korean higher education, through which the five sub-factors of entrepreneurial mindset can be fostered, and the identified factors include innovativeness, need for achievement, risk-taking, autonomy, and proactiveness [35]. Since mindset is not static, but develops over time [16], it can be influenced by the environment and the context in which the individuals are situated [36]. Moreover, it can be trained, learned, and developed via education [15,37]. Thus, the influence of entrepreneurial mindset has been examined in different disciplines, such as the field of engineering [38] and creative and performing arts [39]. Accordingly, enhancing entrepreneurial mindset through entrepreneurship education would benefit individuals, within and outside of the business domain, by helping them to navigate through the challenges of the uncertain world, seeking more opportunities and creating new values.

Since entrepreneurial mindset is understood as cognitive adaptability under uncertain conditions [15], it can be viewed as a universally applicable competence that can be taught and developed [40], and there are studies examining entrepreneurial mindset in relation to other general career-related factors. For instance, Rodriguez and Lieber [41] found that entrepreneurship education led to a significant increase in entrepreneurial mindset, and the entrepreneurial mindset gains were positively associated with perceptions of future career success. Baek and colleagues [18] found a significant effect of entrepreneurship on career decision-making self-efficacy and career maturity. In a study examining the role of entrepreneurship and resilience in Korean college students [42], it was found that, among the sub-factors of entrepreneurship, innovation and risk-taking showed a positive association with challenge-taking behavior while initiative and risk-taking had a positive relation with career preparation, with resilience having a mediating role between risk-taking and challenge-taking behavior and career preparation. Thus far, however, there is no study examining the relation between entrepreneurial mindset and career adaptability.

Career adaptability is a psychosocial construct emphasized in career construction theory that refers to individuals' self-regulatory strengths and competency, which allow them to cope with vocation tasks, transitions, and traumas [19]. It is a multidimensional construct composed of four resources: concern, referring to being interested in and planning for career-related issues and challenges; control, which involves identifying the possible impact one can have on one's own career; curiosity, defined as an exploration of possible selves and career-related information; and confidence, indicating the belief that one is able to tackle career-related challenges [43,44]. Career adaptability has been found to increase job satisfaction and lower job stress [45] and have a positive relation with job search self-efficacy and employment status [46]. It also has a more general influence on wellbeing, such as happiness [47], sense of power, and life satisfaction [48], as well as responses to adversity [49]. Career adaptability has been gaining attention as the labor market becomes more complex and unpredictable, especially in the context of the prolonged COVID-19 pandemic [21]. To empirically investigate the relation between entrepreneurial mindset and career adaptability, the following hypothesis is tested in the current study.

Hypothesis 5 (H5). *Entrepreneurial mindset will have a positive relation with career adaptability.*

1.2.4. The Role of Entrepreneurial Mindset between Intolerance of Uncertainty and Career Adaptability

The mediating role of entrepreneurial mindset in the relationship between intolerance of uncertainty and career adaptability will be examined. Because intolerance of uncertainty is defined as a predisposition, it is usually viewed as a risk factor leading to various pathologies and symptoms of psychological distress, such as anxiety disorder, depression [50–52], and worry [53], and cognitive–behavioral interventions have been emphasized in dealing with intolerance of uncertainty to increase tolerance and to prevent the vicious cycle leading to worry or general anxiety disorder [54]. Although entrepreneurial mindset is not an intervention specifically designed to target intolerance of uncertainty, it is a cognitive ability that allows individuals to adjust and adapt in a precarious context. In order to add evidence to the entrepreneurial mindset research, the present study hypothesized, albeit with limited support, that such cognitive adaptability may play a mediating role, leading intolerance of uncertainty to further adaptability in the context of careers.

Hypothesis 6 (H6). *Entrepreneurial mindset will mediate the relation between prospective anxiety and career adaptability.*

Hypothesis 7 (H7). *Entrepreneurial mindset will mediate the relation between inhibitory anxiety and career adaptability.*

Figure 1 illustrates the proposed model for testing Hypotheses 1 through 7 (H1–H7). The solid lines demonstrate a direct relationship between variables (H1–H5) while the dotted lines depict the mediating effects of entrepreneurial mindset (H6–H7).

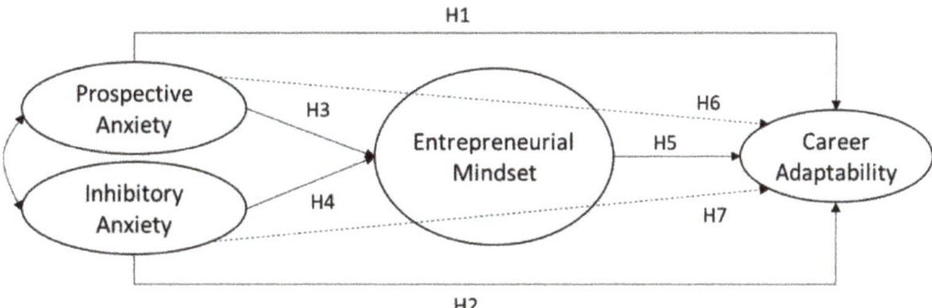

Figure 1. Proposed path analysis model for the intolerance of uncertainty, entrepreneurial mindset, and career adaptability.

2. Materials and Methods

2.1. Participants

In December 2020, when the COVID-19 was at its peak in Korea, we administered an online survey to the college students who are enrolled as the nation-wide college-student panel of an online survey institute in Korea. A screening question was used to identify juniors and seniors, and data from a total of 361 respondents were collected as a part of a large dataset designed to investigate the career development of students in a transition period. In this study, we included only those who responded that they were actively seeking a job, and the resulting number was 274. Table 1 provides the characteristics of the participants. They were 24.21 (S.D. = 1.21) years old on average, 56.6% were female (male: 43.4%) and 72.6% were seniors (juniors: 27.4%). The major composition was 35.8% of liberal arts and social sciences, 27.0% of natural sciences and engineering, 19.7% of economics and business, 9.9% of medical and pharmacology, 6.2% of arts and kinesiology, and 1.5% of undefined areas.

Table 1. Descriptive statistics of the intolerance of uncertainty items.

Subscale	Item	Mean	S.D.	Skew.	Kurt.
Prospective	IU1	2.89	0.70	−0.35	0.19
	IU2	2.94	0.68	−0.25	0.01
	IU3	3.04	0.65	−0.35	0.46
	IU4	2.83	0.73	−0.22	−0.18
	IU5	3.12	0.76	−0.41	−0.57
	IU6	2.52	0.80	0.16	−0.46
	IU7	2.74	0.78	−0.34	−0.18
Inhibitory	IU8	2.59	0.78	−0.21	−0.33
	IU9	2.66	0.75	−0.21	−0.21
	IU10	2.69	0.74	−0.32	−0.05
	IU11	2.45	0.73	−0.10	−0.31
	IU12	2.80	0.70	−0.30	0.11

2.2. Measurement

2.2.1. Intolerance of Uncertainty Scale

Carleton and colleagues [55] devised the Intolerance of Uncertainty Scale-Short Form (IUS-SF), with two factors—prospective and inhibitory anxieties. The Cronbach's αs of the whole scale, the prospective anxiety, and inhibitory anxiety were 0.91, 0.85, and 0.85, respectively [55]. In the current study, we used the Korean version of the IUS-SF, which was measured with a 4-point Likert scale (1: strongly disagree-4: strongly agree) as in Kim's study [56]. Kim reported the Cronbach's α of the whole scale, and it was 0.84.

2.2.2. College Students' Entrepreneurial Mindset Scale

Entrepreneurial mindset was measured by the College Students' Entrepreneurial Mindset Scale (CS-EMS), which was recently developed and validated by Jung and Lee [35]. The CS-EMS includes 19 items and contains 5 sub-factors of innovativeness (6 items), need for achievement (4 items), risk-taking (3 items), autonomy (3 items), and proactiveness (3 items). Each item was measured using a 5-point Likert scale (1 = strongly disagree~5 = strongly agree). Jung and Lee [35] reported that the Cronbach's α reliability coefficients were 0.88, 0.83, 0.88, 0.77, and 0.80 for innovativeness, need for achievement, risk-taking, autonomy, and proactiveness, respectively, while that of the whole scale was 0.94. They also found evidence of construct validity for the five-correlated factor model for the CS-EMS in addition to the predictive validity for start-up intention. Later, Jung and Lee [57] investigated the measurement equivalence of the CS-EMS across the groups based on gender, major, and educational experience, and they found that it holds strict invariance across gender and educational experiences while holding scalar invariance across major.

2.2.3. Career Adapt-Ability Scale (CAAS)

To measure the participants' career adaptability, we used the Career Adapt-Ability Scale [43], which was translated into Korean and validated with Korean college students [44,58]. It consists of 24 items, which were created based on four sub-constructs: concern (6 items), control (6 items), curiosity (6 items), and confidence (6 items). The response options for each item were Likert-type, with five categories (1 = not strong~5 = strongest). In Tak's study [44], the originally supported five-correlated factor model across 13 countries held for 273 Korean college students as well. The Cronbach's α reliability coefficients of the four subscales ranged from 0.80 to 0.93 in Tak's study [44] while those in Jeong's [58] ranged from 0.71 to 0.90.

2.3. Analytic Procedure

In the preliminary analysis phase, we examined the distribution of the item-level and subscale-level data and correlations among the main variables of interest (i.e., prospective anxiety (PA), inhibitory anxiety (IA), entrepreneurial mindset (EM), and career adaptability

(CA)) using Jamovi 1.2.27. In the main analysis phase, we investigated the following: (1) the psychometric properties of the intolerance of uncertainty (IU), EM, and CA in terms of construct validity and internal consistency reliability (i.e., Cronbach's α); (2) the relationships among PA, IA, EM, and CA under a structural equation modeling framework; and (3) the mediating effect of EM between either PA or IA and CA. More detailed procedures for each analysis are presented below.

2.3.1. Psychometric Analysis

Although IU, EM, and CA have been validated before, it is necessary to report sample-specific validity evidence and reliability for each empirical study [59,60]. In the current study, we examined the evidence of the construct validity for both measurements under a confirmatory factor analysis (CFA) framework. Using Jamovi version 1.6.23, we tested the two-correlated factor model for the 12 IU items based on Carleton et al.'s study [55] while testing the five-correlated factor model for EM, which was supported in Jung and Lee's study [35] (detailed information regarding the correlated-five factor model is available in Jung and Lee [57]). For CA, we tested the correlated-four factor model, which was supported by Savickas and Porfeli [43] and Tak [44]. The tested CFA models were evaluated using both chi-square (χ^2) fit statistic and alternative fit indices (CFI: the comparative fit index; RMSE: root mean square of approximation; and SRMR: the standardized root mean squared residual). However, we relied more on the alternative ones than the χ^2 fit statistic, which is prone to reject an acceptable model with minor deviation given a large sample [61,62]. We considered a CFA model adequate given the following criteria: CFI $\geq$ 0.90; RMSEA $\leq$ 0.08; and SRMR $\leq$ 0.08 [61,63,64]. After having examined the acceptability of the CFA models, Cronbach's αs were calculated for the whole scales and every subscale of the CS-EMS and CAAS to investigate internal consistency reliability using Jamovi 1.6.23.

2.3.2. Path Analysis

The relations among IU—specifically the two sub-factors of PA and IA—EM, and CA were investigated using the path analysis model under the structural equation model (SEM) framework. We applied the same criteria for the adequacy of the path model as those for the confirmatory factor analysis models: CFI $\geq$ 0.90; RMSEA $\leq$ 0.08; and SRMR $\leq$ 0.08 [61,63,64]. Then, we investigated the path coefficients in the model and R^2s.

We also tested the mediation effect of EM between either PA or IA and CA using a bootstrap confidence interval following the recommendations of Preacher and Hayes [65]. A statistically significant mediation effect is evidenced by a bootstrap confidence interval that does not include zero at a given confidence level (e.g., 95% confidence interval). All analyses were conducted using MPlus 8.0.

3. Results

3.1. Descriptive Statistics and Correlation Analysis

3.1.1. Item-Level Descriptive Statistics

Table 1 exhibits the mean, standard deviation (S.D.), skewness (Skew.), and Kurtosis (Kurt.) for each of the 12 IU items. The mean scores ranged from 2.45 (S.D. = 0.73) to 3.12 (S.D. = 0.76). For all items, the skewness (range: −0.41~0.16) and kurtosis values (range: −0.57~0.46) were within the criteria for normal distribution suggested by George [66].

Table 2 displays the descriptive statistics of EM items. The item EM14 under the autonomy subscale had the lowest mean score (Mean = 2.96; S.D. = 1.10) while the item EM4 under the innovativeness subscale had the largest mean score (Mean = 3.74; S.D. = 0.92). For EM items, the skewness values and kurtosis values were between −0.54 and 0.08 and between −0.84 and 0.35, respectively, and none of the items appeared to violate normal distribution based on the criteria in George [66].

Table 2. Descriptive statistics of the entrepreneurial mindset items.

Subscale	Item	Mean	S.D.	Skew.	Kurt.
Innovativeness	EM1	3.22	1.05	−0.10	−0.66
	EM2	3.09	1.00	0.08	−0.60
	EM3	3.63	0.80	−0.30	0.35
	EM4	3.74	0.92	−0.41	−0.20
	EM5	3.40	0.90	0.01	−0.49
	EM6	3.16	0.96	−0.01	−0.46
Need for Achievement	EM7	3.60	0.85	−0.35	−0.06
	EM8	3.24	0.94	−0.16	−0.12
	EM9	3.56	0.89	−0.31	0.01
	EM10	3.61	0.86	−0.47	0.07
Risk-taking	EM11	3.38	0.92	−0.21	−0.53
	EM12	3.19	0.93	−0.05	−0.49
	EM13	3.19	0.93	−0.07	−0.43
Autonomy	EM14	2.96	1.10	−0.02	−0.84
	EM15	3.39	1.00	−0.10	−0.73
	EM16	3.63	0.87	−0.22	−0.36
Proactiveness	EM17	3.45	0.86	−0.54	0.33
	EM18	3.25	0.91	−0.07	−0.27
	EM19	3.35	0.91	−0.29	0.04

The descriptive statistics of CA items are presented in Table 3. The item CA7 under the control subscale had the lowest mean score (Mean = 3.47; S.D. = 0.98) while the item CA2 under the concern subscale had the largest mean score (Mean = 4.15; S.D. = 0.85). The skewness and kurtosis values of them were between −0.96 and −0.23 and between −0.15 and 1.28, respectively. All the CA items could be considered normally distributed according to George's [66] criteria.

Table 3. Descriptive statistics of the career adaptability items.

Subscale	Item	Mean	S.D.	Skew.	Kurt.
Concern	CA1	3.99	0.84	−0.74	0.77
	CA2	4.15	0.82	−0.96	1.19
	CA3	3.60	0.94	−0.48	0.13
	CA4	3.68	0.89	−0.59	0.43
	CA5	3.65	0.90	−0.50	0.19
	CA6	3.95	0.88	−0.81	0.74
Control	CA7	3.47	0.98	−0.39	−0.02
	CA8	3.89	0.84	−0.55	0.26
	CA9	3.87	0.80	−0.45	0.23
	CA10	3.81	0.84	−0.48	0.21
	CA11	3.69	1.00	−0.64	0.03
	CA12	3.80	0.84	−0.55	0.43
Curiosity	CA13	3.81	0.78	−0.46	0.55
	CA14	3.73	0.88	−0.54	0.35
	CA15	3.89	0.93	−0.72	0.36
	CA16	3.85	0.82	−0.78	1.28
	CA17	3.71	0.87	−0.35	−0.15
	CA18	3.66	0.88	−0.35	−0.11
Confidence	CA19	3.65	0.82	−0.23	−0.12
	CA20	3.69	0.79	−0.33	0.14
	CA21	3.75	0.86	−0.50	0.07
	CA22	3.76	0.81	−0.47	0.37
	CA23	3.59	0.88	−0.48	0.23
	CA24	3.71	0.79	−0.48	0.50

3.1.2. Subscale-Level Descriptive Statistics and Correlation among the Subscale Scores

Table 4 provides the descriptive statistics of the main variables for the path analysis. The subscale scores were created by averaging out the scores of all items belonging to each of the subscales for IU. The scale scores for EM and CA were calculated by averaging out all the items belonging to each of the scales. The mean scores of the PA and IA were 2.86 (S.D. = 0.45) and 2.60 (S.D. = 0.56), respectively. For the EM and CA, the mean scores were 3.39 (S.D. = 0.57) to 3.79 (S.D. = 0.56), respectively. The skewness values (range: −0.61~−0.06) and the kurtosis values (range: 0.29~1.87) indicated that all variables were reasonably normally distributed [66].

Table 4. Descriptive statistics of the PA, IA, EM, and CA.

Variable	Mean	S.D.	Skew.	Kurt.
Prospective Anxiety	2.86	0.45	−0.11	0.80
Inhibitory Anxiety	2.60	0.56	−0.38	0.29
Entrepreneurial Mindset	3.39	0.57	−0.06	0.72
Career Adaptability	3.79	0.56	−0.61	1.87

The bivariate correlations among PA, IA, EM, and CA are shown in Table 5. PA had a significant bivariate correlation with both EM ($r = 0.13, p < 0.05$) and CA ($r = 0.24, p < 0.001$) in a positive direction, which was not what we had expected. However, IA was negatively correlated with both EM ($r = -0.20, p < 0.01$) and CA ($r = -0.17, p < 0.01$) as we expected. The correlation between EM and CA was statistically significant and in a positive direction ($r = 0.63, p < 0.001$).

Table 5. Bivariate correlations among the PA, IA, EM, and CA.

	Variables	1	2	3	4
1	Prospective Anxiety	-			
2	Inhibitory Anxiety	0.56 ***	-		
3	Entrepreneurial Mindset	0.13 *	−0.20 **	-	
4	Career Adaptability	0.24 ***	−0.17 **	0.63 ***	-

Note. * $p < 0.05$; ** $p < 0.01$; *** $p < 0.001$.

3.2. *Psychometric Analysis*

3.2.1. Confirmatory Factor Analysis

The confirmatory factor analysis results are shown in Table 6. Since all item-level variables were found to be normally distributed, we used the maximum-likelihood estimation method for evaluating the models and estimating the model parameters [61,67]. For the two-correlated factor model for IU, the chi-square fit statistic ($\chi^2_{(df = 51)} = 144.00, p < 0.001$) was statistically significant, whereas the other fit indices (CFI = 0.907; RMSEA = 0.054; SRMR = 0.071) consistently indicated that the model was adequate. Although the chi-square fit statistic ($\chi^2_{(df = 142)} = 439.0, p < 0.001$) indicated that the five-correlated factor model for the EM did not perfectly fit the data, the other fit indices congruently indicated that the model was acceptable (CFI = 0.906; RMSEA = 0.076; SRMR = 0.063). Similarly, the four-correlated factor model for the 24 career adaptability items was considered acceptable based on the alternative fit indices (CFI = 0.923; RMSEA = 0.065; SRMR = 0.045) even though the chi-square fit statistic was significant ($\chi^2_{(df = 244)} = 614.00, p < 0.001$). Hence, the construct validity for the Intolerance of Uncertainty Scale, College Students' Entrepreneurial Mindset Scale, and the Career Adapt-Ability Scale was supported with the sample of the current study.

Table 6. Confirmatory factor analysis results.

Scale	χ^2	df	*p*-Value	CFI	RMSEA	SRMR
Intolerance of Uncertainty	144.00	51	<0.001	0.907	0.054	0.071
Entrepreneurial Mindset	439.00	142	<0.001	0.906	0.076	0.063
Career Adaptability	614.00	244	<0.001	0.923	0.065	0.045

3.2.2. Internal Consistency Reliability

The internal consistency reliability (Cronbach's α) was 0.82 at the whole scale level for the Intolerance of Uncertainty scale, while those of the prospective and inhibitory subscales were 0.71 and 0.74, respectively. The Cronbach's α reliability coefficient of the College Students' Entrepreneurial Mindset Scale as a whole was 0.91, while those for the innovativeness, need for achievement, risk-taking, autonomy, and proactiveness subscales were 0.83, 0.80, 0.81, 0.79, and 0.77, respectively. For the Career Adapt-Ability Scale, the whole scale's Cronbach's α reliability coefficient was 0.95, while those for the concern, control, curiosity, and confidence subscales were, respectively, 0.87, 0.84, 0.82, and 0.87. All Cronbach's α reliability coefficients indicated good to excellent internal consistency and reliability [68,69].

3.3. *Path Analysis*

We used the maximum-likelihood estimation method for the path analysis model because all variables in the model were normally distributed [61,67]. The tested path model was a just-identified model in which the model fit indices were no longer meaningful.

As illustrated in Figure 2, the standardized path coefficient from PA to CA was in a positive direction (β = 0. 30; SE = 0.06, $p < 0.001$), which indicated that Hypothesis 1 was rejected. The standardized path coefficient from IA to CA was in a negative direction ($\beta = -0.23$; SE = 0.06, $p <0.001$), which supported Hypothesis 2. The standardized path coefficient from PA to EM was in a positive direction ($\beta = 0.35$; SE = 0.08, $p < 0.001$), which implied that Hypothesis 3 could not be sustained. The standardized path coefficient from IA to EM was in a negative direction ($\beta = -0.39$; SE = 0.07, $p < 0.001$), which indicated that Hypothesis 4 was supported. Finally, the standardized path coefficient from EM to CA was in a positive direction ($\beta = 0.55$; SE = 0.05, $p < 0.001$), which meant that Hypothesis 5 was supported.

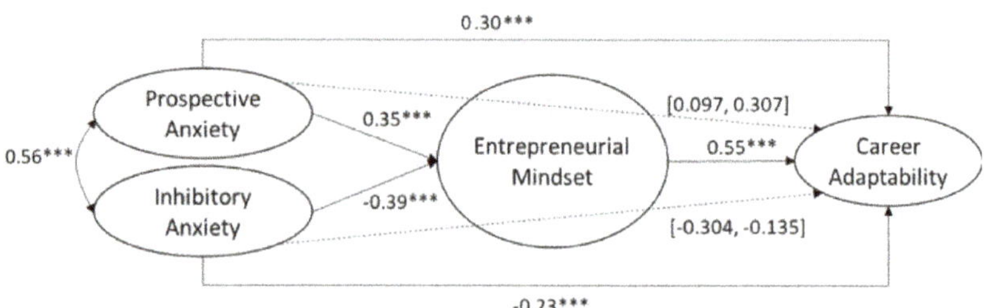

Figure 2. Path analysis results for the PA, IA, EM, and CA. ** $p < 0.01$; *** $p < 0.001$.

The R^2 of CA was 0.46, which implied that approximately 46.0% of the variance of the CA was explained by PA, IA, and EM. The R^2 of EM was 0.12, which indicated that approximately 12.9% of the variability in the EM was accounted for by PA and IA.

The standardized indirect effect of PA through EM to CA was 0.19 (SE = 0.05; $p < 0.001$) with a 95% bootstrapping confidence interval (BS-CI) of [0.097, 0.307], which indicated that EM had a significant partial mediation effect between PA and CA. Thus, Hypothesis 6 was supported. The standardized indirect effect of IA through EM to CA was -0.216 (SE = 0.04; $p < 0.001$) with a 95% bootstrapping confidence interval (BS-CI) of [-0.304,

−0.135], which indicated a significant partial mediation effect of EM between IA and CA, supporting Hypothesis 7.

4. Discussion and Conclusions

4.1. Findings and Implications

The purpose of the current study was to examine the relations among intolerance of uncertainty—consisting of prospective anxiety and inhibitory anxiety—career adaptability, and entrepreneurial mindset, with a specific focus on determining the mediating role of entrepreneurial mindset between intolerance of uncertainty and career adaptability. There are several implications of the study.

One of the most notable results was the positive correlation that prospective anxiety had with career adaptability and entrepreneurial mindset. The authors had hypothesized that the two sub-factors of intolerance of uncertainty, namely prospective anxiety and inhibitory anxiety, would have different degrees of relationship with other variables, albeit with the same directionality. Inhibitory anxiety showed an inverse association with career adaptability and entrepreneurial mindset as hypothesized, but the results for prospective anxiety were contrasting. Previous studies have found that intolerance of uncertainty has a negative association with career adaptability [31,32], as well as other positive variables such as mental wellbeing [23] and positivity [24], but these studies did not examine the sub-factors separately. The seven items of IUS-SF assessing prospective anxiety reflect individuals' inclination to actively seek information to reduce uncertainty and their preference for predictability about the future [28,55], and the responses may also have reflected the participants' level of planning or preparedness.

It should also be noted that individuals' perceived uncertainty was found to be related to the development of entrepreneurial competencies, such as increased uncertainty/ambiguity tolerance and self-efficacy [33,34]. Although these previous studies did not directly link intolerance of uncertainty and entrepreneurial mindset, it may tentatively be suggested that a certain level of fear and anxiety about the future may lead to cognitive adaptability and adaptive resources. Furthermore, in a study conducted by Reuman and colleagues [70], it was found that, when the uncertain aspects of a situation were made more explicit or obvious rather than merely implied, individuals perceived the situation as more anxiety-provoking and were more inclined to perform a safety behavior. Thus, in the current study, the participants may have perceived uncertainty as implied or tacit, perceiving their situation as being less anxiety-provoking. However, these remain the authors' tentative suggestions. Further study is needed to provide evidence and explanation for the results of the current study, regarding the positive relation between prospective anxiety and career adaptability and entrepreneurial mindset.

In the current study, it was found that entrepreneurial mindset was significantly and strongly related to career adaptability. In previous studies, entrepreneurial mindset had been mostly investigated in relation to the outcomes regarding the intention to create a new venture [71,72] or actual venture-creating-related behaviors [73–75]. Although there are few studies that have identified the relation between entrepreneurial mindset and other career-related variables, such as perceived future career success [41], career decision-making self-efficacy and career maturity [18], and career preparation [42], this is the first study, to the best of the authors' knowledge, that has directly identified the positive relationship between entrepreneurial mindset and career adaptability. This result is significant in that it provides the initial groundwork for future studies to determine the role of entrepreneurial mindset in the general career development process.

Furthermore, the mediating effect of entrepreneurial mindset between intolerance of uncertainty and career adaptability also illustrates the importance of cultivating an entrepreneurial mindset in the uncertain era. The prolonged COVID-19 pandemic may continue to impact the level of individuals' intolerance of uncertainty, which has been found to be closely related to generalized anxiety disorder or major depressive disorder [51]. Previous studies have focused on the cognitive–behavioral approach, targeting intolerance of uncertainty as a treatment for generalized anxiety disorder [76] or reduction of anxiety or depressive symptoms [77], and found the treatment to be effective. However, these cognitive–behavioral interventions targeted intolerance of uncertainty in order to treat other emotional disorders. For individuals who are not in the clinical setting, other approaches are needed in order for them to deal with uncertainty and better adapt to the changing environment. The results of the present study initiate a discussion that entrepreneurial mindset and cognitive adaptability may be an approach to enable individuals to tolerate uncertainty and adjust to the precarious world to which they are exposed in the context of their career. Thus, a well-designed entrepreneurship education focusing on cultivating an entrepreneurial mindset may work as an effective cognitive measure for individuals to gain tolerance of uncertainty and increase career adaptability.

In Korea, entrepreneurial education has been gaining interest in higher education as well as in elementary, middle, and high schools. However, education that merely focuses on fostering entrepreneurs and generating venture start-ups has not led to actual changes in the economy [7]. More emphasis is being placed on entrepreneurship education that can develop an entrepreneurial mindset that would lead to positive outcomes in general career development [57]. The current study is significant in that it provides evidence for the extended role of entrepreneurial mindset in relation to career adaptability in the context of uncertainty.

4.2. Limitations and Directions for the Future Research

The present study is limited in that it examined the relations among intolerance of uncertainty, entrepreneurial mindset, and career adaptability using cross-sectional data. In order to further investigate the causal relations among the variables, longitudinal and experimental research is necessary. Moreover, the participants of the study were restricted to Korean college students facing school-to-work transition. Although the study provides a better understanding of the role of entrepreneurial mindset in the educational and economic context of Korea, future study is required to compare and contrast different cultural aspects in order to generalize the results. Moreover, the study did not investigate the relations among the sub-factors of entrepreneurial mindset and career adaptability. Since there is a limited number of studies directly examining the relations among the variables, the initial purpose of the current study was to add to the existing literature by providing supporting evidence for their relations. However, further study should be conducted to identify the specific roles and relations that each sub-factor may have. Finally, the present study could not fully explain why prospective anxiety showed positive relations with entrepreneurial mindset and career adaptability. Although a few assumptions were made by the authors, they should empirically be tested to clarify the effect of prospective anxiety in the context of entrepreneurship education and career development.

Despite the limitations, the current study is significant in that it identified the relationship between entrepreneurial mindset and career adaptability, extending the scope of influence that entrepreneurial mindset has on general career development. In addition, it provided supporting evidence for the mediating effect of entrepreneurial mindset on the relationship between intolerance of uncertainty and career adaptability, providing practical implications for future entrepreneurship education.

Author Contributions: Conceptualization, A.L. and E.J.; funding acquisition, A.L.; methodology, A.L. and E.J.; formal analysis, A.L. and E.J.; writing—original draft preparation, A.L. and E.J. All authors have read and agreed to the published version of the manuscript.

Funding: This work was supported by internal funding for a new faculty at Sejong University, Seoul, Korea, project number 20201114.

Institutional Review Board Statement: Ethical review and approval were waived for this study because the study conducted a secondary analysis of the data already collected under the protocol approved by the Institutional Review Board Committee of Sejong University, Korea (SJU-2020-004 approved on 2 December 2020).

Informed Consent Statement: Not applicable.

Data Availability Statement: The data presented in this study are available on request from the corresponding author.

Conflicts of Interest: The authors declare no conflict of interest.

References

1. Ratten, V.; Usmanij, P. Entrepreneurship education: Time for a change in research direction? *Int. J. Manag. Educ.* **2021**, *19*, 100367. [CrossRef]
2. Schramm, C.J. Building entrepreneurial economies. *Transit. Stud. Rev.* **2005**, *12*, 163–171. [CrossRef]
3. Erikson, T. Towards a taxonomy of entrepreneurial learning experiences among potential entrepreneurs. *J. Small Bus. Enterp. Dev.* **2003**, *10*, 106–112. [CrossRef]
4. Kim, S.Y.; Yoon, S.; Lim, J.Y.; Jang, J.; Kang, M.J.; Park, H.K. Entrepreneurship perception and needs analysis of entrepreneurship education for female engineering students using an im-portance-performance analysis. *J. Eng. Educ. Res.* **2017**, *20*, 43–51.
5. Van Auken, H. Influence of a culture-based entrepreneurship program on student interest in business ownership. *Int. Entrep. Manag. J.* **2013**, *9*, 261–272. [CrossRef]
6. Mok, Y.; Choi, M. A study on curriculum design of entrepreneurship education in undergraduate school. *J. Korea Acad. Ind. Coop. Soc.* **2010**, *11*, 320–334.
7. Kim, Y.S.; Seong, S. On the contents and pedagogy of entrepreneurship education: A way of modular education based on the HAKS model. *J. Bus. Educ.* **2015**, *29*, 1–30. [CrossRef]
8. Im, J.; Kwon, H. A Study on the establishment of an integrated model for the concept of entrepreneurship. *J. Humanit. Soc. Sci. 21* **2020**, *11*, 1395–1410.
9. Kim, M.R.; Eon, W.-y. Research trends on entrepreneurship education in Korea Global creative leader. *Educ. Learn.* **2019**, *9*, 1–20. [CrossRef]
10. Nabi, G.; Liñán, F.; Fayolle, A.; Krueger, N.; Walmsley, A. The impact of entrepreneurship education in higher education: A systematic review and research agenda. *Acad. Manag. Learn. Educ.* **2017**, *16*, 277–299. [CrossRef]
11. Neneh, N.B. An exploratory study on entrepreneurial mindset in the small and medium enterprise (SME) sector: A South African perspective on fostering small and medium enterprise (SME) success. *Afr. J. Bus. Manag.* **2012**, *6*, 3364–3372.
12. Yatu, L.; Bell, R.; Loon, M. Entrepreneurship education research in Nigeria: Current foci and future research agendas. *Afr. J. Econ. Manag. Stud.* **2018**, *9*, 165–177. [CrossRef]
13. Cui, J.; Sun, J.; Bell, R. The impact of entrepreneurship education on the entrepreneurial mindset of college students in China: The mediating role of inspiration and the role of educational attributes. *Int. J. Manag. Educ.* **2019**, *19*, 100296. [CrossRef]
14. Wardana, L.W.; Narmaditya, B.S.; Wibowo, A.; Mahendra, A.M.; Wibowo, N.A.; Harwida, G.; Rohman, A.N. The impact of entrepreneurship education and students' entrepreneurial mindset: The mediating role of attitude and self-efficacy. *Heliyon* **2020**, *6*, e04922. [CrossRef]
15. Haynie, J.M.; Shepherd, D.; Mosakowski, E.; Earley, P.C. A situated metacognitive model of the entrepreneurial mindset. *J. Bus. Ventur.* **2010**, *25*, 217–229. [CrossRef]
16. Bosman, L.; Fernhaber, S. *Teaching the Entrepreneurial Mindset to Engineers*; Springer: Berlin, Germany, 2018.
17. Kim, Y.H.; Kiim, J.K. Impact of NIE on entrepreneurship and career awareness in adolescent: Mediated effect of self-efficacy. *Asian Pac. J. Bus. Ventur. Entrep.* **2016**, *11*, 153–164. [CrossRef]
18. Baek, M.J.; Park, M.S.; Kwon, K.A. A Study on the effect of entrepreneurship upon the career decision-making self-efficacy and career maturity of the entrepreneurial gifted and general students. *J. Gift. Talent. Educ.* **2017**, *27*, 431–449. [CrossRef]
19. Savickas, M.L. Career construction theory and practice. In *Career Development and Counseling: Putting Theory and Research into Work*, 2nd ed.; Lent, R.W., Brown, S.D., Eds.; Wiley: Hoboken, NJ, USA, 2013; pp. 147–183.
20. Daspit, J.J.; Fox, C.J.; Findley, S.K. Entrepreneurial mindset: An integrated definition, a review of current insights, and directions for future research. *J. Small Bus. Manag.* **2021**, 1–33. [CrossRef]
21. Restubog, S.L.D.; Ocampo, A.C.G.; Wang, L. Taking control amidst the chaos: Emotion regulation during the COVID-19 pandemic. *J. Vocat. Behav.* **2020**, *119*, 103440. [CrossRef]
22. Parlapani, E.; Holeva, V.; Nikopoulou, V.A.; Sereslis, K.; Athanasiadou, M.; Godosidis, A.; Stephanou, T.; Diakogiannis, I. Intolerance of uncertainty and loneliness in older adults during the COVID-19 pandemic. *Front. Psychiatry* **2020**, *11*, 842. [CrossRef]
23. Satici, B.; Saricali, M.; Satici, S.A.; Griffiths, M.D. Intolerance of uncertainty and mental wellbeing: Serial mediation by rumination and fear of COVID-19. *Int. J. Ment. Health Addict.* **2020**, 1–12. [CrossRef]
24. Bakioğlu, F.; Korkmaz, O.; Ercan, H. Fear of COVID-19 and positivity: Mediating role of intolerance of uncertainty, depression, anxiety, and stress. *Int. J. Ment. Health Addict.* **2020**, 1–14. [CrossRef] [PubMed]
25. Ladouceur, R.; Gosselin, P.; Dugas, M.J. Experimental manipulation of intolerance of uncertainty: A study of a theoretical model of worry. *Behav. Res. Ther.* **2000**, *38*, 933–941. [CrossRef]
26. Freeston, M.H.; Rhéaume, J.; Letarte, H.; Dugas, M.J.; Ladouceur, R. Why do people worry? *Pers. Indiv. Differ.* **1994**, *17*, 791–802. [CrossRef]

27. Carleton, R.N.; Sharpe, D.; Asmundson, G.J. Anxiety sensitivity and intolerance of uncertainty: Requisites of the fundamental fears? *Behav. Res. Ther.* **2007**, *45*, 2307–2316. [CrossRef]
28. Hale, W.; Richmond, M.; Bennett, J.; Berzins, T.; Fields, A.; Weber, D.; Beck, M.; Osman, A. Resolving uncertainty about the Intolerance of Uncertainty Scale-12: Application of modern psychometric strategies. *J. Personal. Assess.* **2016**, *98*, 200–208. [CrossRef]
29. Carleton, R.N.; Duranceau, S.; Shulman, E.P.; Zerff, M.; Gonzales, J.; Mishra, S. Self-reported intolerance of uncertainty and behavioural decisions. *J. Behav. Ther. Exp. Psychiatry* **2016**, *51*, 58–65. [CrossRef]
30. Nelson, B.D.; Shankman, S.A. Does intolerance of uncertainty predict anticipatory startle responses to uncertain threat? *Int. J. Psychophysiol.* **2011**, *81*, 107–115. [CrossRef]
31. Hwang, J.N.; Kim, E.H. The relationship between intolerance of uncertainty and career adaptability among master's-level counseling students: A mediated moderation effect of career search self-efficacy through self-reflection. *Korea J. Couns.* **2016**, *17*, 289–312.
32. Kim, S.Y.; Kwon, K. The Effects of intolerance of uncertainty on the retirement anxiety of middle-aged men: The mediating effects of emotional awareness clarity and career adaptability. *CNU J. Educ. Stud.* **2020**, *41*, 101–130. [CrossRef]
33. Lackéus, M. An emotion based approach to assessing entrepreneurial education. *Int. J. Manag. Educ.* **2014**, *12*, 374–396. [CrossRef]
34. Kubberød, E.; Pettersen, I.B. Exploring situated ambiguity in students' entrepreneurial learning. *Educ. Train.* **2017**, *59*, 265–279. [CrossRef]
35. Jung, E.; Lee, Y. Development and validation of indicators to measure entrepreneurial mindset and competency scales for college students. *Korea Educ. Rev.* **2019**, *25*, 259–287. [CrossRef]
36. Mathisen, J.-E.; Arnulf, J.K. Entrepreneurial mindsets: Theoretical foundations and empirical properties of a mindset scale. *Int. J. Manag. Bus.* **2014**, *5*, 84–107. [CrossRef]
37. Bellotti, F.; Berta, R.; de Gloria, A.; Lavagnino, E.; Antonaci, A.; Dagnino, F.; Ott, M.; Romero, M.; Usart, M.; Mayer, I.S. Serious games and the development of an entrepreneurial mindset in higher education engineering students. *Entertain. Comput.* **2014**, *5*, 357–366. [CrossRef]
38. Rae, D.; Melton, D.E. Developing an entrepreneurial mindset in US engineering education: An international view of the KEEN project. *J. Eng. Entrep.* **2017**, *7*, 1–16.
39. Pollard, V.; Wilson, E. The "entrepreneurial mindset" in creative and performing arts higher education in Australia. *Artivate* **2014**, *3*, 3–22.
40. Zupan, B.; Cankar, F.; Setnikar Cankar, S. The development of an entrepreneurial mindset in primary education. *Eur. J. Educ.* **2018**, *53*, 427–439. [CrossRef]
41. Rodriguez, S.; Lieber, H. Relationship between entrepreneurship education, entrepreneurial mindset, and career readiness in secondary students. *J. Exp. Educ.* **2020**, *43*, 277–298. [CrossRef]
42. Kim, D.S.; Yoon, J.; Lee, D.S. A mediating effects of resilience among entrepreneurship and challenge taking behavior and career preparation. *J. Bus. Educ.* **2017**, *31*, 201–236. [CrossRef]
43. Savickas, M.L.; Porfeli, E.J. Career Adapt-Abilities Scale: Construction, reliability, and measurement equivalence across 13 countries. *J. Vocat. Behav.* **2012**, *80*, 661–673. [CrossRef]
44. Tak, J. Career Adapt-Abilities Scale—Korea Form: Psychometric properties and construct validity. *J. Vocat. Behav.* **2012**, *80*, 712–715. [CrossRef]
45. Fiori, M.; Bollmann, G.; Rossier, J. Exploring the path through which career adaptability increases job satisfaction and lowers job stress: The role of affect. *J. Vocat. Behav.* **2015**, *91*, 113–121. [CrossRef]
46. Guan, Y.J.; Deng, H.; Sun, J.Q.; Wang, Y.N.; Cai, Z.J.; Ye, L.H.; Fu, R.C.Y.; Wang, Y.; Zhang, S.; Li, Y.H. Career adaptability, job search self-efficacy and outcomes: A three-wave investigation among Chinese university graduates. *J. Vocat. Behav.* **2013**, *83*, 561–570. [CrossRef]
47. Johnston, C.S.; Luciano, E.C.; Maggiori, C.; Ruch, W.; Rossier, J. Validation of the German version of the Career Adapt-Abilities Scale and its relation to orientations to happiness and work stress. *J. Vocat. Behav.* **2013**, *83*, 295–304. [CrossRef]
48. Hirschi, A. Career adaptability development in adolescence: Multiple predictors and effect on sense of power and life satisfaction. *J. Vocat. Behav.* **2009**, *74*, 145–155. [CrossRef]
49. Tian, Y.; Fan, X. Adversity quotients, environmental variables and career adaptability in student nurses. *J. Vocat. Behav.* **2014**, *85*, 251–257. [CrossRef]
50. Carleton, R.N.; Mulvogue, M.K.; Thibodeau, M.A.; McCabe, R.E.; Antony, M.M.; Asmundson, G.J. Increasingly certain about uncertainty: Intolerance of uncertainty across anxiety and depression. *J. Anxiety Disord.* **2012**, *26*, 468–479. [CrossRef]
51. Gentes, E.L.; Ruscio, A.M. A meta-analysis of the relation of intolerance of uncertainty to symptoms of generalized anxiety disorder, major depressive disorder, and obsessive-compulsive disorder. *Clin. Psychol. Rev.* **2011**, *31*, 923–933. [CrossRef]
52. Norton, P.J. A psychometric analysis of the Intolerance of Uncertainty Scale among four racial groups. *J. Anxiety Disord.* **2005**, *19*, 699–707. [CrossRef]
53. Dugas, M.J.; Freeston, M.H.; Ladouceur, R. Intolerance of uncertainty and problem orientation in worry. *Cogn. Ther. Res.* **1997**, *21*, 593–606. [CrossRef]
54. Robichaud, M.; Dugas, M.J. A cognitive-behavioral treatment targeting intolerance of uncertainty. In *Worry and Its Psychological Disorders: Theory, Assessment and Treatment*; Wiley Publishing: Hoboken, NJ, USA, 2006; pp. 289–304.

55. Carleton, R.N.; Norton, M.P.J.; Asmundson, G.J. Fearing the unknown: A short version of the Intolerance of Uncertainty Scale. *J. Anxiety Disord.* **2007**, *21*, 105–117. [CrossRef]
56. Kim, S. The Relationship of Fear of Negative and Positive Evaluation, Intolerance of Uncertainty, and Social Anxiety. Master's Thesis, Ewha University, Seoul, Korea, 2010.
57. Jung, E.; Lee, Y. College students' entrepreneurial mindset: Educational experiences override gender and major. *Sustainability* **2020**, *12*, 8272. [CrossRef]
58. Jeong, J. *The Casual Relationship of Undergraduate Students' Career Adaptability Social Support, Self-Esteem and Self-Directed Learning*; Seoul National University: Seoul, Korea, 2013.
59. Barry, A.E.; Chaney, B.; Piazza-Gardner, A.K.; Chavarria, E.A. Validity and reliability reporting practices in the field of health education and behavior: A review of seven journals. *Health Educ. Behav.* **2014**, *41*, 12–18. [CrossRef]
60. Thompson, B. ALPHAMAX: A program that maximizes coefficient alpha by selective item deletion. *Educ. Psychol. Meas.* **1990**, *50*, 585–589. [CrossRef]
61. Brown, T.A. *Confirmatory Factor Analysis for Applied Research*; Guilford Publications: New York, NY, USA, 2015.
62. Lee, K.; Lee, H. The effects of career self-efficacy in predicting the level of career attitude maturity of college students. *Korean J. Couns. Psychother.* **2000**, *12*, 127–136.
63. Barrett, P. Structural equation modelling: Adjudging model fit. *Pers. Indiv. Differ.* **2007**, *42*, 815–824. [CrossRef]
64. Hu, L.t.; Bentler, P.M. Cutoff criteria for fit indexes in covariance structure analysis: Conventional criteria versus new alternatives. *Struct. Equ. Modeling* **1999**, *6*, 1–55. [CrossRef]
65. Preacher, K.J.; Hayes, A.F. SPSS and SAS procedures for estimating indirect effects in simple mediation models. *Behav. Res. Methods Instrum. Comput.* **2004**, *36*, 717–731. [CrossRef]
66. George, D. *SPSS for Windows Step by Step: A Simple Study Guide and Reference, 17.0 Update*, 10th ed.; Pearson Education India: Noida, India, 2011.
67. Kline, R.B. *Principles and Practice of Structural Equation Modeling*; Guilford Publications: New York, NY, USA, 2015.
68. Davidshofer, K.; Murphy, C.O. *Psychological Testing: Principles and Applications*; Pearson: London, UK, 2005.
69. Nunnally, J.C. *Psychometric Theory 3E*; Tata McGraw-Hill Education: New York, NY, USA, 1994.
70. Reuman, L.; Jacoby, R.J.; Fabricant, L.E.; Herring, B.; Abramowitz, J.S. Uncertainty as an anxiety cue at high and low levels of threat. *J. Behav. Ther. Exp. Psychiatry* **2015**, *47*, 111–119. [CrossRef]
71. Pfeifer, S.; Šarlija, N.; Zekić Sušac, M. Shaping the entrepreneurial mindset: Entrepreneurial intentions of business students in Croatia. *J. Small Bus. Manag.* **2016**, *54*, 102–117. [CrossRef]
72. Mukhtar, S.; Wardana, L.W.; Wibowo, A.; Narmaditya, B.S. Does entrepreneurship education and culture promote students' entrepreneurial intention? The mediating role of entrepreneurial mindset. *Cogent Educ.* **2021**, *8*, 1918849. [CrossRef]
73. Pidduck, R.J.; Clark, D.R.; Lumpkin, G. Entrepreneurial mindset: Dispositional beliefs, opportunity beliefs, and entrepreneurial behavior. *J. Small Bus. Manag.* **2021**, 1–35. [CrossRef]
74. Saptono, A.; Wibowo, A.; Narmaditya, B.S.; Karyaningsih, R.P.D.; Yanto, H. Does entrepreneurial education matter for Indonesian students' entrepreneurial preparation: The mediating role of entrepreneurial mindset and knowledge. *Cogent Educ.* **2020**, *7*, 1836728. [CrossRef]
75. Eftekhari, N.; Bogers, M. Open for entrepreneurship: How open innovation can foster new venture creation. *Creat. Innov. Manag.* **2015**, *24*, 574–584. [CrossRef]
76. Dugas, M.J.; Ladouceur, R. Treatment of GAD: Targeting intolerance of uncertainty in two types of worry. *Behav. Modif.* **2000**, *24*, 635–657. [CrossRef]
77. Boswell, J.F.; Thompson-Hollands, J.; Farchione, T.J.; Barlow, D.H. Intolerance of uncertainty: A common factor in the treatment of emotional disorders. *J. Clin. Psychol.* **2013**, *69*, 630–645. [CrossRef]

MDPI

Article

Team Learning as a Model for Facilitating Entrepreneurial Competences in Higher Education: The Case of Proakatemia

Timo Nevalainen [1,*], Jaana Seikkula-Leino [1,2] and Maria Salomaa [1,3]

1 TAMK Business, Tampere University of Applied Sciences, 33520 Tampere, Finland; jaana.seikkula-leino@tuni.fi (J.S.-L.); maria.salomaa@tuni.fi (M.S.)
2 Department of Education, Mid-Sweden University, SE-891 22 Östersund, Sweden
3 Lincoln International Business School, University of Lincoln, Lincoln LN6 7TS, UK
* Correspondence: timo.nevalainen@tuni.fi

Abstract: In the past decades, there has been a growing interest in entrepreneurship education, and many higher education institutions have developed specific programs and courses to support entrepreneurial competencies. However, there have been significant changes in how universities train competences related to business skills and entrepreneurship in practice. Whereas entrepreneurship courses used to focus on the different forms of businesses and drafting business plans, the overall perception of entrepreneurship and entrepreneurial competences has shifted this toward a more holistic educational approach to develop students' entrepreneurial competencies. In this comparative quantitative case study, we investigate the university students' perception of the development of their entrepreneurial competencies in the case of Proakatemia (Tampere University of Applied Sciences). The aim was to examine how the entrepreneurial competencies are reflected and strengthened in their thinking and everyday functions through the concept of team learning. The survey involved, altogether, 64 students, of which 21 studied in Proakatemia. The results of this study indicate that the team learning concept of Proakatemia facilitates learning entrepreneurial competencies. Therefore, these results provide insights for universities aiming to develop their curricula, programs and pedagogy, thus promoting sustainable societal development. However, we recommend further studies, e.g., from a qualitative point of view, to assess the effective of the concept in other learning environments.

Keywords: team learning; entrepreneurship; entrepreneurship education; entrepreneurial competencies; innovation

Citation: Nevalainen, T.; Seikkula-Leino, J.; Salomaa, M. Team Learning as a Model for Facilitating Entrepreneurial Competences in Higher Education: The Case of Proakatemia. *Sustainability* **2021**, *13*, 7373. https://doi.org/10.3390/su13137373

Academic Editor: Isabel Novo-Corti

Received: 17 May 2021
Accepted: 24 June 2021
Published: 1 July 2021

Publisher's Note: MDPI stays neutral with regard to jurisdictional claims in published maps and institutional affiliations.

1. Introduction

In the past decades, there has been a growing interest towards entrepreneurship education as part of academic discussion related to the 'entrepreneurial university' (e.g., [1,2]). While the policy makers consider entrepreneurial education to be a tool to support job creation and economic growth [3], the education institutions, especially post-secondary level, have been busy developing a large range of entrepreneurship programs in practice [4]. Subsequently, the range of offered educational opportunities have increased dramatically [5,6]. More recently, there have been significant changes in how universities train competences related to business skills and entrepreneurship in practice. Whereas the entrepreneurship courses offered as part of higher education study paths used to focus on the different forms of businesses and drafting business plans—with the idea that the main task for a new entrepreneur is to utilize the business plan in applying for funds for the start-up—the overall perception on entrepreneurship and entrepreneurial competences has shifted from business ownership and stewardship towards providing students strong entrepreneurial competencies through more holistic educational approaches [6–10].

There is some previous evidence that entrepreneurship programs can indeed be effective in reinforcing entrepreneurial interests among higher education students (e.g., [8,11–13]).

However, the knowledge on the competences gained through entrepreneurship education is rather limited, and should be further investigated based on solid theoretical foundations (e.g., [3,14]). Instead of the volume of provided entrepreneurship courses, the focus of the research should be on 'holistic' and 'integrated' approaches toward entrepreneurship in both curriculum design and delivery, including exposure to role models, peer examples and applied learning [15]. Moreover, as Fayolle and Linan [16] state, the role of the university institutions and their operational (business) environment—the context—in supporting entrepreneurial behavior should be further studied as part of entrepreneurship education. This is also emphasized in the sense that societies, including education, must change in terms of sustainable development [17]. For example, how do we produce new operating models and solutions for, e.g., climate change or pandemic prevention? This requires people to have the will and ability to think and act entrepreneurially. Therefore, traditional higher education programs and delivery methods need further consideration: What do we educate and how?

As an attempt to contribute to the rather limited theoretical discussion on entrepreneurship education [3], this paper investigates team learning as a conceptual framework for supporting students' entrepreneurial competencies within higher education institutions. More specifically, we explore the efficacy of the team learning method in entrepreneurship education as adapted in the case of entrepreneurship unit entitled 'Proakatemia' located in Tampere University of Applied Sciences (TAMK) in Finland. The team learning method is based on a holistic concept perceiving human being as the basis of education; instead of more traditional education delivery models, team learning builds on a strict requirement to support the students' ownership of their own shared learning processes [18], and the goals of the team enterprise in their study paths. The quantitative study collected from Proakatemia and non-Proakatemia students at the Tampere Higher Education Community builds on Seikkula-Leino's & Salomaa's [19], and Ruskovaara et al.'s [20] previous works.

Previously, Seikkula-Leino's & Salomaa's [19] framework for entrepreneurial competencies has been used to assess entrepreneurial competencies of university staff members. The competence areas presented in the framework form the theoretical basis of our research. In this study, they are applied to the university student's perception of the development of their entrepreneurial competencies, based on the idea of how these entrepreneurial competencies are reflected and strengthened in their thinking and everyday functions through team learning within the chosen case university. By utilizing this new framework in comparing the effects of a team learning approach on entrepreneurship education to the effects of other more individually oriented approaches in the university context, this study contributes to research on pedagogical and communal factors shaping the development of entrepreneurial competencies during university studies. Previous recent research has already pointed towards the importance of several features of the learning community that are present in entrepreneurial team learning, such as entrepreneurial culture that has been found to impact entrepreneurial intentions (mediated by entrepreneurial attitude) [21], as well as psychological empowerment [22] that has an impact on the students' intrinsic motivation (e.g., [23]) and knowledge sharing behavior. Typically, entrepreneurship education within universities has focused on entrepreneurial knowledge instead of becoming an owner of a team enterprise in a learning community. Entrepreneurial knowledge has been found to have an impact on the learners' entrepreneurial intentions, but less so on their entrepreneurial mindset [24].

The paper is structured as follows: first, the theoretical framework for assessing entrepreneurial competencies [19] of higher education students is discussed in detail in the literature review together with the concept of team learning. Then, we introduce the chosen case study, and the selected methods, after which we present the main results from the survey followed by discussion, conclusion and limitations of the study. The results indicate that utilizing team learning is an effective way to foster the development of entrepreneurial competencies of higher education students. Finally, we elaborate on the theoretical and practical implications of the key findings, as well as suggesting further avenues for research.

2. Literature Review

2.1. Entrepreneurial Competencies Beyond Business Ventures

Bosman, Grard and Roegiers [25] argue that an individual, competency-based approach is among the most common structures to deliver entrepreneurship education training programs and courses. This approach digresses from what entrepreneurs are towards what they do, and what are the key competencies needed in the society. As Chandler and Jansen [26] found out in their study, the entrepreneurial competencies are essential skills for performing and succeeding well in life and work. Since then, entrepreneurship research has focused also on the psychological aspects having an impact on the entrepreneurial and enterprising processes (e.g., [27]) of individuals, and subsequently, several intention models have been proposed to shape the development of entrepreneurial competencies, such as combining personal and contextual factors as well as self-efficacy [28–31]. One of the rather widely used frameworks for analyzing the impact of entrepreneurship education is Ajzen's 'Theory of Planned Behaviour', which focuses on individual entrepreneurial intentions [32] and positive effect on the desirability and feasibility of starting a business [33]. However, currently entrepreneurship education can also have other goals beyond starting up a company. As an example, Miller & Breton-Miller [34] argue that the focus could be on the role of entrepreneurial courage and imagination fostered through entrepreneurship education, although these aspects have been underemphasized in the research literature.

Recently, new elements have been introduced to entrepreneurship education, such as integrating the concept of 'competence' to entrepreneurial learning processes (e.g., [19,35–38]). In this approach, the focus is on the process; entrepreneurial competencies should not be viewed as inputs or outputs, but as context-dependent processes of learning. Entrepreneurial processes are often iterative rather than linear [39], which means that also entrepreneurial attitudes, intentions and behavior become dynamically interrelated [40] and they may vary in different performances [41]. This dynamic runs through experimental learning, thus transforming entrepreneurial learning into a process in which knowledge is created through the transformation of experiences [42] and, according to the socio-constructivist approach, new knowledge is then created and revised (collectively) in a social context.

According to Man and Chan [43], entrepreneurial competencies consist of personal attributes, knowledge, and skills. Considering that entrepreneurship competences are highly diversified, Bacigalupo et al. [44] build an entrepreneurial competency framework that includes opportunity identification, entrepreneurial skills that represent resources, action areas, and a list of 15 competencies. Gianesini et al. [45] compared models and classifications of entrepreneurial competencies, arguing that entrepreneurial competencies include personality traits, entrepreneurial knowledge and skills. Each classification states that personal qualities are an inherent part of entrepreneurial competencies, which are the focus of this article.

According previous research [19], the underpinnings of entrepreneurial thinking and behavior involves a range of competencies, such as, (1) trust, (2) getting to know yourself, (3) cooperation, (4) learning to set goals, (5) practicing success and (6) creating pathways to future studies and working life. The model is based on Borba's [46,47] psychological and educational work of creating self-empowerment, which can also be formed in group activities (see [48–50] and through experiential learning, e.g., [42]). At first, an individual's self-esteem builds on three key elements, which are basic security, selfhood and affiliation (see [1–3]). Moreover, the environment, which involves opportunities for cooperation (see [3,6]), has an important role in their development in which one can form a more specific and realistic picture of themselves. As a result, goal setting (see 4) and success (see 5) of the individual will improve. Thus, the importance of external control gradually decreases, and the individual does not need to rely on others' opinions, but he/she becomes internally driven through self-empowerment.

All six modules of entrepreneurial learning concern the aspects of an entrepreneurial way of thinking as well as behavior, which fosters creativity, taking risks and solving problems. As an example, in terms of cooperation, a student may find new creative ways to develop friendship. In terms of developing self-trust, a student will create solutions for a problem which may arise when presenting new ideas aloud. In terms of goal-oriented behavior, students also set goals for themselves and accomplish them without being discouraged by problems and hardships. The individual is also able to find alternative solutions and minimize problems in challenging situations. Therefore, entrepreneurial behavior is a way of thinking and acting, not just about establishing a company, although this approach is also supported (e.g., working life and entrepreneurship), and can be seen as a career opportunity among other options. This is supported by, e.g., [37,51,52]. approaches to entrepreneurial learning by raising academic debate on holistic understanding of the work-life, in which individuals can find their own meaningful ways to contribute after having developed an entrepreneurial mindset.

2.2. *Team Learning Model Employed in TAMK Proakatemia*

Next, we describe team learning as a pedagogical concept and method to empower entrepreneurial learning within higher education students. Team learning is based on holistic and situated concept of human beings as the basis of education (e.g., [53,54]), in which team coaches support the students' ownership of their own learning processes and the goals of the team enterprise throughout their study path. One major difference between individual coaching and team coaching, is that in the latter, the coach will always have to encounter each student as a unique human being. At the same time, they help the students in guiding their attention and intention to relationships, and how they affect the 'good of the whole'—the team and the community. Where individual goals clash with the goals of the team or appear irrelevant in relation to those goals, the coach must also be ready to engage in dialogue with the student about his or her priorities and, if necessary, to challenge them.

The role of the team coach in professional higher education (for example, in Finnish universities of applied sciences) differs significantly from the traditional role of a lecturer in a conventional university setting. The primary task of the team coach is not to act as an expert source of information on a specific topic, but to assist the team and individual team entrepreneurs in guiding their attention and energy to the team learning processes and the goals set by the team and the individuals for themselves through dialogue [55,56]. Team coaches also give feedback with the aim of encouraging critical reflection and transformative or double-loop learning [57] where their focus is on critically examining the habituated ways of thinking and action instead of focusing solely on the student behavior or assumed knowledge.

Entrepreneurial team learning approach in TAMK Proakatemia, also referred to as 'experiential action learning' [58] is originally based on 'Tiimiakatemia' team learning methodology developed in the beginning of 1990's by Johannes Partanen in the Jyväskylä University of Applied Sciences. Team learning approach differs from other forms of action learning in that it highlights the relationship between the individual and the team, as well as the whole learning community, and the role of this relationship in the transformative learning process. In TAMK Proakatemia, the original Tiimiakatemia model has been developed further to facilitate greater student autonomy and participation in the strategic leadership of the community and the degree programme. Team enterprise in TAMK Proakatemia is a holistic and relational framework [59] in which the learning process takes place in a platform of collaborative learning and experimentation for the students. Team coaching in Proakatemia focuses on both collaborative team learning and the personal growth of individual students. Team learning environment and coaching in Proakatemia is expected to facilitate the development of skills needed in teamwork, collaborative product development and problem-solving, leadership and management, as well as sales and marketing and to help the students develop a clearer vision of their own strengths and motivations as entrepreneurs and business professionals.

2.3. Team Learning and the Development of Entrepreneurial Competencies

Team learning model implements Seikkula-Leino's [52] and Seikkula-Leino's and Salomaa's [19] approach to fostering and assessing entrepreneurial competencies in practice, focusing on both collaborative team learning and the personal growth of individual students. Team learning is a pedagogical concept and method utilized in our case study, while entrepreneurship is a unit entitled Proakatemia within Tampere University of Applied Sciences, in which students can accomplish Bachelor's Degree in Business Administration. In annual graduate feedback for universities of applied sciences, collected by Finnish Ministry of Culture and Education, Proakatemia has received consistently higher marks from its current students than any other degree program in the university [60,61]. When interviewed about their experience in Proakatemia, and how it has helped them succeed as entrepreneurs, the members of the alumni who are business owners or leaders often refer to the team as a practice environment and the development of their ability to learn new things that they need in developing their business independently through, for example, the habit of reading books which they developed in Proakatemia. Many also indicate the coaching in Proakatemia as a transformational influence in their lives. Qualitative study of Proakatemia alumni views from 2017 [62] showed that entrepreneurs graduated from Proakatemia do not necessarily view economic success as the self-evident goal of entrepreneurial activity. They may also equate success with ability to provide a living for themselves and employ others, ability to focus on a particular professional field or to create a 'workplace of dreams' for themselves and others. For the interviewed entrepreneurs and managers, money is often viewed as a means of livelihood and a reward for a job well done and taking responsibility.

For the interviewed alumni members, success as an entrepreneur has required a positive attitude toward experimentation and trying out new ideas, as well as reasonable tolerance for risk and uncertainty. They viewed the trust and encouragement by other team members as a major influence in developing their entrepreneurial attitude. The role of the coach as someone who encouraged, supported and challenged them to think for themselves and leave their 'comfort zone' was seen as instrumental in their choosing to start their own enterprises after graduation. Alongside the important influences of the team and the coach for later success as entrepreneurs and managers, the alumni members mentioned the importance of own office facilities in Proakatemia and the significance of the wider community during the studies.

Based on these experiences, we conducted a quantitative case study in order to examine if the key entrepreneurial competence areas (see Table 1) can be developed through the above-mentioned concepts and practices. In the following section, we will introduce the research setting and the methodology of our study.

Table 1. Description of entrepreneurial competencies in university context. After [19,46,47].

Competence Area	Description
Trust and Respect	There is trust between the students, academia, staff, and in the organization (university) as a whole. There is trust enough to allow mistakes that may lead to new solutions or ideas.
Everyone is Special	Students have an understanding of individual respect, and students are given the space and opportunity to act individually. Individuals are open to express their ideas and thoughts. This also promotes new, innovative ways to study and work.
Open Collaboration	A collaborative approach is encouraged in studies. Students are proud of the team spirit. Ideas are shared. Furthermore, the university does not cooperate only internally. Students are developing their external networks and communication.
Towards Goals	The achievement of personal and group goals is supported at the university. Students are encouraged to seek out new opportunities and ways of doing things to achieve goals. The community participates in decision making. Meaningful changes in a working and learning community bring improvements to the studies.
Competence and Pleasure	Students' skills are recognized, and they have an opportunity to leverage their strengths at the university. There is a feeling that students are able to positively influence one another's results. Students evaluate whether objectives have led to results. Continuing evaluation supports reaching the goals during studies and promotes the feeling of satisfaction.
Working Life and Entrepreneurship	The university supports the development of understanding of different fields and professions, and networking and partnerships with working life and the society around that. The university encourages the development/further development of ideas, solutions, services, or business ideas for customers or other target groups. Moreover, understanding and interest in entrepreneurship is shared within the university students.

3. Materials and Methods

3.1. Case Study Overview

The Finnish higher education landscape is based on a dual model consisting of research universities and universities of applied sciences as high-level vocational institutions. The Universities of Applied Sciences (UAS) are especially active in collaborative RDI activities and they collaborate with a range of different stakeholders. The Finnish UASs are considered to be significant promoters of innovation, particularly through their strong business and work-life connections. As Seikkula-Leino and Salomaa [19] observed, strong entrepreneurial competences of both staff and students would further strengthen the establishment of linkages with external partners and other collaborative initiatives.

The selected case unit Proakatemia is part of Tampere University of Applied Sciences (TAMK), which is among the largest UASs in Finland. TAMK is a multidisciplinary higher education institution, and it provides many BA and MA degree programs, especially in health and wellbeing, business studies, and technology, and has over 13,000 students and ca. 800 staff members. TAMK has strong working life connections, and the management has committed towards becoming an entrepreneurial organization [63]. TAMK's mission statement emphasizes the importance of developing collaboration as well as the higher education's societal role: 'Our strong orientation towards working life ensures the best learning possibilities for our students. Furthermore, we are involved in research, development and innovation which specifically target the development needs of working life.' In 2019, TAMK became a member of the newly established Tampere Higher Education Community, after the merger of the former University of Tampere and Tampere University of Technology in 2019. Currently, there are ca. 150 students and 12 team coaches in Proakatemia. Since being founded in 1999, Proakatemia students have started 46 team enterprises. During recent years, the overall turnover of companies operating in Proakatemia has been about EUR 1.2 million per year: For example, the community has also had a set goals for increasing the overall revenue of the team enterprises by 20% yearly from 2014 (EUR ~400,000) until 2019 (EUR ~1.2 million). However, not only business development is emphasized, but the students are encouraged to develop their literacy knowledge and reading competencies while focusing on the psychological wellbeing of the team entrepreneur students.

Daily work in TAMK Proakatemia is led by the board of leaders which consists of one team entrepreneur student from each team enterprise, the Head Coach of Proakatemia who is the only member of the teaching faculty in the board of leaders, and the Assistant Coach, a team entrepreneur student hired by TAMK to assist the coaches in managing the community and to work with the leaders of team enterprises, the Marketing and Communications Team, the International Team and the more recently formed Data Team. The strategic leadership roles of the Assistant Coach, as well as the leaders of the Marketing and Communications Team and the International Team have become more visible during the last few years. This has significantly helped to expand Proakatemia's visibility and reach both in locally and internationally.

The main forums for dialogue in Proakatemia are training sessions where whole teams participate two times three to four hours each week and which are usually planned, led, and facilitated by the team entrepreneur students themselves. The training sessions usually focus on a theme or topic that the team entrepreneur students have identified as an important one. There is a formal curriculum in Proakatemia but rather than listing the required content and topics for teaching, it outlines the entrepreneurial team learning process (based on core values of Proakatemia) and the expected learning outcomes. This way, the control over content and topics to be learned remains almost entirely with the learning community, the team enterprises and the students themselves.

3.2. Research Design and Target Group

Previous studies imply that the development of entrepreneurship by entrepreneurship education is not straightforward in an academic education [64–66], nor creating 'real entrepreneurship' [67,68]. Furthermore, there is no clear understanding of what kind of entrepreneurial competencies is needed to empower entrepreneurship by entrepreneurship education (e.g., [69]). Reconfirming past reviews and meta-analyses, it has been seen that entrepreneurship education impact research still predominantly focuses on short-term and subjective outcome measures, and also tends to severely under describe the actual pedagogies being tested [70].

These findings provided a profitable starting point for our study, allowing us to build on existing viewpoints related to entrepreneurial competencies in the context of higher education. Thus, we wanted to further investigate how different students studying in different entrepreneurial related courses or programs perceive entrepreneurialism within the TAMK university, and how these students perceive the development of their individual entrepreneurialism. It is also interesting to see how pedagogically successful entrepreneurship education programs, such as Proakatemia, supports the development of entrepreneurial competencies. All students in Proakatemia establish a company (registered as a co-operative) together with their peers, and about 40% of those students continue as entrepreneurs after their studies. Therefore, we compare how the entrepreneurial competencies have developed during university studies of students from different study programs. We have selected two groups for the comparison: (1) Proakatemia students (N 21) in their final years of their studies, and (2) non-Proakatemia students (N 43) coming from various disciplines, and having a slight interest towards entrepreneurship, since they attended a mini-course of entrepreneurship education called 'Summer Challenge', in May–June 2020. The collection of the data was implemented in May–June 2020 by two sets of SKILLOON assessment tools focusing on (1) the entrepreneurialism of university, and (2) the self-assessment of entrepreneurial competencies.

3.3. Research Questions and Assessment Tools

The research questions are following:

1. How do students assess the entrepreneurialism of their university in Tampere Higher Education Community?
    1.1. How do Proakatemia students of the entrepreneurialism of their university?
    1.2. How do other, not Proakatemia students, in the Tampere Higher Education Community assess the entrepreneurialism of their university?

   1.3. How do the assessments of Proakatemia and other students in the Tampere Higher Education Community differ?
2. How do students self-assess their entrepreneurial competencies?
   2.1. How do Proakatemia students in Tampere Higher Education Community self-assess their entrepreneurial competencies?
   2.2. How do other, not Proakatemia, students in the Tampere Higher Education Community self-assess their entrepreneurial competencies?
   2.3. How do the self-assessments of the entrepreneurial competencies of Proakatemia and other Tampere Higher Education Community students differ?

Overall, our study continues previous 'entrepreneurial organization research' implemented by SKILLOON (www.skilloon.com, accessed on 10 December 2021), which is an official education concept of Education Finland supported by the Finnish National Board of Education. SKILLOON involves assessment tools of entrepreneurial activities and a mentoring program for learners. It is created in research cooperation with schools and universities, and it is used for education and research purposes. Thus, this study builds on a series of entrepreneurial organization research, which initially took place in August–May 2020. In the first part, entrepreneurial staff competencies were studied by using Seikkula-Leino's theoretical approach as in this study, thus also applying SKILLOON assessment tools for university staff. In the first research setting, the staff assessment tools were analyzed to be reliable and valid, and the phenomenon has been examined through a multidisciplinary approach, and with a range of different assessment tools and two different respondent groups. In general, Cronbach's alpha levels varied from 0.60 to 0.95. In addition, in our previous studies, the SKILLOON assessment tools have been successfully used in the corporate world (e.g., Wihuri Group, Property Management Association, Raisio, pharmacies etc.) between 2012–2015. These individual studies confirm the reliability of the assessment tools; as an example, Cronbach's alpha levels varied in different categories between 0.67–0.96 [19]. However, there is still room for further development of the assessment tools and research design for other target groups, which is our focus in this study.

The SKILLOON assessment tool, targeted to students, has two sets of different assessment tools, each of which included six sets of research questions. The first assessment tool contained an evaluation of the different (entrepreneurial characteristics) of the organization. The second assessment tool focused on self-assessment of the students. Each of these two assessment tools contained between five to seven claims. The respondents specified their level of agreement or disagreement on a symmetric agree/disagree scale between 1–10, where 1 meant that the respondent fully disagreed with the claim, and 10 that the respondent fully agreed. Each competence area forms an individual summation notation, by calculating each respondent's mean for each set of questions. The tables below show examples of survey questions for evaluating the school (Table 2) and self-evaluation (Table 3).

Table 2. Examples of survey questions and statements.

Competence Area	Evaluation of the School (The 1st Assessment Tool)
Trust and Respect	1. Common rules are characterized by a mutual understanding between the school staff and the students. 2. There is open communication between students and the entire staff, which makes, for example, the presentation of "crazy" ideas possible. 3. A climate of mutual trust prevails between the students and the teaching staff. 4. Students can rely on promises made by the teaching staff. 5. The procedures applying to students are clear. 6. In our view, mistakes made lead to new solutions or ideas.
Everyone is Special	1. There are opportunities to point out students' knowledge or appreciation. 2. The unique features of individual students are valued and taken into account. 3. Teachers and/or friends pay attention to students' personal life (birthdays, hobbies etc.) 4. Students feel that they are valued as individuals. 5. In school, we experience the feeling that the entire community is valued. 6. In our school students have the opportunity to take risks, and there is no need to be afraid of failure.
Open Collaboration	1. Pride in the school's team spirit is clearly visible among staff and students. 2. In our school, we encourage a collaborative approach. 3. The atmosphere in our school suggests that we keep our ideas to ourselves. 4. The school staff and students want to work for the benefit of the whole school—not just for their own benefit. 5. There is a team spirit among students. 6. We actively develop cooperation with other people and organizations outside the school.

Table 3. Examples of survey questions and statements in students' self-evaluation.

Competence Area	Students' Self-Evaluation (The 2nd Assessment Tool)
Towards goals and new opportunities	1. I try to look for alternative solutions to problems. 2. I set goals for my studies. 3. I try to create an encouraging atmosphere with my friends in order to achieve better results. 4. New things related to school operations make me interested in school. 5. I think about new ways of making my studies more effective.
Competence and Pleasure	1. I am also happy to try absolutely impossible things. 2. I can also take advantage of my weaknesses. 3. After failures, I know how to direct my focus forward, towards new goals. 4. I have evaluated how effectively my set objectives have guided me towards my results. 5. I also discuss the objectives I have set with others.

3.4. The Analysis of the Data and Reliability

Analysis of variance (ANOVA) was used to examine the differences between the means of different groups. The necessary assumptions concerning the normality and uniformity of deviations were appropriately addressed. After this, a suitable method (parametric vs. non-parametric) was chosen. The reliability of every assessment tool was examined by calculating the Cronbach's alpha coefficients. Internal consistency of the assessment tools is measured with Cronbach's alpha. All the alphas are either good or excellent (Table 1), ranging from 0.63 to 0.90. The table below (Table 4) shows the measurements for the consistency of the assessment tools by Cronbach's alpha.

Table 4. Measuring the consistency of the assessment tools by Cronbach's alpha.

Student's Self-Evaluation	Cronbach's Alpha
1. Trust and respect	0.80
2. Everyone is special	0.63
3. Open collaboration	0.88
4. Towards goals	0.86
5. Competence and pleasure	0.82
6. Working life and entrepreneurship	0.83
Evaluation of the School	
1. Trust and respect	0.86
2. Everyone is special	0.90
3. Open collaboration	0.82
4. Towards goals	0.89
5. Competence and pleasure	0.90
6. Working life and entrepreneurship	0.86

3.5. *Generalization*

TAMK provides an interesting case of HEI, as it has a strategic aim to strengthen entrepreneurial skills and competencies on an organizational level. Our current case study, as a second part of 'entrepreneurial organization research' targeted to students, together with the previous case study of staff, provides a suitable platform for investigating how these organizational goals can be detected in different individual members', as staff's and students', attitudes, beliefs, and behavior. According to Cohen, Manion, and Morrison [71], the generalizability of single experiments, such as case and pilot studies, can be further extended through replication or multiple experiment strategies [72]. This allows individual case studies to contribute to the development of a growing pool of data for eventually achieving a wider generalizability. Therefore, the results obtained from this study contribute to 'analytic' rather than 'statistical' generalization to build on further studies.

4. Results

In this section, we will summarize our main research results according to each research question.

4.1. *How Do Students Assess the Entrepreneurialism of Their University in Tampere University Higher Education Community?*

Students in the Tampere Universities assess the entrepreneurialism of their university to be high with an overall score 3.19 (1 = Poor, ... , 4 = Excellent). Within the six assessment tools averages are Trust and respect 3.29, Everyone is special 3.23, Open collaboration 3.22, Towards goals 3.25, Competence and pleasure 3.14 and Working life and entrepreneurship with the lowest score 3.02.

4.1.1. How Do Proakatemia Students of Tampere Higher Education Community Assess the Entrepreneurialism of Their University?

In Table 5 and Figure 1, it can be seen that the students of Proakatemia consider the level of entrepreneurialism of their university to be very high. Within the six assessment tools averages are Trust and respect 3.63, Everyone is special 3.64, Open collaboration 3.54, Towards goals 3.61, Competence and pleasure 3.52 and Working life and entrepreneurship with the lowest score 3.27.

Table 5. Assessment of entrepreneurialism within Tampere Higher Education Community.

	Proakatemia Students, Mean	Non-Proakatemia Students, Mean	Sig.
Trust and respect	3.63	3.13	8.147×10^{-6} ***
Everyone is special	3.64	3.03	1.727×10^{-6} ***
Open collaboration	3.54	3.07	1.135×10^{-5} ***
Towards goals	3.61	3.07	1.028×10^{-5} ***
Competence and pleasure	3.52	2.96	9.3×10^{-6} ***
Working life and entrepreneurship	3.27	2.90	0.002729 **

Note: ** $p < 0.01$. *** $p < 0.001$.

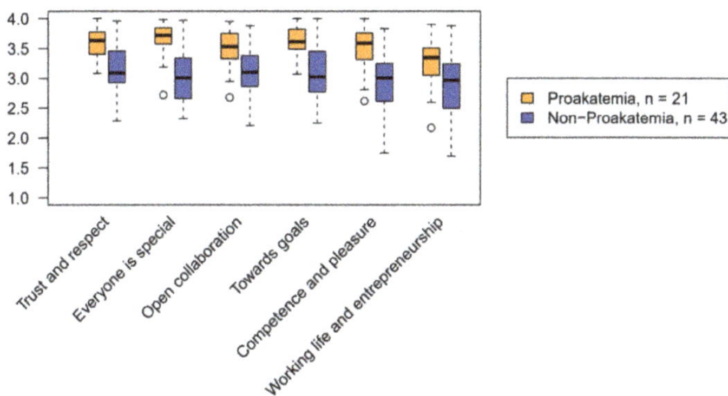

Figure 1. Evaluation of the school ($n = 64$).

4.1.2. How Do Other, Non-Proakatemia Students, in the Tampere Higher Education Community Assess the Entrepreneurialism of Their University (TAMK or Tampere University)?

Furthermore, non-Proakatemia based students assess the entrepreneurialism of their university to be rather high. In the Table 5 can be seen that the averages of each assessment tool are 3.13, 3.03, 3.07, 3.07, 2.96 and 2.90, respectively.

4.1.3. How Do the Assessments of Proakatemia and Other Students in the Tampere Higher Education Community Differ?

When assessing the entrepreneurship of the Tampere Higher Education Community, the scores depend a lot on whether the respondent is studying at the Proakatemia or not. In all the assessment tools, the differences between groups are statistically significant (Table 5). It was examined by Mann-Whitney U-test. Proakatemia students rate university's entrepreneurship at a higher level than non-Proakatemia students in all six assessment tools, although both rate university entrepreneurship as very high. The lowest mean in both groups is in the assessment tool working life and entrepreneurship (Figure 1). Proakatemia students have the average of 3.27 and non-Proakatemia students have the average of 2.90. Within Proakatemia students the highest mean (3.64) is in the assessment tool Everyone is special and within non-Proakatemia students the highest mean (3.13) is in the assessment tool Trust and respect.

When examining individual questions inside the assessment tools, altogether in two questions the averages of the two groups are almost the same ("We actively develop cooperation with other people and organizations outside the school" Proakatemia average is 3.23 and non-Proakatemia average is 3.21, and "In our school we are guided to develop our own CVs or portfolios which can be used, for example, in looking for a job.)" Proakatemia average is 3.11 and non-Proakatemia average is 3.17. Only in question "In our school we are guided to look for a job (for example jobs during weekends/holidays etc.)" the students of Proakatemia have a very low score (2.08), lower than those studying in non-Proakatemia (2.91). The biggest difference (1 or over) in averages is in questions "In our school we do brainstorming about our business ideas and/or create jobs for ourselves." Proakatemia average is 3.93 (very high score) and non-Proakatemia average is 2.72, "Together we think about new solutions and/or policies for the development of our school." Proakatemia mean 3.73 and non-Proakatemia mean 2.73 and "In our school we practice job hunting." Proakatemia average is 3.89 and non-Proakatemia average is 2.89. All of these questions mentioned above are in assessment tool Working life and entrepreneurship except the question "We actively develop cooperation with other people and organizations outside the school" is in assessment tool Open collaboration.

The result in the assessment of *Trust and respect* reflects the values that are explicitly promoted in Proakatemia. Trust is put forward in the "Value path" of Proakatemia as the basic building block of entrepreneurial team learning and the community that supports their development. The measure for "Everyone is special" appears to reflect the focus on coaching encounters with individual students, as well as the interplay of the whole team and each individual team member. Open collaboration measures could reflect the open leadership structure of Proakatemia, as well as the intention of the coaches and the students to build and maintain a safe and collaborative environment.

4.2. How Do Students Self-Assess Their Entrepreneurial Competencies?

The sum variables were formed from the responses of 64 students: Proakatemia students and non-Proakatemia students.

4.2.1. How do Proakatemia Students in Tampere Higher Education Community Self-Assess Their Entrepreneurial Competencies?

The averages of each sum variable are very high as we can see from Table 6—they all are clearly above the mid-range 2.5. The highest average is in the measurement tool *Trust and respect* and the lowest average is in measurement tool *competence and pleasure*. We conclude that students in Proakatemia assess their own entrepreneurial competencies very highly.

Table 6. Self-assessment of the entrepreneurial competencies of students (n = 64).

	Proakatemia, Mean	Non-Proakatemia, Mean	Sig
1. Trust and respect	3.47	3.16	0.0142 *
2. Everyone is special	3.29	3.21	0.3639
3. Open collaboration	3.31	3.13	0.1398
4. Towards goals	3.33	3.19	0.07621
5. Competence and pleasure	2.93	2.88	0.7443
6. Working life and entrepreneurship	3.11	3.01	0.5046

Note: * $p < 0.05$.

4.2.2. How Do Other, non-Proakatemia Students in the Tampere Higher Education Community Self-Assess Their Entrepreneurial Competencies?

In this case, the averages of each sum variable are high and above the mid-range. The highest average is in the measurement tool 'Everyone is special' and the lowest average is in the measurement tool 'Competence and pleasure'.

4.2.3. How Do the Self-Assessments of the Entrepreneurial Competencies of Proakatemia and Other Tampere Higher Education Community Students Differ?

It was investigated with analysis of variance (ANOVA) whether there were any differences between the means of Proakatemia students and non-Proakatemia students. Only statistically significant difference was in the assessment tool Trust and respect. From Table 6 and Figure 2, we can indeed agree with these results. Although the averages of Proakatemia students are slightly higher than the averages of non-Proakatemia students there is more dispersion in the answers of students not in Proakatemia which can be seen from Figure 2. This dispersion could be explained by the fact that there are significantly more respondents among the students not in Proakatemia than the students in Proakatemia.

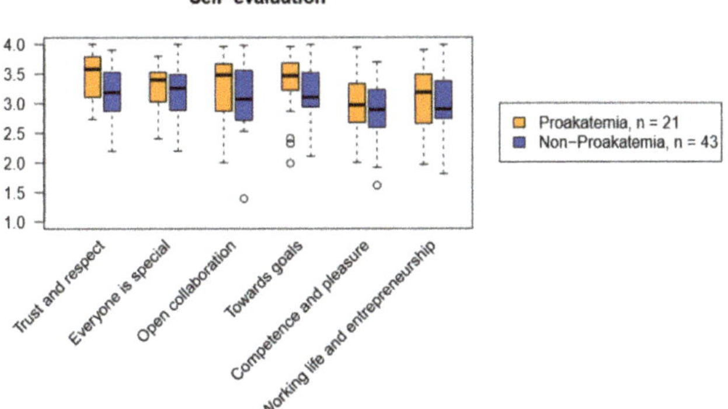

Figure 2. Self-evaluation (n = 64).

Self-evaluation assessments appear to confirm that the team learning and coaching methodology, the integrated curriculum and the open leadership structure of Proakatemia provide a solid basis for developing entrepreneurial capabilities. It would seem that those measures which are most emphasized also by the structures of the learning community (trust and respect, open collaboration and goal-orientation) are also assessed highest in the students' self-evaluations. In most measures for the non-Proakatemia group of respondents, the near opposite was true and the students in this group more often rated themselves higher than their institution in terms of entrepreneurialism.

However, the applicability of the results in other contexts may be limited because the participants of the study have a strong interest in entrepreneurship. All the participants have taken part in entrepreneurship education courses offered by Tampere Higher Education Community, and may thus be expected to be more oriented towards entrepreneurship than Finnish university students in general. Moreover, students in Proakatemia commit themselves to work in an entrepreneurial team and as an actual co-operative team enterprise for the duration of their studies (~3.5 years) which in itself indicates a significant entrepreneurial intention on their part. Therefore, we suggest further studies comparing the competencies of new students in both Proakatemia and other programs in the Tampere Higher Education Community with recent graduates from both groups. Another matter that requires further studies is the applicability of the results in multicultural groups and in other cultural contexts.

5. Discussion and Conclusions

The results indicate that the chosen case university has strong working life connections, and the management has committed towards becoming an entrepreneurial university. Our results are thus in line with our previous study, in which we investigated the entrepreneurial staff competencies at Tampere University of Applied Sciences (Seikkula-Leino & Salomaa 2020). It also suggests that the Proakatemia program is efficient in educating the students to become entrepreneurs, and to think and act entrepreneurially. Proakatemia has been running for over 15 years and the success of the program has been evident. In Proakatemia, all students establish their company, and about 40% of those students continue as an entrepreneur. Furthermore, the concept has been disseminated to Europe, Latin America and China. Even though Proakatemia evidence is "strong" in practice, we still have had concerns about its quality from the research point of view. Previous studies imply that the outcomes of entrepreneurship education are not straightforward [64–68]. Furthermore, recent studies emphasize the need of studying used pedagogy in the field to understand the phenomenon of entrepreneurship education and learning (e.g., [70]). Therefore, there was a definite need for our study; do we do "the right things" in Proakatemia entrepreneurship education (or do we only think so)? What kind of entrepreneurial competencies do we develop having team learning as our pedagogy? Then, how these Proakatemia students might differ compared to other students joining entrepreneurship education courses provided by Tampere Higher Education Community.

These questions provided a profitable starting point for our study, allowing us to build on existing viewpoints related to entrepreneurial competencies in the context of higher education. Thus, we wanted to further investigate how different students studying in different entrepreneurial related courses or programs perceive entrepreneurialism within the TAMK university, and how these students perceive their individual entrepreneurial competencies. Proakatemia curriculum and team learning method both allow the students high degree of freedom in following their own interests within the studies and enable them to dedicate time and effort on developing capabilities in the areas they feel they need to develop. The primary constraints or guiding factors in this, such as in most aspects of studies in Proakatemia, are the pressure coming from the team for each student to contribute to its success (often measured in terms of both, economic success and the development of the team and its members) and the Proakatemia curriculum which translates into personal study plan for each student. Students in Proakatemia are connected with working life organizations and entrepreneurs mainly through commercial projects where those organizations are their paying customers. This allows them to build real-life business relationships already during their studies. Proakatemia also has a very active alumni network of entrepreneurs and business professionals which provides events for dialogue and knowledge sharing, mentoring for current Proakatemia teams and individual students, as well as new business opportunities. The degree to which the current students utilize these structural affordances depends on the individual interests and personality.

Even though Proakatemia students' entrepreneurial assessments are generally very positive, it is interesting to note that Proakatemia students' individual assessments are lower compared to the other students. Given the nature of the degree program this somehow understandable; as previously described, at Proakatemia students have to challenge themselves genuinely as entrepreneurs in the business world. In this way, perhaps they do not share too idealistic perceptions of entrepreneurship change, and students look at themselves more critically. On the other hand, the added value of Proakatemia is highly evident in the results, demonstrated by the consistency in the way in which the students assessed the entrepreneurialism of the community. This is also highlighted in open collaboration measure that could reflect the open leadership structure of Proakatemia, as well as the intention of the coaches and the students to build and maintain a safe and collaborative environment. Proakatemia also has an explicit process for setting the renewed vision for the whole community every five years, and each year the coaches and the leadership team together select a specific development theme for the community.

These results are also in line with the yearly feedback from recent graduates collected on a national level also suggests that team learning model in Proakatemia provides a holistic learning environment where different elements that contribute to students' personal growth are well-balanced [60,61], while some degree programs may attain higher marks from their graduates on some aspects of the learning process, guidance and the environment, Proakatemia gets consistently high marks on measured satisfaction on learning process guidance and the environment, including the use of process and peer feedback, which supports the utilization of team learning as a driver of entrepreneurial competencies.

This is evident in the survey results, which stress developing trust and respect in cooperation within teams. This reflects the values that are explicitly promoted by Proakatemia. As previously described, trust is promoted in the 'value path' of Proakatemia as the basic building block of entrepreneurial team learning, in which the community supports its development. These findings suggest that the Proakatemia concept could be utilized not only in the development of higher education curricula, but also in other levels of education, such as leadership training and continuous education. The various educational concepts of the Proakatemia are already being utilized internationally, such as "From Teacher to Coach Training", in which traditional academic teacher role shifts towards a coach role enhancing team learning. However, further studies will be needed to validate the utilized conceptual framework of entrepreneurial learning

Based on these results, we conclude that the Proakatemia pedagogical concept facilitates the learning of the entrepreneurial competencies (e.g., [19,30,31,33,36–38,40]). Therefore, these results provide insights for universities aiming to develop their curricula, programs and pedagogy. However, we recommend further studies; for example, the conceptual framework used in this study and its evaluation could be approached from a qualitative point of view to increase knowledge on the effectivity of its' adaptations in different educational contexts.

In conclusion, this study contributes to the development of theory-based framework of entrepreneurial competencies and their pedagogical operationalization within the context of higher education. As an implication for practice, we encourage to adapt Proakatemia pedagogy and further develop it, e.g., with the higher education's own needs in mind. Furthermore, we have introduced novel assessment tools for higher education students' entrepreneurial competencies by introducing a new framework for self-evaluation, which has previously been applied to assess the development of higher education staff's entrepreneurial competencies. Overall, the research has increased our understanding of the development of entrepreneurial competencies, their realization, and their assessment and validation towards sustainable transformation within societies.

The results of the study also indicate the need for further comparative studies, including new students and recent graduates, as well as studies conducted using the framework with multicultural groups and in other cultural contexts.

Author Contributions: Conceptualization, J.S.-L., T.N. & M.S.; methodology, J.S.-L.; software, J.S.-L.; validation, J.S.-L.; formal analysis, J.S.-L.; writing—original draft preparation, T.N., J.S.-L. & M.S.; writing—review and editing, T.N. & M.S. All authors have read and agreed to the published version of the manuscript.

Funding: This research received no external funding.

Data Availability Statement: The dataset generated for this study will not be made publicly available because of the sensitive nature of the questions. All study participants were assured that the data will remain confidential and will not be shared. Therefore, all requests concerning the access to the dataset should be directed to the corresponding author.

Conflicts of Interest: The authors declare no conflict of interest.

References

1. Fayolle, A. *Handbook of Research in Entrepreneurship Education*; International Perspectives; Edward Elgar: Cheltenham, UK, 2010.
2. Gianiodis, P.T.; Meek, W.R. Entrepreneurial education for the entrepreneurial university: A stakeholder perspective. *J. Technol. Transf.* **2020**, *45*, 1167–1195. [CrossRef]
3. Rideout, E.C.; Gray, D.O. Does entrepreneurship education really work? A review and methodological critique of the empirical literature on the effects of University-Based entrepreneurship education. *J. Small Bus. Manag.* **2013**, *51*, 329–351. [CrossRef]
4. Morris, M.H.; Webb, J.W.; Fu, J.; Singhal, S. A competency-based perspective on entrepreneurship education: Conceptual and empirical insights. *J. Small Bus. Manag.* **2013**, *51*, 352–369. [CrossRef]
5. Seikkula-Leino, J.; Ruskovaara, E.; Pihkala, T.; Rodríguez, I.D.; Delfino, J. Developing entrepreneurship education in Europe: Teachers' commitment to entrepreneurship education in the UK, Finland, and Spain. In *The Role and Impact of Entrepreneurship Education*; Fayolle, A., Kariv, D., Matlay, H., Eds.; Methods, Teachers and Innovative Programmes; Edvard Elgar: Cheltenham, UK, 2019; pp. 130–145.
6. Hägg, G.; Kurczewska, A. Towards a learning philosophy based on experience in entrepreneurship education. *Entrep. Educ. Pedagog.* **2020**, *3*, 129–153. [CrossRef]
7. Ciappei, C.; Pellegrini, M.M.; Giacomo, M. Philosophy Theory Into Entrepreneurial Education Practice: A Holistic Model. *Proceedings* **2014**, *2014*, 17160.
8. Guelish, U.; Bosma, N. Youth Entrepreneurship in Asia and the Pasific 2018–2019. *GEM Global Entrepreneurship Monitor*. 2018. Available online: https://www.gemconsortium.org/file/open?fileId=50262 (accessed on 1 May 2021).

9. Guerrero, M.; Urbano, D.; Fayolle, A. Entrepreneurial activity and regional competitiveness: Evidence from European entrepreneurial universities. *J. Technol. Transf.* **2016**, *41*, 105–131. [CrossRef]
10. Hytti, U.; Stenholm, P.; Heinonen, J.; Seikkula-Leino, J. Perceived learning outcomes in entrepreneurship education-the impact of student motivation and team behaviour. *Educ. Train.* **2010**, *8*, 587–606. [CrossRef]
11. OECD. *Developing Entrepreneurship Competencies. Parallel Session 3. Policy Note*; OECD: Mexico City, Mexico, 2018. Available online: https://www.oecd.org/cfe/smes/ministerial/documents/2018-SME-Ministerial-Conference-Parallel-Session-3.pdf (accessed on 1 May 2021).
12. Vanevenhoven, J.; Ligouri, E. The impact of entrepreneurship education: Introducing the entrepreneurship education project. *J. Small Bus. Manag.* **2013**, *51*, 315–328. [CrossRef]
13. Wei, X.; Liu, X.; Sha, J. How does the entrepreneurship education influence the students' innovation? Testing on the multiple mediation model. *Front. Psychol. Educ. Psychol.* **2019**, *10*. [CrossRef]
14. Sánchez, J.C. The impact of an entrepreneurship education program on entrepreneurial competencies and intention. *J. Small Bus. Manag.* **2013**, *51*, 447–465. [CrossRef]
15. Shirokova, G.; Tsukanova, Y.; Morris, M.H. The moderating role of national culture in the relationship between university entrepreneurship offerings and student start-Up activity: An embeddedness perspective. *J. Small Bus. Manag.* **2018**, *56*, 103–130. [CrossRef]
16. Fayolle, A.; Linan, F. The future of research on entrepreneurial intentions. *J. Bus. Res.* **2014**, *67*, 663–666. [CrossRef]
17. Times Higher Education. Impact Ranking. Times Higher Education (THE). 2020. Available online: https://www.timeshighereducation.com/rankings/impact/2020/overall (accessed on 14 May 2021).
18. Nevalainen, T. Core Principles of Team Coaching in TAMK Proakatemia. *Tamk Journal*. 2020. Available online: https://tamkjournal-en.tamk.fi/core-principles-of-team-coaching-in-tamk-proakatemia/ (accessed on 1 May 2021).
19. Seikkula-Leino, J.; Salomaa, M. Entrepreneurial competencies and organisational change—Assessing entrepreneurial staff competencies within higher education institutions. *Sustainability* **2020**, *12*, 7323. [CrossRef]
20. Ruskovaara, E.; Rytkölä, T.; Seikkula-Leino, J.; Pihkala, T. *Building a Measurement Tool for Entrepreneurship Education: A Participatory Development Approach*; Edward Elgar: Cheltenham, UK, 2015.
21. Wardana, L.W.; Narmaditya, B.S.; Wibowo, A.; Fitriana; Saraswati, T.T.; Indriani, R. Drivers of entrepreneurial intention among economics students in Indonesia. *Entrep. Bus. Econ. Rev.* **2021**, *9*, 61–74. [CrossRef]
22. Fauzi, M.A.; Martin, T.; Ravesangar, K. The influence of transformational leadership on Malaysian students' entrepreneurial behaviour. *Entrep. Bus. Econ. Rev.* **2021**, *9*, 89–103. [CrossRef]
23. Ryan, R.M.; Deci, E.L. Intrinsic and extrinsic motivation from a self-determination theory perspective: Definitions, theory, practices, and future directions. *Contemp. Educ. Psychol.* **2020**, *61*, 101860. [CrossRef]
24. Karyaningsih, R.P.D.; Wibowo, A.; Saptono, A.; Narmaditya, B.S. Does entrepreneurial knowledge influence vocational students' intention? Lessons from Indonesia. *Entrep. Bus. Econ. Rev.* **2020**, *8*, 138–155. [CrossRef]
25. Bosman, C.; Grard, F.-M.; Roegiers, R. *Quel Avenir Pour Les Compétences?* De Boeck: Brussels, Belgium, 2000.
26. Chandler, G.N.; Jansen, E. The founder's self- assessed competence and venture performance. *J. Bus. Ventur.* **1992**, *7*, 223–236. [CrossRef]
27. Davidsson, P.A.W.J. Levels of analysis in entrepreneurship research: Current research practice and suggestions for the future. *Entrep. Theory Pract.* **2001**, *25*, 81–100. [CrossRef]
28. Ajzen, I. *Attitudes, Personality, and Behavior*; Dorsey: Chicago, IL, USA, 1988.
29. Ajzen, I. The theory of planned behaviour. *Organ. Behav. Human Decis. Process.* **1991**, *50*, 179–211. [CrossRef]
30. Karali, S. The Impact of Entrepreneurship Education Programs on Entrepreneurial Intentions: An Application of the Theory of Planned Behaviour. Master's Thesis, Erasmus University of Rotterdam, Rotterdam, The Netherlands, 2013.
31. Küttim, M.; Kallastea, M.; Venesaar, U.; Kiisk, A. Entrepreneurship education at university level and students' entrepreneurial intentions. *Procedia Soc. Behav. Sci.* **2014**, *110*, 658–668. [CrossRef]
32. Krueger, N.F.; Reilly, D.; Carsrud, A.L. Competing models of entrepreneurial intensions. *J. Bus. Ventur.* **2000**, *5*, 41–432.
33. Peterman, N.E.; Kennedy, J. Enterprise education: Influencing students' perceptions of entrepreneurship. *Entrep. Theory Pract.* **2003**, *28*, 129–145. [CrossRef]
34. Miller, D.; le Breton-Miller, I. Sources of entrepreneurial courage and imagination: Three perspectives, three contexts. *Entrep. Theory Pract.* **2017**, *41*, 667–675. [CrossRef]
35. Cope, J. Toward a dynamic learning perspective of entrepreneurship. *Entrep. Theory Pract.* **2005**, *29*, 373–398. [CrossRef]
36. Hayton, J.C.; Kelley, D.J. A competency- based framework for promoting corporate entrepreneurship. *Hum. Resour. Manag.* **2006**, *45*, 407–427. [CrossRef]
37. Kyrö, P.; Seikkula-Leino, J.; Mylläri, J. Meta processes of entrepreneurial and enterprising learning-the dialogue between cognitive, conative and affective constructs. In *European Research in Entrepreneurship*; Borch, J.B., Fayolle, A., Kyrö, P., Ljungren, E., Eds.; Edward Elgar: Cheltemham, UK, 2011; pp. 56–84.
38. Kyguolienė, A.; Švipa, L. Personal entrepreneurial competencies of participants in experiential entrepreneurship education. *Manag. Organ. Syst. Res.* **2020**, *82*. [CrossRef]
39. Sarasvathy, S.D.; Dew, N. Entrepreneurial logics for a technology of foolishness. *Scand. J. Manag.* **2005**, *21*, 385–406. [CrossRef]
40. Lackéus, M. *Entrepreneurship in Education-What, Why, When, How*; OECD Publishing: Paris, France, 2015.

41. Rauch, A.; Wiklund, J.; Lumpkin, G.T.; Frese, M. Entrepreneurial orientation and business performance: An assessment of past research and suggestions for the future. *Entrep. Theory Pract.* **2009**, 33, 761–784. [CrossRef]
42. Kolb, D. *Experiential Learning: Experience as the Source of Learning and Development.Englewood Cliffs*; Prentice Hall: Englewood Cliffs, NJ, USA, 1984.
43. Man, T.; Chan, T. The competitiveness of small and medium enterprises: A conceptualization with focus on entrepreneurial competencies. *J. Bus. Ventur.* **2002**, *17*, 123–142. [CrossRef]
44. Bacigalupo, M.; Kampylis, P.; Punie, Y.; Brande, G. *EntreComp: The Entrepreneurship Competence Framework*; Publication Office of the European Union: Luxembourg, 2016. [CrossRef]
45. Gianesini, G.; Cubico, S.; Favretto, G.; Leitão, G.C.C. *Entrepreneurial Competences: Comparing and Contrasting Models and Taxonomies" in Entrepreneurship and the Industry Life Cycle*; Springer International Publishing: Cham, Switzerland, 2018; Volume 6. [CrossRef]
46. Borba, M. *A K-8 Self-Esteem Curriculum for Improving Student Achievement, Bahavior and School Climate*; Jalmar Press: Torrance, CA, USA, 1989.
47. Borba, M. *Home Esteem Builders: Activities Designed to Strengthen the Partnership between Home and School*; Jalmar Press: Torrance, CA, USA, 1994.
48. Bandura, A. Self-efficacy: Toward a unifying theory of behavioral change. *Psychol. Rev.* **1977**, *84*, 191. [CrossRef]
49. Bandura, A. Self-efficacy beliefs in adolescents. In *Guide for Constructing Self-Efficacy Scales*; Pajares, F., Urdan, T., Eds.; Information Age Publishing: Greenwich, UK, 2005; pp. 307–338.
50. Bandura, A. Perceived self-efficacy in cognitive development and functioning. *Educ. Psychol.* **2010**, *28*, 117–148. [CrossRef]
51. Seikkula-Leino, J. The implementation of entrepreneurship education through curriculum reform in Finnish comprehensive schools. *J. Curric. Stud.* **2011**, *43*, 69–85. [CrossRef]
52. Seikkula-Leino, J. Developing Theory and Practice for Entrepreneurial Learning–Focus on Self-Esteem and Self-Efficacy. *Teach. Teach. Educ.* Submitted.
53. Purjo, T. *The Spiritual Capabilities of Viktor Frankl's Logotheory: Guidelines for Successful Application*; BoD-Books on Demand: Helsinki, Finland, 2020.
54. Rauhala, L. *Ihmiskäsitys Ihmistyössä*; Gaudeamus: Helsinki, Finland, 2014.
55. Isaacs, W. *Dialogue and the Art of Thinking Together: A Pioneering Approach to Communicating in Business and in Life*; Currency: New York, NY, USA, 1999.
56. Senge, P.M. *The Fifth Discipline: The Art & Practice of the Learning Organization*, Revised&Updated ed.; Doubleday: New York, NY, USA, 2006.
57. Argyris, C.; Schon, D.A. *Theory in Practice: Increasing Professional Effectiveness*, 1st ed.; Jossey-Bass: San Francisco, CA, USA, 1992.
58. Arpiainen, R.-L. *Tiimityötä, Toimintaa ja Tunteita–Näkökulmia Yrittäjyyskasvatuksesta ja Yrittäjämäisestä Oppimisesta Kokemuksellisen Toimintaoppimisen Viitekehyksessä*; Aalto University: Espoo, Finland, 2019. Available online: https://aaltodoc.aalto.fi:443/handle/123456789/37108 (accessed on 10 December 2020).
59. Gergen, K.J. *Relational Being: Beyond Self and Community*; Oxford University Press: Oxford, UK; New York, NY, USA, 2009.
60. Ministry of Ecucation and Culture. *AVOP: Ammattikorkeakoulujen Valmistumisvaiheen Opiskelijapalautekysely*; Ministry of Ecucation and Culture: Helsinki, Finland, 2019.
61. Ministry of Ecucation and Culture. *AVOP: Ammattikorkeakoulujen Valmistumisvaiheen Opiskelijapalautekysely*; Ministry of Ecucation and Culture: Helsinki, Finland, 2020.
62. Nevalainen, T.; Hautamäki, P. Proakatemian alumnien käsitykset yrittäjänä menestymisestä. In *TAMK-Konferenssi-TAMK Conference 2017*; Tampereen Ammattikorkeakoulun Julkaisuja: Tampere, Finland, 2017; pp. 210–218.
63. Tampere University of Applied Sciences. TAMK Strategy 2030. 2020. Available online: https://www.tuni.fi/sites/default/files/2020-04/tamk_strategy_2030_en.pdf (accessed on 7 April 2021).
64. Seikkula-Leino, J.; Ruskovaara, E.; Hannula, H.; Saarivirta, T. Facing the changing demands of Europe: Integrating entrepreneurship education in finnish teacher training curricula. *Eur. Educ. Res. J.* **2012**, *11*, 382–399. [CrossRef]
65. Seikkula-Leino, J.; Satuvuori, T.; Ruskovaara, E.; Hannula, H. How do finnish teacher educators implement entrepreneurship education? *Educ. Train.* **2015**, *57*, 392–404. [CrossRef]
66. Deveci, I.; Seikkula-Leino, J. A review of entrepreneurship education in teacher education. *Malays. J. Learn. Instr.* **2018**, *15*, 105–148. [CrossRef]
67. Martin, B.C.; McNally, J.J.; Kay, M.J. Examining the formation of human capital in entrepreneurship: A meta-analysis of entrepreneurship education outcomes. *J. Bus. Ventur.* **2013**, *28*, 211–2224. [CrossRef]
68. Pittaway, L.; Cope, J. Entrepreneurship education: A systematic review of the evidence. *Soc. Sci. Electron. Publ.* **2016**, *25*, 479–510. [CrossRef]
69. Graevenitz, G.; Harhoff, D.; Weber, R. The effects of entrepreneurship education. *J. Econ. Behav. Organ.* **2010**, *76*, 90–112. [CrossRef]
70. Nabi, G.; Liñán, F.; Fayolle, A.; Krueger, N.; Walmsley, A. The impact of entrepreneurship education in higher education: A systematic review and research agenda. *Acad. Manag. Learn. Educ.* **2017**, *16*. [CrossRef]
71. Cohen, L.; Manion, L.; Morrison, K. *Research Methods in Education*; Routledge: Abington, UK, 2017.
72. Yin, R.K. *Case Study Research: Design and Methods*; Sage: Thousand Oaks, CA, USA, 2003.

Article

Entrepreneurship and Sustainable Development Goals: A Multigroup Analysis of the Moderating Effects of Entrepreneurship Education on Entrepreneurial Intention

Hasbullah Ashari [1], Iffat Abbas [1,*], Asmat-Nizam Abdul-Talib [2,*] and Siti Norhasmaedayu Mohd Zamani [3]

[1] Department of Management and Humanities, Universiti Teknologi Petronas, Seri Iskandar 32610, Perak, Malaysia; hasbullah.ashari@utp.edu.my
[2] School of International Studies, Universiti Utara Malaysia, Sintok 06010, Kedah, Malaysia
[3] School of Technology Management & Logistic, Universiti Utara Malaysia, Sintok 06010, Kedah, Malaysia; edayu@uum.edu.my
* Correspondence: iffatabbas24@gmail.com (I.A.); asmat@uum.edu.my (A.-N.A.-T.); Tel.: +60-16-111-677-7961 (I.A.); +60-19-479-3664 (A.-N.A.-T.)

Abstract: The role of entrepreneurs in attaining Sustainable Development Goals (SDGs) is paramount. Entrepreneurs with strong awareness and commitment to sustainable development help to attain almost all SDGs, as they create businesses that will help employment, eliminate poverty, provide decent work and economic growth, help to reduce hunger, assist in attaining good health and wellbeing, help to achieve affordable and clean energy, and enhance their industries. Realizing the importance of entrepreneurs and entrepreneurship, the government of Malaysia has taken proactive actions to develop and inculcate the entrepreneurial mindset through entrepreneurship education at higher education. This study aims to apply the Theory of Planned Behavior (TPB) to analyze the effect of an entrepreneurship course on entrepreneurial intentions of the engineering students at Universiti Teknologi Petronas, as entrepreneurial intention is effective in predicting behavior. A quantitative technique and descriptive cross-sectional study have been employed to collect data. The result of this study indicates that the TPB explains and predicts the entrepreneurial intention. However, the Multigroup Analysis (MGA) results show that attending the entrepreneurship course does not increase the strength of the relationship between the exogenous and endogenous construct compared to those who do not attend the course. The results of this study raise a positive implication toward the improvement of the course curriculum and the teaching pedagogy. An in-depth qualitative study to understand the issue will help to improve the curriculum and pedagogy of entrepreneurship education, and eventually enable a realization of the government's aspirations.

Keywords: entrepreneurship education; entrepreneurial intention; Theory of Planned Behavior; multigroup analysis; Sustainable Development Goals

Citation: Ashari, H.; Abbas, I.; Abdul-Talib, A.-N.; Mohd Zamani, S.N. Entrepreneurship and Sustainable Development Goals: A Multigroup Analysis of the Moderating Effects of Entrepreneurship Education on Entrepreneurial Intention. *Sustainability* **2022**, *14*, 431. https://doi.org/10.3390/su14010431

Academic Editor: Jaana Seikkula-Leino

Received: 19 November 2021
Accepted: 28 December 2021
Published: 31 December 2021

1. Introduction

The paradigm of education is shifting from traditional towards entrepreneurial university due to the importance of education as one of the main promoters of Sustainable Development (SD) [1,2]. The importance of entrepreneurs and entrepreneurship to influence the overall wellbeing of humans has attracted the attention of governments, and they have taken proactive actions to develop and inculcate entrepreneurial thought, particularly through education [3,4]. The interest is evident in the rapid development of entrepreneurship curricula and education programs since the early 80s [5]. In the case of Malaysia, which is the focus of this study, the government has included in its 2015–2025 Malaysia Education Blueprint a requirement for academic programs in higher institutions to include an entrepreneurship course or education as part of the curriculum. The intention is to produce graduates who can create jobs, instead of graduates who can only search for jobs [6,7].

Since Entrepreneurship Education (EE) was introduced into higher education in Malaysia, many studies have been done to measure the effectiveness of the entrepreneurship programs. Although the themes of most studies are on the effectiveness of the EE, the scopes and the emphases are different. They emphasize different aspects, such as the effects of demographics [8] or specific students' characteristics (need for achievement, locus of control, propensity to take risk, self-confidence, tolerance of ambiguity and uncertainty, and leadership) [9]; emotional intelligence dynamics [9]; risk thinking and self-efficacy [10,11]; and proactive personality [10,11]. There are also previous studies on the effect of EE on entrepreneurial intention [12–17]. However, there is a fundamental issue that has not been very much discussed in the literature related to entrepreneurial education which is the effect of entrepreneurial knowledge on the development of the entrepreneurial behavior of the students. Does knowledge influence behavior? Studies by researchers, including Krueger [18], Piperopoulos and Dimov [19], and Sabah [20], on the effect of knowledge on behavior in studies related to the "Theory of Planned Behavior" (TPB) find that knowledge does not influence the behavior [18,20]. Ajzen [21,22] developed the TPB, and it has been broadly used in multiple academic perspectives for predicting and understanding human behavior [22,23].

The fundamental question raised is significant since it will determine the right approach to the teaching of entrepreneurship courses in universities. Shall the course syllabus focus on imparting the entrepreneurial knowledge only, or shall it use other approaches that will enhance the development of the entrepreneurial behavior? These questions motivate this study to analyze the effect of the entrepreneurship course on entrepreneurial behavior. Since behavior is not something that can be observed immediately, this study focusses on the development of entrepreneurial intention among the students. Intention has been shown to consistently be able to predict behavior [22,23]. The performance of a behavior is determined by the individual's intention to engage in it [22,23].

The effectiveness of the entrepreneurship course is measured through the intention of the students to be an entrepreneur or to be self-employed. This is in alignment with the aspiration of the government to produce graduates that can create job opportunities. In doing so, this study employs the TPB model to investigate the influence of the entrepreneurship course towards the entrepreneurial intention. The entrepreneurship course acts as a moderator in this study, since, according to the TPB, the external variable does not influence directly, except through the attitudinal variables—the entrepreneurial attitude, the subjective norm, and the perceived behavior control [5,24]. Therefore, the current study has employed the TPB to the effect of the entrepreneurial course towards the entrepreneurial intention. Multigroup analysis (MGA) has been employed to analyze the effects of the moderator on the entrepreneurial intention. MGA is suggested by. Hair et al. [24] for the categorical moderators that influence the relationship of all the independent variables and the dependent variable simultaneously.

The study is conducted on the students that have taken the entrepreneurship course at Universiti Teknologi Petronas, Malaysia Universiti Teknologi Petronas is one of the universities in Malaysia that has introduced the entrepreneurship course as a compulsory course in its academic curriculum. The introduction is in alignment with the government aspiration. Since the introduction of the course in this university, there has not been any study performed to analyze the effectiveness of the course in enhancing students' inclination or intention to set-up a business or to venture into a business upon their graduations.

This study, with its focus on the role of entrepreneurial knowledge, contributes to the enrichment of knowledge in the study of the effectiveness of entrepreneurship education. The outline of the remaining paper is as follows: the proceeding section explains the literature review, which includes the theoretical framework and hypotheses development, and the conceptual framework of this study. The following section explains the methodology part, followed by the results and discussion. Furthermore, the last part shows the conclusions, significance of this study, and future recommendations.

2. Literature Review

2.1. Previous Studies on Entrepreneurship Education in Higher Education in Malaysia

Malaysia, like other developed and developing nations, realized the importance of entrepreneurship education, and started to introduce entrepreneurship courses in the middle of the 1990s. Serious initiatives have been taken to foster entrepreneurship at all levels of education. Malaysian higher education institutions offer entrepreneurship as a course, as a specialization under the undergraduate business program, or as a degree program. There is a growing trend in Malaysia to blend or amalgamate different programs [24,25]. Entrepreneurship is one of the key engines of economic growth, as described in the Malaysia Plans and the New Economic Model Strategy. This focus on entrepreneurship as a core factor of development has contributed to a rising in entrepreneurship study. Cheng et al. [16] state that despite growing interest in entrepreneurship decisions to become entrepreneurs, motives for establishing a new venture are very low in students after graduation, and there is a need to study the effectiveness of the subject and teaching methods [24,25].

There have been many studies performed in this country to analyze the effectiveness of entrepreneurship education in higher education. Although the theme of the studies is the effectiveness of entrepreneurship education, the scope and focus of most of the studies vary. The scope of the studies is mostly confined to a university or group of universities. For example, researchers Din et al. [13] performed a study on Malaysian students in Universiti Utara Malaysia to gauge the effectiveness of entrepreneurship education programs. They examined the effectiveness of the study concerning business planning, risk thinking, and self-efficacy [25–27]. Similarly, Cheng et al. [16] measured effectiveness as an intention to start a business. Similarly Othman et al. [12] examined the perception of public university students towards entrepreneurship education. Understanding the role of entrepreneurship towards entrepreneurship intention is crucial for universities and policy-makers, as entrepreneurship intention is the best predictor of entrepreneurial behavior [27,28]. Table 1 below lists past literature related to the study of the effectiveness of EE.

Worldwide, many studies have employed using the TPB to study entrepreneurial intention in students in different fields. The TPB has been used to study entrepreneurial intention in nursing students in South Korea [28]. Similarly, another study has been performed in Greek universities to study the entrepreneurial intention among business students [28,29]. The current study has also employed the TPB to study entrepreneurial intention among students at University Technology PETRONAS, Malaysia. The following section has discussed the theoretical foundation of the current study.

2.2. Theoretical Framework

Theory of Planned Behavior (TPB)

Ajzen [21,29] developed the "Theory of Planned Behavior" (TPB), and, since then, it has been used broadly in multiple academic perspectives for predicting and understanding human behavior [23,29]. For example, marketing researchers and social psychologists have effectively utilized TPB-based models for practical applications and fundamental studies [7,29]. According to the theory, human behavior is determined by intentions. Intentions, in turn, are influenced by three attitudinal constructs: "Attitude", "Subjective Norm", and "Perceived Behavioral Control" [21,29].

Instead of the actual rate of venture creation, which is the more accurate measure of the effectiveness of the EE, intentions have been used widely to measure the effectiveness [5,29]. This is because, first, it is not practical to measure the actual rate of venture creation, since a long timelapse between the students graduating and becoming entrepreneurs could weaken the reliability of the study. Many other factors could affect the decision, and, in the end, weaken the relationship between education and entrepreneurial behavior [5,29]. Second, the entrepreneurship course affects intentions, as well as behavior, even implicitly, through attitude changes [21,29]. The TPB provides a complete model of the relationship of the attitudinal constructs to intention. Thus, through the TPB, the effect of entrepreneurship education on intention could be determined.

Table 1. Summary of previous studies on the effectiveness of entrepreneurship education in Malaysia.

Authors	Summary
Jumaatbin Mahajar [7]	This study explores the inclination towards entrepreneurship among university Pendidikan Sultan Idris students. It studies the relationship of two demographic factors—qualification and program of study, and their relationship with the inclination towards entrepreneurship.
Keat [8]	This study examines the relationship between entrepreneurship education and inclination towards entrepreneurship. The influence of demographic characteristics and family business background on university students' inclination towards entrepreneurship is also examined. The study is on three public universities.
Mohamed [9]	This paper evaluates the effectiveness of the Basic Student Entrepreneurial Program (BSEP) among local university graduates who have undergone the training program in entrepreneurship development. Three variables under the study are the origin of the participant, the presence of family members already involved in entrepreneurial activities, and educational background. The students are from all over Malaysia.
Othman [11]	This study focuses on the emotional intelligence dynamics that foresee the choice of entrepreneurship as a career. Entrepreneurs with high emotional intelligence typically manage their emotions efficiently, and make decisions and implement actions wisely. The samples are from 21 Public Universities in Selangor.
Din [12]	The study evaluates the effectiveness of entrepreneurship education programs on Malaysian university students. It stresses on the importance of risk thinking and self-efficacy. The sample is from a public university in Malaysia.
Mustafa [13]	The study analyses the effects of students' proactive personality, and the university support environment (education support, concept development support, and business development support) on entrepreneurial intentions. The population is a private university in Malaysia.
Ambad [14]	This study uses the TPB to analyze the relationships of perceived educational support, perceived relational support, perceived structural support, personal attitude, and perceived behavioral control with entrepreneurial intention. The sample is a public university.
Che Embi [17]	The study explores the effects of students' entrepreneurial characteristics (need for achievement, locus of control, propensity to take risk, self-confidence, tolerance of ambiguity and uncertainty, and leadership) on their propensity to become entrepreneurs in Malaysia. The samples are students from the International Islamic University of Malaysia (IIUM).
Mohd Ariff [18]	This study analyzes the relationship of entrepreneurial attitudes with the entrepreneurial intention through the TPB.

In this study, entrepreneurial attitude refers to the personal desirability in becoming an entrepreneur [5,29]. It is manifested in the high expectations and beliefs towards self-employment. Generally, the intention to perform a behavior depends on the personal attitude towards that behavior [20,30]. On the other hand, the subjective norm is the belief that is based on the perception of whether people will approve or disapprove of a behavior. Specifically, it is an individual's perception of what most people of importance think of them performing the behavior [30,31]. Perceived behavior control is the belief of the ease or difficulty of performing the behavior of interest. This reflects the perceived feasibility of performing the behavior. The belief is based on the perceptions of the ability to have the required resources, opportunities, and skills to perform the behavior. The effects of entrepreneurial attitude, subjective norm, and perceived behavior lead to the formation of intention. In this study, the entrepreneurial intention is defined as a cognitive condition that will direct individual attention and action towards self-employment [31–35].

3. Conceptual Model and Hypotheses

The whole discussion of the TPB, as well as the relationship between the entrepreneurship course and entrepreneurial intention can be summarized into the theoretical framework seen in Figure 1. The entrepreneurship course is an exogenous variable that acts as a moderator. It influences entrepreneurial intention through the three attitudinal variables: entrepreneurial attitude, subjective norm, and perceived behavior.

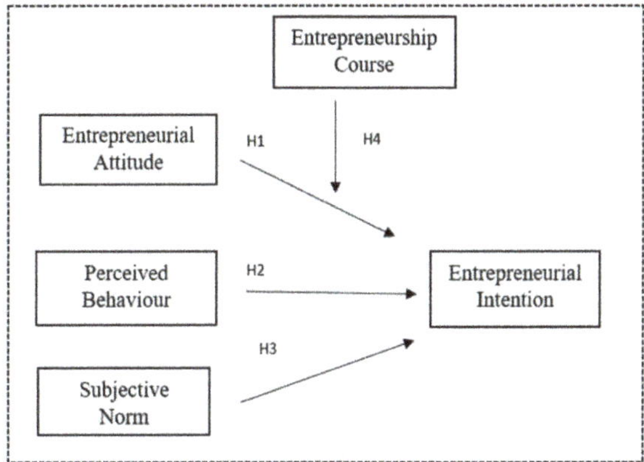

Figure 1. The theoretical framework.

Since the entrepreneurship course affects the intention only indirectly through the three attitudes (attitude towards self-employment, subjective norm, and perceived behavioral control), it is essential and fundamental to confirm or to establish the relationship of the said constructs in the TPB model and theory. Hence, the following hypotheses are suggested for this purpose:

Hypothesis 1 (H1). *Students' high attitude towards self-employment positively influences their entrepreneurial intention.*

Hypothesis 2 (H2). *Students' high subjective norm towards self-employment positively influences their entrepreneurial intention.*

Hypothesis 3 (H3). *Students' high perceived control towards self-employment positively influences their entrepreneurial intention.*

Hypothesis 4 (H4). *The effect of entrepreneurship education on attitude, subjective norm, and perceived behavioral control is stronger for the group of students that have attended the entrepreneurship course than for those who have not attended.*

4. Methodology

A descriptive cross-sectional study has been employed to collect data to answer the research questions, as well as to fulfill the purpose of the study. The quantitative technique has been chosen for this study to collect data and to analyze them, and to explain the phenomena of this study (e.g., descriptive, correlation, and inferential statistics) [36]. This study focused on UTP students that have taken the entrepreneurship course, and those who have not taken the course. Most of the students (83%) are from various fields of engineering. The remaining are information science, management and humanities, and geoscience undergraduate students. They must take the entrepreneurship course as a compulsory university course. This study employs a purposive sampling technique, since it fits the purpose of this research, to determine the effect of the entrepreneurship course on entrepreneurial intention. GPower is used to calculate the minimum sample size needed, and the test suggested a minimum sample size of 111 to achieve the power of 0.95. The same minimum sample size also applies to the two groups for the Multigroup Analysis (MGA). The survey has been administrated online, including e-mail and WhatsApp. The respondents have been informed about the purpose of the research. They have been given informed consent prior to their responses. Many students haven't shown interest to fill their response. This approach ensures the method is less sensitive to biases, as no interviewers are involved in the process, and it gives a sense of privacy and flexible time to the respondents [33,34,36]. The data analysis process has been done using SMART-PLS 3.0 software, and interpreted with descriptive and correlative statistics. In this study, the total sample size is 230 (after removal of the outliers). For the Multigroup Analysis (MGA), the total number of respondents that have attended the entrepreneurship course is 111, and 121 respondents have not attended the course. All these numbers equal and surpass the suggested minimum sample size. There are more male students (60%) compared to female students (40%). This, however, reflects the population of the university, which has more male students compared to female students [37,38]. For the students' academic programs, the highest percentage of the respondents come from mechanical engineering (23%) and chemical engineering (22%). They are followed by electrical and electronic engineering (16%), civil engineering (14%), computer and information science (12%), and petroleum engineering (8%). The lowest percentages are from geoscience (1%), and management and humanities (4%). These distributions of the respondents also reflect the actual composition of the number of students for each program. Management and humanities has just started its undergraduate degree program, and the total number of students are less than 30 students. Meanwhile, geoscience has the fewest undergraduate students among the established undergraduate program in UTP.

Questionnaire and Measure of the Constructs

The questions for entrepreneurial attitude are adopted from Heuer and Kolvereid [5], which consists of four items, measured by a seven-point Likert-type scale ranging from 1 = completely disagree, to 7 = totally agree. Subjective norm questions are adopted from Heuer and Kolvereid [5], which consists of six items, measured by a seven-point Likert-type scale ranging from 1 = completely disagree, to 7 = totally agree. The items for perceived behavioral control and entrepreneurial intention have been adopted from Heuer and Kolvereid [5], which consists of a seven-point Likert-type scale and a five-point Likert-type scale, respectively, ranging from 1 = completely disagree, to 7 = totally agree. Entrepreneurship course construct is a categorical variable that asks the question of "yes" or "no". The categorical question allows the construct to test its effect on all the attitudinal variables at once to understand its indirect effect on the entrepreneurial intention. Relevant demographics of the respondents are asked to reflect the characteristics of the UTP students, such as the respondent's gender, race, age, the discipline of study, and marital status, along with several control questions, such as whether the respondents have parents as entrepreneurs, the respondents' experience as entrepreneurs, and their close relatives as entrepreneurs.

5. Analysis, Findings, and Discussion

This section deals with the analysis of the measurement and the structural model. The measurement model deals with the indicators and their relationship with the latent variables [39–42]. The discussion then proceeds with the findings of the structural model assessment. The structural model assessment examines the relationship of the inner model or, in other words, the relationship of the exogenous variables and the endogenous variables.

5.1. Measurement Model

The measurement model deals with the indicators and their relationship with the latent variables. The fundamental purpose of this stage is to evaluate the assumptions related to reliability and validity. Reliability is measured through internal consistency or reliability (Cronbach's Alpha (CA) and Composite Reliability (CR)). The convergent validity is measured through the Average of Variance Extracted (AVE). Table 2 below summarized the results of the assessment of all the above measurements. According to the results in Table 2, all the factor loadings, AVE, CR, and CA, for the complete model, and the groups "Attend Class" and "Did not Attend Class", are above the threshold values. "Attend Class" refers to the respondents that have attended the entrepreneurship course, and "Did not Attend Class" refers to respondents who have not attended the class. The cross-loadings indicate that all the items load more strongly on their constructs in the model (refer to Table 2).

Table 2. Construct reliability and validity.

		Full Sample (230)				Attend Class (109)				Did Not Attend Class (122)			
Construct	Items	Loadings	AVE	CR	CA	Loadings	AVE	CR	CA	Loadings	AVE	CR	CA
Entrepreneurial Attitude	EA1	0.867	0.77	0.93	0.90	0.891	0.74	0.92	0.88	0.823	0.79	0.94	0.91
	EA2	0.895				0.911				0.869			
	EA3	0.884				0.882				0.885			
	EA4	0.858				0.867				0.852			
Entrepreneurial Intention	EI1	0.891	0.85	0.95	0.91	0.902	0.84	0.94	0.91	0.884	0.85	0.95	0.91
	EI2	0.935				0.942				0.925			
	EI3	0.934				0.925				0.943			
Perceived Behavior Control	BC1	0.923	0.81	0.93	0.88	0.935	0.76	0.91	0.84	0.899	0.83	0.94	0.90
	BC2	0.911				0.918				0.889			
	BC3	0.86				0.876				0.825			
Subjective Norm	SN1	0.912	0.82	0.97	0.96	0.941	0.74	0.94	0.93	0.865	0.87	0.98	0.97
	SN2	0.928				0.932				0.915			
	SN3	0.937				0.944				0.93			
	SN4	0.879				0.927				0.789			
	SN5	0.888				0.914				0.834			
	SN6	0.891				0.934				0.803			

5.2. Structural Model and Hypothesis Testing

The structural model, or the inner model, is based on the relationship between the latent variables in the form of exogeneous and endogenous variables. The structural model path coefficients for the direct model have been assessed to determine the strength and the significance of the relationship of the exogenous and endogenous variables. Estimated path coefficients close to +1 or −1 indicate strong positive or negative relationships accordingly. Results for this study indicate that all the relationships are significant, and variable attitude has the largest positive relationship with the endogenous variable, intention, followed by behavioral control, and finally by the subjective norm. Hence, H1, H2, and H3 are all supported. The path coefficients with the *p*-values are as in Table 3.

Table 3. Structural model path coefficients for the complete model.

Hypotheses		Std Beta	STDEV	t-Value	Decision
H1	Attitude -> Intention	0.447	0.071	6.291	Accept
H2	Behavioral Control -> Intention	0.301	0.065	4.643	Accept
H3	Subjective Norm -> Intention	0.189	0.071	2.661	Accept

5.3. The Multigroup Analysis for the Moderator Effects

Multigroup analysis (MGA) has been performed to analyze the path coefficient between the groups, and, eventually, in testing the hypothesis. MGA is suggested by Hair et al. [43] for the categorical moderators that influence the relationship of all the independent variables and the dependent variables simultaneously. However, before MGA can be employed, the Measurement Invariance of Composites (MICOM) test needs to be run first. This is because group comparisons can be inaccurate unless researchers establish the invariance of their measures. The variations in the structural relationships between latent variables could be because of different interpretations or understandings of a phenomenon being measured by different groups rather than due to differences in structural relationships [44]. The following sections discuss the procedures of MICOM and MGA.

Measurement Invariance of Composites (MICOM)

The MICOM procedure consists of three step-by-step tests. They are the configural invariance assessment, compositional invariance assessment, and the assessment of equal mean value and variance across groups. If the configural invariance cannot be established in step one, then step two cannot be proceeded. Similarly, if, in step two, the compositional invariance assessment cannot be established, then step three cannot be proceeded.

Step 1: Configural Invariance

Configural invariance involves qualitative assessment of the composites' specification across all the groups. The objectives are to ensure identical indicators per measurement model, identical data treatment, and identical algorithm settings or optimization criteria. In this study, similar measurement model indicators have been employed to each group ("Attend Class" and "Did not Attend Class"), as can be observed in the above discussion of the measurement model. Identical data treatment has been observed, as data for both groups have been treated similarly. The coding process, and data handling, such as the treatment of the missing value and the determination of the outliers, have been treated similarly for each group. Finally, identical algorithm settings or optimization criteria has been observed, since both groups are going through similar path model estimation methods and similar structural assessment methods as have been discussed in the previous sections. As a result of MICOM's Step 1, it can be concluded that configural invariance has been established.

Step 2: Compositional Invariance

The next step is to assess the compositional invariance, which focuses on analyzing whether a composite is formed equally across the groups. In Step 2, a permutation test has been performed. The result of the 5000 permutations is in the table below. Compositional invariance exists if the original correlation is greater than or equal to five (5) percent quantiles. In this case, as shown in Table 4 below, the compositional invariance has been established, since all original correlations are greater or equal to 5% quantiles.

Step 3: Assessment of the Equality of Composite Mean Values and Variances.

In Step 3, the composites' equality of mean values and variances across groups has been assessed. There are two conditions that need to be met. The first is the mean original difference must fall between 5% boundaries and 95% boundaries. The second is the variance original difference must fall between the 5% and 95% boundaries. If each condition is met, then there is a full invariance. However, if only one of the above conditions is met, then there is a partial invariance. Finally, if none of the conditions above are met, then there is no invariance. Results of Step 3 show that the above conditions have not been met. The results, shown in Table 5 below, do not support the invariance condition. Hence, MGA will be run to analyze the path coefficient between the two groups.

Table 4. Compositional invariance.

	Original Correlation	Correlation Permutation Mean	5.00%	Permutation p-Values
Attitude	1	1	0.999	0.172
Behavioral Control	1	1	0.998	0.535
Intention	1	1	1	0.974
Subjective Norm	1	1	0.999	0.614

Table 5. Composites' equality of mean values and variance.

	Mean Original Difference (Attend Course (1.0)–Did Not Attend (0.0)	Mean Permutation Mean Difference (Attend Course (1.0)–Did Not Attend (0.0)	5.00%	95.00%	Permutation p-Values	Variance Original Difference (Attend Course (1.0) Did Not Attend (0.0))	Variance Permutation Mean Difference (Attend Course (1.0) Did Not Attend (0.0))	5.00%	95.00%	Permutation p-Values
Attitude	0.178	0.001	−0.215	0.215	0.086	−0.32	−0.003	−0.291	0.29	0.033
Behavioural Control	0.371	0.001	−0.22	0.214	0.002	−0.509	−0.001	−0.282	0.28	0.001
Intention	0.307	0	−0.221	0.219	0.01	−0.178	−0.004	−0.266	0.243	0.131
Subjective Norm	0.418	0	−0.215	0.214	0	−0.441	0.001	−0.28	0.288	0.005

Finally, Multigroup Analysis (MGA) has been performed to analyze the path coefficient between the groups, and, eventually, in testing the hypothesis. The results are shown in Table 6 below.

Table 6. Multigroup analysis.

Hypotheses		Path Coefficients-Diff (Attend Course (1.0)–Did Not Attend (0.0))	p-Value Original 1-Tailed (Attend Course (1.0) vs. Did Not Attend (0.0))	p-Value New (Attend Course (1.0) vs. Did Not Attend (0.0))	Decisions
	Attitude -> Intention	−0.308	0.981	0.038	
	Behavioral Control -> Intention	0.112	0.193	0.385	Rejected
	Subjective Norm -> Intention	0.158	0.165	0.33	

From the results, it can be concluded, in general, that attending the entrepreneurship course ("Attend Course") does not increase the magnitude of the positive relationship of the independent variables on dependent variables. In the first relationship of entrepreneurial attitude towards entrepreneurial intention, the path coefficient difference between the two groups ("Attend Course" and "Did Not Attend Course") is negative (−0.308) and significant (p-value = 0.038). For the second relationship between behavioral control and intention, the path difference is positive (0.112), but not significant (p-value = 0.385). In the last relationship between subjective norm and intention, the path coefficient difference between the two groups is also not significant (p-value = 0.33). Thus, H4 is rejected.

6. Conclusions

This research aims to analyze the effect of an entrepreneurship course on entrepreneurial intention through exogenous attitudinal constructs. The respondents are generally second year students, and the majority of them are engineering undergraduate students. This is a quantitative study that employs structural equation modeling (SEM) to analyze the results. Based on the results of the analysis, a few conclusions can be drawn. Firstly, consistent with many other studies from various disciplines, the results of this study also show that the Theory of Planned Behavior (TPB) explains and predicts entrepreneurial intention. The results of the effects of entrepreneurial attitude, perceived behavior control, and subjective norms on entrepreneurial intention are also consistent with most of the previous studies [43–47]. All the variables have significant direct effects on the intention. Based on the results, personal attitude is the main factor affecting entrepreneurial intention, followed by behavior control, and the least influential is the subjective norms, and this is consistent with previous studies [47,48]. Similarly, subjective norms, in general, show small or insignificant effects, and provide inconsistent results.

Secondly, the MGA results show that attending the entrepreneurship course does not increase the strength of the relationship between the exogenous and endogenous construct compared to those who do not attend the course. In other words, the entrepreneurship course does not moderate the relationship between the exogenous and endogenous constructs. Furthermore, the group that has not attended the class ("Did Not Attend Course") shows a stronger positive relationship between attitude and entrepreneurial intention compared to the group that has attended the course ("Attend Course"). The findings of the current study are consistent with a study published in Sweden by Gabriel Linton and Markus Klinton [48]. They stated that an entrepreneurship course in a structured classroom environment cannot inculcate entrepreneurial behavior in students. Further, they argued that the entrepreneurship process is nonlinear, and can not be inculcated through a structured process. The current study also states that entrepreneurship behavior can not be inculcated by traditional entrepreneurship courses. Similarly, Santhosh and Dinesh [49] have also stated that despite taking entrepreneurship courses, students in India are not willing to opt into startups. Students who graduated from Indian institutions have claimed that they lack the required set of skills to start their own business. They have shown dissatisfaction towards entrepreneurial education and entrepreneurial courses. They feel that the entrepreneurship course taught during their degree program is not enough to inculcate entrepreneurship behavior in Indian youth. Hence, it can be concluded from the current study and the available literature that entrepreneurship is a creative and nonlinear process which requires more than a formal entrepreneurship course.

A study by Ismail et al. [50] shows that different teaching pedagogy results in different effects of the entrepreneurship education. The didactic or teacher-centric approach (e.g., relying on learning materials, such as PowerPoint slides, notes, textbooks, or online learning platforms) is effective in enhancing students' understanding of certain topics. However, this approach has been criticized for its ineffectiveness in developing entrepreneurial behavior, knowledge, and skills [46,48]. Ismail et al. [50] also found that the student-centric approach in teaching the entrepreneurial course is more effective in the development of entrepreneurial behaviors among students. A student-centric or experiential learning model is a learning process by which knowledge is created through the transformation of experience [49,50]. The entrepreneurship course in UTP, although it does have experiential learning aspects (such as developing their business ideas, and testing the ideas through a feasibility study), still focuses on the didactic approach to ensure the students obtain the understanding and knowledge. The result of this study manifests this approach. Additionally, the teaching method should be tailored towards the engineering-based product or service that is related to the students' field of study. This will help to make them appreciate the course, and not feel that the course is only for the business degree students.

The result of the study implicates the improvement of the curriculum and teaching pedagogy of the course. Besides the practical contribution, this study also contributes to the methodology in terms of the application of MGA in the field of entrepreneurship education. The MGA helps this study to analyze whether different estimates occur for each group. The effect of the entrepreneurship course on the intention could falsely be understood without understanding the heterogeneity of the data. The application of the MGA allows the understanding of group-specific effects that facilitate obtaining further differentiated findings [50]. Several constraints need to be addressed and considered in this study. Although the quantitative approach has achieved its objectives, a qualitative approach will enhance our understanding on various interactions among students, and between them and their lecturers to further understand this phenomenon. Further study on the relationship of the education on intention, focusing on the engineering students in Malaysia, could help to uncover more information that will help to improve the effectiveness of this course or program. Besides, it is suggested for the future researchers in this area in the university to take up qualitative research to provide a deeper understanding of the issue of the effectiveness of the course. This understanding will help the further enhancement of the teaching of this course.

Limitations

The current study, like other studies, is no exception to limitations. One of the limitations is the cross-sectional nature of the study. Usually, cross-sectional studies look for data from the population at a specific time, whereas the current study employed the Theory of Planned Behavior (TPB) to analyze the effect of an entrepreneurship course on entrepreneurial intention and behavior of the students. It is difficult to measure the behavior or intention at a specific period. Ideally, longitudinal studies can better explain intention and behavior, as change in behavior is a long process that can be influenced by specific incidents. Additionally, this research is purposive, and is meant to study the behavior of engineering students towards entrepreneurial education, so the current study is not generalizable to the entire student population.

The current study has included data from engineering students, as Universiti Teknologi Petronas (UTP) mostly offer engineering courses, and the number of students taking engineering courses is more than other disciplines. Therefore, the results of this study cannot be generalized, and do not represent the entire population, or the entire UTP student population.

UTP offers entrepreneurship courses to all undergraduate second year students, and the average age of second-year students is 22 years. The student sample is homogenous, with slight variation in their gender. Therefore, mean and standard deviation are not taken into account for the current study. However, the current study has used PLS, which has the capability to handle non-normal data. Despite the ability of PLS to handle non-normal data, the current study has skewness or kurtosis, and all are within -1 and 1 (means distribution is normal). Furthermore, UTP has more male students enrolled in comparison to female students. This unequal proportion of males and females is another limitation of the current study.

Author Contributions: Conceptualization, H.A. and I.A.; methodology, H.A.; review and editing, I.A.; supervision, A.-N.A.-T. and S.N.M.Z. All authors have read and agreed to the published version of the manuscript.

Funding: We would like to express our gratitude to Universiti Teknologi Petronas (UTP) for supporting this study under the STIRF grant (grant number: 015LA0-019).

Institutional Review Board Statement: Not applicable.

Informed Consent Statement: Not applicable.

Data Availability Statement: Not applicable.

Conflicts of Interest: The authors declare no conflict of interest.

References

1. Del Carmen Pegalajar-Palomino, M.; Burgos-García, A.; Martinez-Valdivia, E. What does education for sustainable development offer in initial teacher training? A systematic review. *J. Teach. Educ. Sustain.* **2021**, *23*, 99–114.
2. UNESCO. *Progress on Education for Sustainable Development and Global*; UNESCO: Paris, France, 2018.
3. Rae, D.; Carswell, M. Towards a conceptual understanding of entrepreneurial learning. *Dev. J. Small Bus. Enterp.* **2001**, *8*, 150–158. [CrossRef]
4. Politis, D. The Process of Entrepreneurial Learning: A Conceptual Framework. *Entrep. Theory Pract.* **2005**, *29*, 399–424. [CrossRef]
5. Ministry of Education Malaysia. *Malaysia Education Blueprint 2015–2025*; Kementerian Pendidikan Malaysia: Putrajaya, Malaysia, 2015.
6. Krueger, A.L.; Reilly, N.F.; Carsrud, M.D. Competing models of entrepreneurial intentions. *J. Bus. Ventur.* **2000**, *15*, 411–443. [CrossRef]
7. Jumaatbin Mahajar, J.; Yunus Mohd binti, A. Inclination Towards Entrepreneurship Among Universiti Pendidikan Sultan Idris Students. *J. Glob. Bus. Manag.* **2012**, *8*, 248–256.
8. Keat, D.; Selvarajah, O.Y.; Meyer, C. Inclination towards entrepreneurship among university students: An empirical study of Malaysian university students. *Int. J. Bus. Soc. Sci.* **2015**, *2*, 206–220.
9. Mohamad, N.; Lim, H.E.; Yusof, N.; Soon, J.J. Estimating the effect of entrepreneur education on graduates' intention to be entrepreneurs. *Educ. Train.* **2015**, *57*, 874–889. [CrossRef]
10. Embi, N.A.C.; Jaiyeoba, H.B.; Yussof, S.A. The effects of students' entrepreneurial characteristics on their propensity to become entrepreneurs in Malaysi. *Educ. Train.* **2019**, *61*, 1020–1037. [CrossRef]
11. Othman, N.; Muda, T.N.A.A.T. Emotional intelligence towards entrepreneurial career choice behaviours. *Educ. Train.* **2018**, *60*, 953–970. [CrossRef]
12. Din, B.H.; Anuar, A.R.; Usman, M. The Effectiveness of the Entrepreneurship Education Program in Upgrading Entrepreneurial Skills among Public University Students. *Procedia-Soc. Behav. Sci.* **2014**, *224*, 117–124. [CrossRef]
13. Mustafa, M.J.; Hernandez, E.; Mahon, C.; Chee, L.K. Entrepreneurial intentions of university students in an emerging economy: The influence of university support and proactive personality on students' entrepreneurial intention. *J. Entrep. Emerg. Econ.* **2016**, *8*, 162–179. [CrossRef]
14. Ambad, S.N.A.; Damit, D.H.D.A. Determinants of Entrepreneurial Intention Among Undergraduate Students in Malaysia. *Procedia Econ. Financ.* **2016**, *37*, 108–114. [CrossRef]
15. Cheng, M.Y.; Chan, W.S.; Mahmood, A. The effectiveness of entrepreneurship education in Malaysia. *Educ. Train.* **2009**, *51*, 555–566. [CrossRef]
16. Ariff, A.H.M.; Bidin, Z.; Sharif, Z.; Ahmad, A. Predicting entrepreneurship intention among Malay University Accounting students in Malaysia. *UNITAR E J.* **2010**, *6*, 1–10.
17. Krueger, N.F. The cognitive infrastructure of opportunity emergence. *Entrep. Theory Pract.* **2000**, *24*, 5–22. [CrossRef]
18. Piperopoulos, D.; Dimov, P. Burst bubbles or build steam? Entrepreneurship education, entrepreneurial self-efficacy, and entrepreneurial intentions. *J. Small Bus. Manag.* **2015**, *53*, 970–985. [CrossRef]
19. Sabah, S. Entrepreneurial Intention: Theory of Planned Behaviour and the Moderation Effect of Start-Up Experience. In *Entrepreneurship-Practice-Oriented Perspectives*; InTechOpen: London, UK, 2016.
20. Ajzen, I. The theory of planned behavior. *Orgnizational Behav. Hum. Decis. Processes* **1991**, *50*, 179–211. [CrossRef]
21. Ajzen, I.; Czasch, C.; Flood, M.G. From intentions to behavior: Implementation intention, commitment, and conscientiousness. *J. Appl. Soc. Psychol.* **2009**, *39*, 1356–1372. [CrossRef]
22. Armitage, M.; Conner, C.J. Efficacy of the theory of planned behaviour: A meta- analytic review. *Br. J. Soc. Psychol.* **2001**, *40*, 471–499. [CrossRef]
23. Mason, C. Entrepreneurship Education and Research: Emerging Trends and Concerns Entrepreneurship Education and Research: Emerging Trends and Concerns International School of Entrepreneurship. *J. Glob.* **2011**, *1*, 13–22.
24. Othman, N.; Othman, S.H. The Perceptions of Public University Students of Entrepreneurship Education in Malaysia. *Int. Bus. Manag.* **2017**, *11*, 865–873.
25. Shah, I.A.; Amjed, S.; Jaboob, S. The moderating role of entrepreneurship education in shaping entrepreneurial intentions. *J. Econ. Struct.* **2020**, *9*, 2–15. [CrossRef]
26. Kim, J. Predictors of Entrepreneurial Intention of Nursing Students Based on Theory of Planned Behavior. *J. Multidiscip. Healthc.* **2021**, *14*, 533–543.
27. Tsordia, C. The Role of Theory of Planned Behavior on Entrepreneurial Intention of Greek Business Students. *Int. J. Synerg. Res.* **2015**, *4*, 1–23. [CrossRef]
28. Tornikoski, E.; Maalaoui, A. Critical reflections–The Theory of Planned Behaviour: An interview with Icek Ajzen with implications for entrepreneurship research. *Int. Small Bus. J.* **2019**, *37*, 536–550. [CrossRef]
29. Bird, B. Implementing Entrepreneurial Ideas: The Case for Intention. *Acad. Manag. Rev.* **1988**, *13*, 442–453. [CrossRef]

30. Creswell, J.W.; Creswell, J.D. *Research Design: Qualitative, Quantitative, and Mixed Methods Approach*; Sage Publications: Thousand Oaks, CA, USA, 2013.
31. Abdul-Latif, S.A.; Abdul-Talib, A.N. Consumer racism: A scale modification. *Asia Pacific J. Mark. Logist.* **2016**, *29*, 616–633. [CrossRef]
32. Sekaran, U.; Bougie, R. *Research Method for Business A Skill Building Approach*, 4th ed.; John Wiley & Sons: New York, NY, USA, 2017.
33. Abu Bakar, A.R.; Abdul Talib, A.N.; Hashim, F. Restoring service quality, satisfaction and loyalty in higher education institutions through market orientation. *J. Glob. Bus. Adv.* **2014**, *7*, 88–108. [CrossRef]
34. Heuer, A.; Kolvereid, L. Education in entrepreneurship and the theory of planned behaviour. *Eur. J. Train. Dev.* **2014**, *38*, 506–523. [CrossRef]
35. Ashari, H.; Yusof, Y.; Mohd Zamani, S.N.; Abdul Talib, A.N. A study of the effect of market orientation on Malaysian automotive industry supply chain performance. *Int. J. Technol.* **2018**, *9*, 1651–1657. [CrossRef]
36. Zamani, S.N.M.; Abdul-Talib, A.N.; Ashari, H. Strategic orientations and new product success: The mediating impact of innovation speed. *Int. Inf. Inst. (Tokyo) Inf.* **2016**, *19*, 2785–2790.
37. Hair, M.; Hult, J.; Ringle, G.T.M.; Sarstedt, C.M. *A Primer on Partial Least Squares Structural Equation Modeling (PLS-SEM)*; Sage Publications: Thousand Oaks, CA, USA, 2014.
38. Hair, J.F., Jr.; Sarstedt, M.; Ringle, C.M.; Gudergan, S.P. *Advanced Issues in Partial Least Squares Structural Equation Modeling*; Sage: Thousand Oaks, CA, USA, 2017.
39. Utami, W.C. Attitude, Subjective Norms, Perceived Behavior, Entrepreneurship Education and Self-efficacy toward Entrepreneurial Intention University Student in Indonesia. *Eur. Res. Stud. J.* **2017**, *20*, 475–495.
40. Dinc, M.S.; Budic, S. The Impact of Personal Attitude, Subjective Norm, and Perceived Behavioural Control on Entrepreneurial Intentions of Women. *Eurasian J. Bus. Econ.* **2016**, *9*, 23–35. [CrossRef]
41. Mohammed, B.S.; Fethi, A.; Djaoued, O.B. The Influence of Attitude, Subjective Norms and Perceived Behavior Control on Entrepreneurial Intentions: Case of Algerian Students. *Am. J. Econ.* **2017**, *7*, 274–282. [CrossRef]
42. Mawardi, M.K.; Baihaqi, A.I. Impact of Attitudes Towards Entrepreneurship, Subjective Norms and Perceived Behavioral Control in Creating Entrepreneurial Intention. In Proceedings of the 2nd Annual International Conference on Business and Public Administration (AICoBPA 2019), Malang, Indonesia, 23–24 October 2019; pp. 53–56.
43. Al-Jubari, I. College Students' Entrepreneurial Intention: Testing an Integrated Model of SDT and TPB. *SAGE Open* **2019**, *9*, 1–15. [CrossRef]
44. Alam, M.; Kousar, S.; Rehman, C.A. Role of entrepreneurial motivation on entrepreneurial intentions and behaviour: Theory of planned behaviour extension on engineering students in Pakistan. *J. Glob. Entrepr. Res.* **2019**, *9*, 50. [CrossRef]
45. Helweg-Larsen, M.; Collins, B.E. A Social Psychological Perspective on the Role of Knowledge About AIDS in AIDS Prevention. *Curr. Dir. Psychol. Sci.* **1997**, *6*, 23–25. [CrossRef]
46. Linton, G.; Klinton, M. University entrepreneurship education: A design thinking approach to learning. *J. Innov. Entrep.* **2019**, *8*, 3. [CrossRef]
47. Kumar Santhosh, A.V.; Dinesh, N. Inculcating Entrepreneurial Spirit in Students Through Entrepreneurship Education. *Indian J. Appl. Res.* **2019**, *9*, 54–56. [CrossRef]
48. Ismail, A.B.; Sawang, S.; Zolin, R. Entrepreneurship education pedagogy: Teacher-student- centred paradox. *Educ. Train.* **2018**, *60*, 168–184. [CrossRef]
49. Kolb, A.Y. *The Kolb Learning Style Inventory–Version 3*; Hay Resources Direct: Boston MA, USA, 2015; Volume 200, pp. 166–171.
50. Mustapha, M.; Selvaraju, M. Personal attributes, family influences, entrepreneurship education and entrepreneurship inclination among university students. *Kaji. Malays. J. Malays. Stud.* **2016**, *33*, 155–172.

Review

Cooperation and Environmental Responsibility as Positive Factors for Entrepreneurial Resilience

Tancredi Pascucci *, Brizeida Raquel Hernández-Sánchez and José Carlos Sánchez-García

Department of Social Psychology and Anthropology, Salamanca University, 37005 Salamanca, Spain; brizeida@usal.es (B.R.H.-S.); jsanchez@usal.es (J.C.S.-G.)
* Correspondence: tancredipascucci@usal.es

Abstract: In this review, we study the state of entrepreneurial education as it applies to business resilience. We consider records over the last 20 years about entrepreneurial resilience that consider their social impact and focus on sustainability. The aim of the study was to determine whether an enterprise that stresses social impact and sustainability rather than profits could reinforce entrepreneurial resilience. The importance of this study is that it offers a more complex description of entrepreneurial resilience by connecting social and environmental sensitivity with a profit-oriented logic. We found a mild incremental rise in, first, the years of the 2000s and a jump by 2010. We then used VosViewer to create a cluster map from the record list of WOS, creating three clusters of: "education and sustainability", "entrepreneurship and social impact" and "innovation", and these three clusters were related to superior entrepreneurial resilience. This approach should be adopted in real time to be able to adapt to socio-economic crises, adopting a functional approach based on cooperativeness and awareness of complexity.

Keywords: sustainability; resilience; social impact; empowerment

Citation: Pascucci, T.; Hernández-Sánchez, B.R.; Sánchez-García, J.C. Cooperation and Environmental Responsibility as Positive Factors for Entrepreneurial Resilience. *Sustainability* **2022**, *14*, 424. https://doi.org/10.3390/su14010424

Academic Editors: Jaana Seikkula-Leino, Mats Westerberg, Priti Verma and Maria Salomaa

Received: 22 November 2021
Accepted: 29 December 2021
Published: 31 December 2021

1. Introduction

Crises in the last 20 years and throughout the 20th century have reached international proportions, often based on economic triggers. For example, two world wars occurred as a consequence of political and economic expansion, the Great Depression followed the Wall Street (NY, USA) crash of 1929, the 1973 petroleum crisis, the capitalist re-invention of former Soviet republics following the fall of the Berlin Wall, the Great Recession of 2007–2009 and, more recently, the COVID-19 recession as a consequence of the pre-existing vulnerability of socio-economic systems around the world, which led to the chaotic management of the flow of goods and people around the world [1,2]. These events stressed the need for Entrepreneurship Education (EE) to equip new and existing entrepreneurs with the managerial and entrepreneurial skills to manage similar difficulties and prevent similar crises in the future. A firm's survival depends on its ability to withstand difficulties, and it can be defined as "resilient" if it can adapt positively without altering its mission. [3–6]. "Resilience" is a term borrowed from Civil Engineering, which defines a material that has good resistance under pressure, is also used in Individual Psychology to define good adaptation during difficulties and has similarly been adopted in Management Science to define a "resistant" organization that can survive without significant impairment during international crises [7–9]. Not every business organization is resilient, and those that are are not at risk of being eliminated by a sort of economic Darwinian selection. EE is a discipline that began in 1947 to train new entrepreneurs to rebuild world economies after the war and received increasing attention during the 1980s, when universities began offering courses to train future entrepreneurs [10,11] and create entrepreneurial research in the U.S. and Europe and then also in Asia [12].

International markets are prone to unpredictable events that can negatively influence a business, be they political, financial, environmental, technological, health-related or cultural. These can significantly affect consumer behavior, reducing the enterprise's earnings [13–19], but we cannot adopt a fatalistic view of the economy, whereby we renounce the responsibility to prevent similar, unexpected events or, at least, to buffer their negative consequences on markets and economic activity. Following a liberal logic, especially after the fall of the Berlin Wall and the conflict between capitalist and communist countries, many entrepreneurs followed an aggressive business strategy based on saving resources and maximizing profits without considering workers' rights, ecosystem balance or community needs [20–23]. This has impaired societies and the environment. For example, an entrepreneur who is entirely oriented toward profit maximization is not motivated to create a bond with the area where the enterprise operates; instead, they exploit the community's workforce, raw materials or strategic position [24], and the capital generated is sent elsewhere, leaving the community that invested in this activity impoverished. Sometimes, the environment in which these communities live become polluted, and they suffer socio-economic distress [25–27]. In contrast, some projects offer an alternative entrepreneurial model based not only on economics, but also on innovative strategies and social aspects of the area in which they operate [28–31], also involving some integrated models of the stakeholder theory [32]. An entrepreneurial organization cannot consider itself to be an isolated institution, considering that it has a precise community context, even if it operates across different regions [33]. This aggressive and hypercompetitive strategy does not consider the importance of cooperation [34,35], which requires a coordinated approach, even in Entrepreneurship, where different institutions and organizations have a functional approach in order to reach a common goal. The approach of cooperativeness first emerged at the end of the 1980s [36], and there are some interesting studies concerning this approach [37–39]. We considered the importance of sensibility for environmental responsibility where an enterprise, even a small business, adopts an approach aiming to reduce the impact of its activity in terms of pollution or territorial alteration. In this case, we cite ecological intelligence [40] and community psychology [41,42], both of which must be considered so as to improve entrepreneurial performance. Future entrepreneurs must also be trained to consider these factors, as well as the social impact in terms of community wellness, including terms of employment, social cohesion, a sense of community and community empowerment [43–45]. It is not just an ethical question because an enterprise that acts responsibly will be appreciated by the community, which may lead to stronger partnerships [3,5,25,28,37]. In the next section, we describe our hypothesis for conducting a literature review in relation to the coexistence of multiple factors, such as earnings and environmental and social sustainability, to reinforce entrepreneurship organization and then describe this scientific literature with state-of-the-art cluster mapping that defines its various components.

This study, designed to overturn Fisher's Separation Theorem [46,47], sought records in which entrepreneurial organizations merged their profit motive with both social and environmental aspects to become more resilient and robust [4,5,8,26]. Entrepreneurship Education should equip entrepreneurs with not only the right skills, but should also motivate them to improve the world by extending beyond simple profit accrual. In this case, it is important to reinforce the social function, and in this work, we define this as a pro-social and collaborative attitude characterized by a sustainable strategy, positive social impact and a cooperative entrepreneurial approach that reinforces the empowerment of communities in which the organization operates [6,7,26,36,38].

2. Materials and Methods

For this review, we studied records from the last 20 years on entrepreneurial education that reinforce entrepreneurial resilience and survival, expecting that most would focus on sustainability and social sensitivity. To conduct this analysis, we used the following Boolean string: "(entrepreneur and education)" AND "(social and impact or resilience)" AND "(sustainability)". We searched the literature between 2000 and 2020, without limitations in relation to area, type of record or language; however, the most prevalent language was

English. We decided on a wide selection because we noticed that this was a recent and uncrowded line of research, and we wanted to select a significant number of records in order to generate a satisfying review and cluster analysis. For this review, we stated an ambitious social function that involved all three aspects. We verified that there would be a more abundant record selection if we considered EE only from the sustainability, social impact or empowerment points of view. This is not just a choice governed by a practical need, but has the goal of evidencing that social, sustainability and entrepreneurial performance and resistance are not mutually exclusive domains.

Due to the fact that we opted for a restricted definition of our interest area, including different aspects contributing to a "virtuous" entrepreneur that aims to satisfy all three missions of social, ecological and economic goals, our record selection was poor, but specific, with just 16 excluded records that were defined as not pertinent. These records mostly involved a type of organization that is not dedicated to a sustainable and/or a social mission.

We used three databases on 26 August 2021—SCOPUS (https://www-scopus-com.ezproxy.usal.es/search/form.uri?display=basic&zone=header&origin=#basic), WOS (https://www.webofscience.com/wos/alldb/basic-search) and EBSCO (https://web-s-ebscohost-com.ezproxy.usal.es/ehost/search/advanced?vid=2&sid=b763d2d8-f1f6-4ffa-90af-8eb7241e75a8%40redis)—for record mapping and then VosViewer to analyze a list of records extrapolated from WOS, the platform from which most records were excluded due to a lack of relevance. We used the PRISMA Statement [48] to represent our records, as demonstrated in the selection chart provided in Figure 1.

Following the PRISMA checklist, we catalogued the title, abstract, keywords and type of study for each record. After the duplicates were removed, we excluded another group of records that mainly focused on financial aspects, history, university educational strategies, medical treatments, entrepreneurial orientation, philosophy, pedagogical strategies and blended education. This selection process aimed to be highly specific, uniting different domains for a holistic overview, instead of separating aspects related to, for example, sustainability, social impacts or Entrepreneurship Education, thereby differentiating itself from other reviews that are broader and more difficult to analyze than this record selection process. After this strict selection process, we chose the database with the most records for cluster mapping, which was WOS.

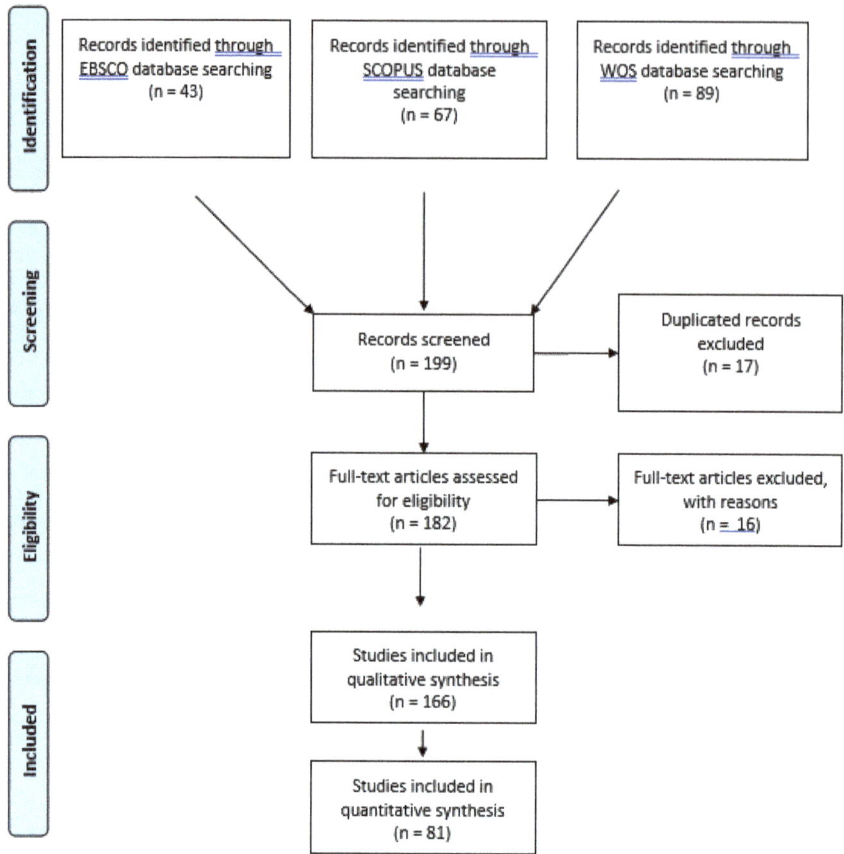

Figure 1. PRISMA Statement chart representation.

3. Results

Interest in the topic of entrepreneurial education to foster organizational resilience has emerged recently, judging from the evolution of the number of records in the last 20 years (Figure 2). Beginning in 2000, there was discontinuous and poor production of work on the subject, with a mild increase during the economic crisis of 2007–2009 and a jump after that. We hypothesized that this world crisis provided an important lesson to entrepreneurial organizations: that an approach that was totally focused on profits was dysfunctional and dangerous for economic stability [13,17,18,20].

Most of the contributions originated from the U.S. and UK, confirming a primacy trend found in many other research lines. We appreciate, as illustrated in Figure 3, that, unexpectedly, the third most active country was Spain, followed by India, Canada and Malaysia, demonstrating that there was also interest from European and Asian countries. We think that that the need to reinforce research on Entrepreneurship Resilience is related to the actual economic crisis unfolding across the globe. We could hypothesize that this urgency is changing the research trend, passing from developed and English-speaking countries to a new group of developed and non-English speaking countries.

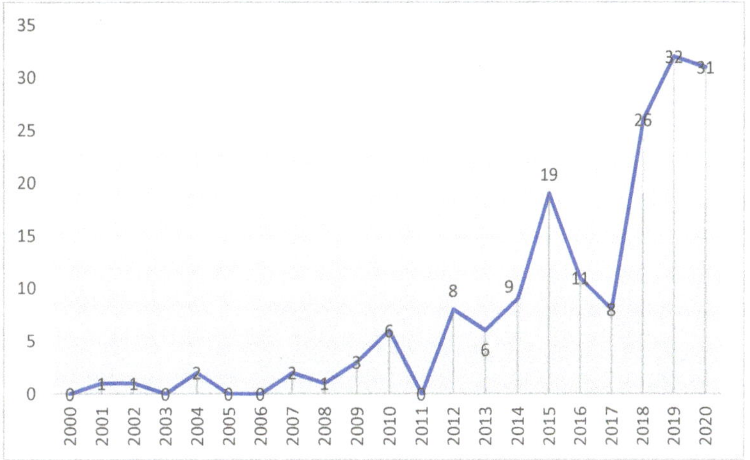

Figure 2. Publication progression during the last 20 years.

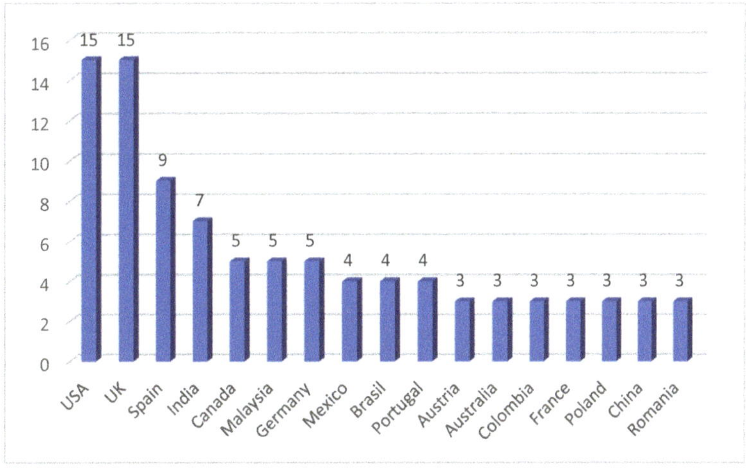

Figure 3. Representation of most productive countries for record publishing.

As previously stated, this is an emerging topic, and only a small number of authors have published papers on it. As represented in Table 1 they come from both developed and developing countries that have an h-index between 4 and 20. We do not yet have a large enough number of publications to hypothesize that the results are significantly generalizable, but we can observe that most of these authors are from developing countries. Some of them work together, forming research lines concerning sustainable entrepreneurship [49,50].

Table 1. Representation of authors with more than 2 publications in our record.

No.	Author	Actual Affiliation	H-Index	Prevailing Research Area
2	Matzenbacher D.E.	Universidade Federal do Rio Grande do Sul, Porto Alegre, Brazil	4	Environmental Sciences, Social Sciences, Business Management and accounting
2	Mets T.	Tartu Ulikool, Tartu, Estonia	9	Business Management and Accounting, Social Sciences, Computer Sciences
2	Raudsaar M.	Tartu Elikool, Tartu, Estonia	3	Business Management and Accounting, Social Sciences, Environmental Sciences
2	De Barcellos M.D.	Universidade Federal do Rio Grande do Sul, Porto Alegre, Brazil	20	Business Management and Accounting, Environmental Sciences, Social Sciences
2	Iyer V.G.	University of Louisville, Louisville, KY, USA	14	Neurosciences, Nursing, Psychology

Table 2 represents the most active journals. The first is Sustainability, which specializes in this area (in particular with regard to Environmental Sciences), but a variety of journals in countries, such as the U.S., UK, the Netherlands and Germany, have research areas that are quite varied, from social science to engineering.

Table 2. Representation of most productive journals.

No.	Journals	H-Index	Research Area
16	Sustainability	85 (Q1)	Environmental Sciences, Social Sciences, Energy
6	International Journal of Entrepreneurial Behaviour and Research	67 (Q1)	Business, Management and Accounting
3	International Journal of Sustainability in Higher Education	59 (Q2)	Social Sciences, Education, Human factors and Ergonomics
2	Smart Innovation Systems and Technologies	22 (Q4)	Computer Sciences, Decision Sciences
2	International Journal of Entrepreneurial Venturing	17 (Q3)	Business and International Management, Management of Technology and Innovation, Strategy and Management
2	Journal of Cleaner Production	200 (Q1)	Strategy Management, Renewanable Energy, Industrial and Manufacturing Engineering
2	Journal of Rural Studies	104 (Q1)	Forestry, Development, Sociology

4. Cluster Analysis

If we had conducted a brief study on Entrepreneurship Education with just one component, such as "entrepreneur and education" AND "social and impact or resilience" or "entrepreneur and education" AND "sustainability", we would have had a much larger group of records; for example:

- "Entrepreneur * AND Education" AND "Social AND impact OR resilience" has 809 records on EBSCO, 700 on SCOPUS and 1229 on WOS;
- "Entrepreneur * AND Education" AND "Sustainability" has 578 records on EBSCO, 508 on SCOPUS and 552 on WOS.

Once we refined the list of records on WOS, which included all previous dimensions, we used VosViewer [51] to load this list to create an analysis that clustered different research areas related to this research line. We decided to use VosViewer for its graphical intuitive representation, whereby the most important keywords are located in the representation area. Cluster mapping is an important analysis technique that provides a graphical representation of research lines, where similar topics, summarized by their tracking keywords, are regrouped into wider categorizations in each thematic cluster. A map of 3 clusters with 11 items is represented in Figure 4. The most powerful keywords identified in this case are Sustainability (Link strength = 25; Occurrence = 20), Education (Link strength = 24;

Occurrence = 18), Innovation (Link strength = 23; Occurrence = 12), Social Entrepreneurship (Link strength = 13; Occurrence = 9) and Impact (Link strength = 11; Occurrence = 7), with a relationship between them that reinforces the others in a holistic conception, whereby, for example, a Sustainability-centered approach is the goal for improving an Organization but can also act as a method by which to improve it.

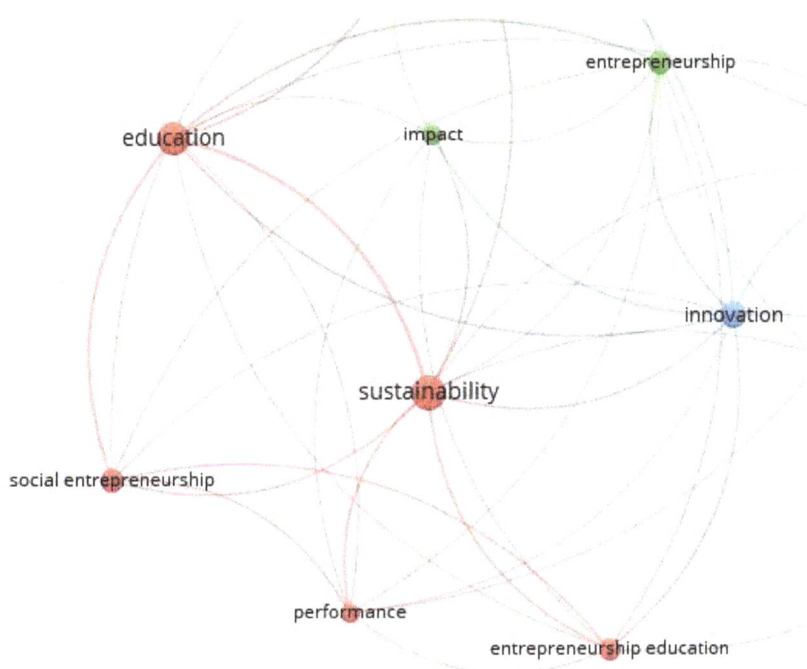

Figure 4. Cluster map of our research topic.

4.1. Cluster 1: Education and Sustainability (5 Items)

The first and most numerous cluster on Figure 5 (Education with link strength = 24 and occurrence = 18; Sustainability with link strength = 25 and occurrence = 20) occurs around two keywords concerning the training of future entrepreneurs with a sensitive, pro-environmental attitude [26,52–57]. In this cluster, entrepreneurial education [23,52–62] and performance appear because a well-formed entrepreneur has a higher performance level [6,63–65], which reinforces his or her empowerment [21,66]. This cluster also includes the social entrepreneurship keyword [31,67–69], an area distinguished from sustainability [24,44,52,53,70,71], although it has some common points regarding socio-economic interdependence [54,72].

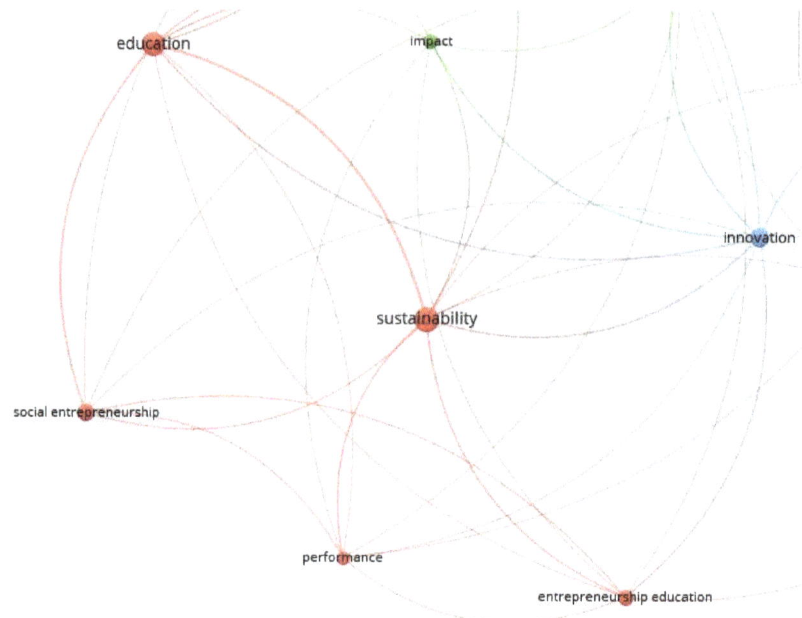

Figure 5. Representation of Cluster 1 map.

4.2. *Cluster 2: Entrepreneurship and Social Impact (3 Items)*

This cluster on Figure 6 focuses more on the management [17,20,26,28,68,73–75] of this kind of enterprise [75]. Following the previously cited stakeholder theory [7,58,76], these kinds of organizations pursue an entrepreneurial strategy to consider the wider social impact [49,58,70,75,77,78], thereby reinforcing entrepreneurial resilience through stronger community approval (Entrepreneurship with link strength = 16 and occurrence = 10; Social Impact with link strength = 11 and occurrence = 7), which would help it to survive in a crisis.

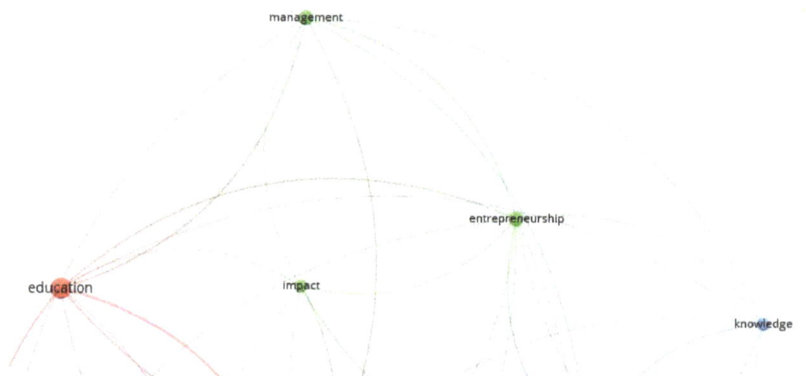

Figure 6. Representation of Cluster 2 map.

4.3. *Cluster 3: Innovation (3 Items)*

An important aspect in Figure 7 of a functional education is the incorporation of innovative points of view and protocols (Innovation with link strength = 23 and occurrence = 12;

Knowledge with link strength = 9 and occurrence = 7) into a mindful entrepreneurial strategy [34,79–84]. This can be realized through an exchange of knowledge [11,54,60] with university institutions [11,59,62,85], which creates the right combination between theory and practice.

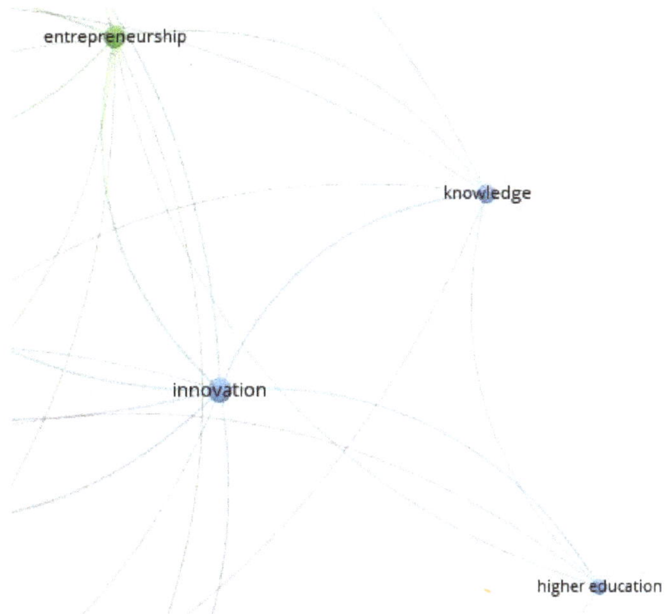

Figure 7. Representation of Cluster 3 map.

Figure 8 clearly shows the main point of view of this review. The keyword "Sustainability" is, in this case, the most powerful keyword (with a total link of 25 and 20 co-occurrences) that recalls the other cluster keywords. This means that, over the last 20 years, Sustainability has become a core theme, considering the relatively recent formation of the Kyoto Protocol and other green initiatives that have called attention to the pollution emergency and the need for sustainable development [86–89]. This need does not find appreciation in countries that have so far stressed the urgency of a solid economy without considering the environmental cost [90,91]; however, in countries trying to integrate environmental approaches within economic and social planning [91–95], the sustainable approach is not seen as antithetical to economic and social development.

Figure 9 regroups the main items that have been characterized for their cluster strength and co-occurrences. Social Impact, Innovation and Sustainability are the most important keywords in this study, which is founded on intersections between these three domains in which they merge and combine to create a new area of research. The main area of this study is defined by Entrepreneurship Education, which connects these three aspects, namely, Innovation, Sustainability and Social Impact; there are some intersections between the Innovation and Social Impact areas, as defined by Social Entrepreneurship, which combine a managerial approach of traditional enterprise with the need to satisfy the social function [31,46,47,61,96]; the intersection between Innovation and Sustainability is located in Higher Education due to a proper use of knowledge [7,11,54,59,60,62,85]; Entrepreneurship Resilience is related to a combination between Social Impact and Sustainability, where the Organization is strategy oriented so as to consider the interdependence of environmental and social factors in relation to a business [5,7,10,17].

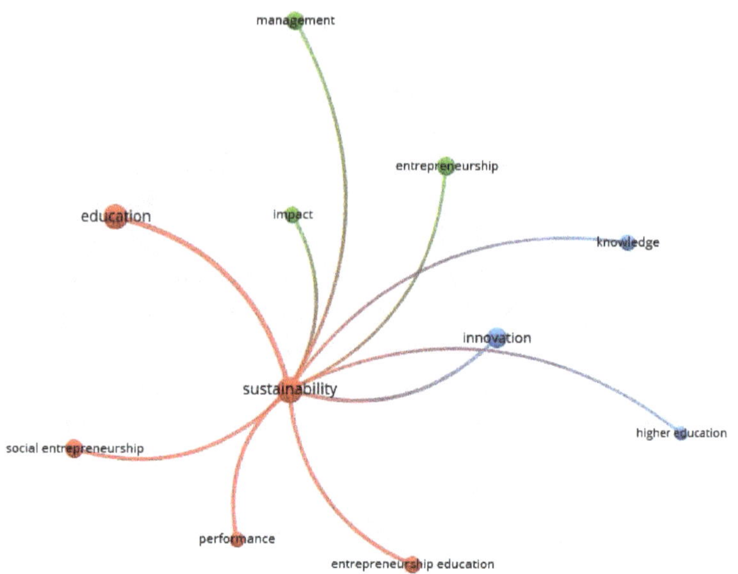

Figure 8. Representation of a cluster focused on sustainability.

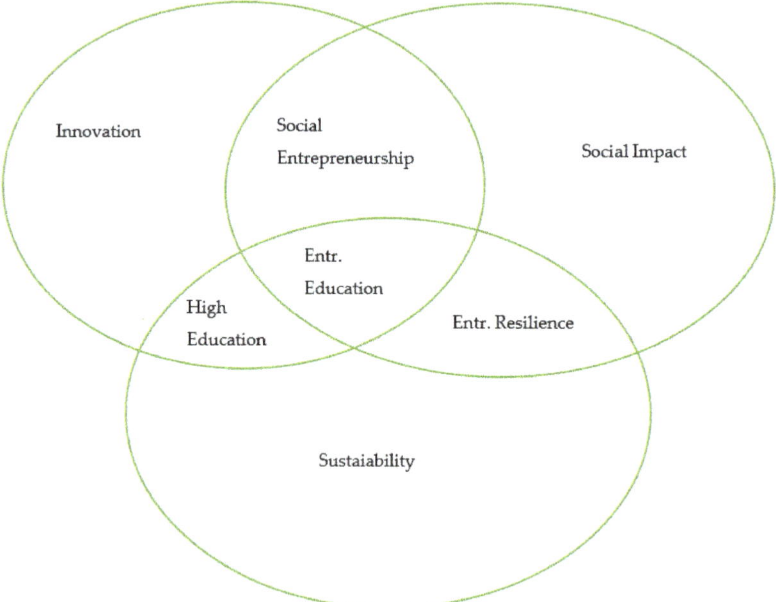

Figure 9. Representation of main keywords.

5. Discussion and Conclusions

A quick database search for entrepreneurial education "AND" resilience "OR" social impact "OR" sustainability produced many records, with a total of just under 1000 on SCOPUS. We chose to consider these keywords together because they are often considered independently. For example, an enterprise that focuses on societal change alone may neglect the environmental aspects, as was observed during a Boolean search of entrepreneurial education "AND" resilience "OR" social change "AND NOT" sustainability and vice versa when social change OR resilience was excluded. This trend of neglecting certain aspects is often encouraged, especially for ideological reasons. This was verified in some districts that refused to shut down their industrial structures because they feared the loss of jobs [6,11,75,81,97]. Despite this, innovations in technology are now making possible an effective industrial conversion that saves jobs and worker identities and preserves a sense of community as well as the environment [26,60,98–101]. From this point of view, entrepreneurial resilience must be considered as the result of different components. Mutual interaction reinforces the organization, in contrast to the traditional entrepreneurial philosophy in which a firm must maximize earnings to avoid failure [4,5,15,19], act as an individual [21,22,102,103] and avoid cooperation [74,75,80,87,88,93,94,96].

Recently, COVID-19 has exposed the illusion of medical and institutional invulnerability in the most privileged countries as social disparity, individualism, mental problems, economic instability and social injustice have been exacerbated. Consequently, humankind has had to rediscover the values of honesty, generosity, courage and foresight. The rejection of neo-liberal management provides the possibility of understanding the interdependence between world and market events [104–112]; adopting this mode of entrepreneurship, we will live in a better place—one in which an organization gains trust from the community and the entrepreneurial ecosystem in which it operates and receives help in return [6,36].

In the future, Entrepreneurship Education will have to negotiate some fundamental strategic challenges, such as training new entrepreneurs to use innovative strategies based on the skilled use of technology [29,99]; promoting managerial competences [59,68,89] to consider social [25,27–31,33,40,113] and environmental aspects [3,52,113]; and using electronic communication to facilitate learning [99]. EE has to adapt to different economic areas, including developing countries such as China, which is a complex and populous country with a high level of economic activity, consumption and pollution [84,95,114], but also in countries currently managing their economic transition [113,115,116]. There is a need for entrepreneurs to use wisdom in management strategies despite their fear of failure [33,116] and the risk of losing profits. [4,5,25,30]. A sustainable entrepreneurial strategy can assist in sectors such as "slow food" or agriculture [94,95], but also in those that have slow growth, and can provide a level of stability that can help them to resist a crisis [4,5,18,26,33,74]. The stakeholder theory underlines how important an ethical approach is for management, not only for business interest, but also for an interdependent socio-economic network, especially during world crises such as pandemics [9,19,109]. With this work, we state the urgent need for a "wealthy" entrepreneurial ecosystem [6,110].

This study suffers from some limitations, such as the use of a cluster analysis using just the WOS database, and the lack of precise restriction criteria for record selection. Perhaps it is too early to define a precise research line due to the significant dispersion among authors' contributions in this area, but we are fairly certain that it is a promising and growing topic for future research, especially after the end of the pandemic, as there will be a clear need to rebuild and re-organize interactions among people, organizations and communities, starting with the resilient organizations that survive the crisis.

It is tempting and easier to employ a reductive approach and focus on just one or two objectives when starting a business. This focus could just be to make money while neglecting civil rights and exploiting the environment, creating social distress and pollution as a result. Furthermore, it is important to underline, in this case, the relevance of the stakeholder theory, in which a responsible act performed by a restricted group of people encourages collective action to improve the world within and outside of an institutional framework [7,63,92]. We can also set a double objective, combining economic and social goals, economic and green goals, or social and green goals, while neglecting the third aspect. Even if the Gross Domestic Product (GDP) is sometimes considered to be an incomplete criterion to evaluate a country's economic performance [15], the World

Bank (N.H., U.S.A.) shows that the annual growth of GDP for all countries in the world and—with the exception of China, a large country experiencing continuous growth—of most economic superpowers is decreasing, and we hypothesize that the current economic strategy, based primarily on an individualistic short-term planning strategy, should be reconsidered [103]. These approaches are often encouraged by ideology, but this can be a superficial approach that does not appreciate the entrepreneurial ecosystem complexity. In this case, the enterprise will fail, lose its resilience and collapse because it will not have a functional, long-term strategy. There is a need for entrepreneurial education to avoid the superficial, short-sighted approach. In this case, it is important to consider recent contributions from Nobel Prize researchers, which encourage the consideration of emotional triggers in economic behaviors [117,118], restructuring a dysfunctional belief about economic–rational infallibility.

In line with the Community Psychology Paradigm [43–45], Entrepreneurship Education could reinforce concern for Sustainability and Social Impacts with regard to the territory, developing a sense of empowerment among citizens and Entrepreneurial Organizations, which could foster a functional attitude with a spontaneous initiative and/or through Institutional Intervention provided by the Government, which could encourage people, services and communities to adopt social functions, as represented in Figure 10. It suggests that organizational change for entrepreneurs comes from the top, via direct Statal–Institutional intervention, combined with change at the bottom. This requires the modification of the personal attitudes of entrepreneurs so that they are not just led by Institutions, but so that they also have a genuine, intrinsic motivation for creating a business organization that has a social function. Entrepreneurs should also be well informed about the interdependence of these worlds and their events and actions. [7,25,28,31]

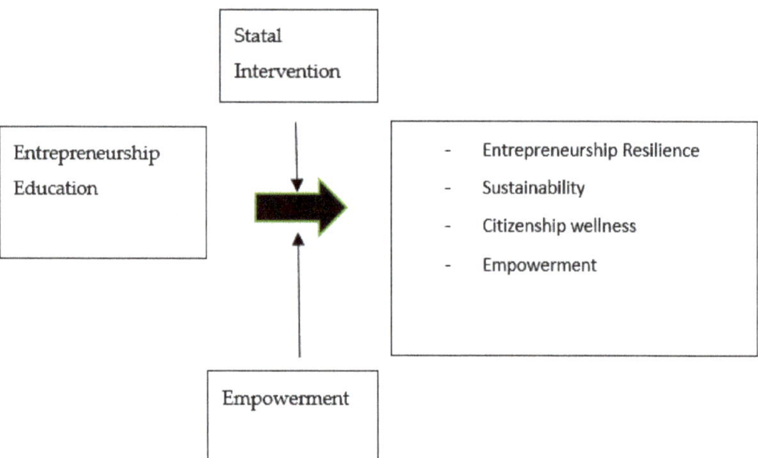

Figure 10. Representation of positive factors for Sustainability and Organizational Resilience.

The empowerment of a community could be considered in this case both as a result of and a positive contributor to providing resilience, wellness and sustainability within communities [21,50,55]. In the future, we hope to use similar instruments for cluster mapping, such as SciMAT, CitNetExplorer and Sci2Tool [119–121] and databases such as SSCI [122] or EBSCO, following the example of other papers [123], with a different approach regarding co-occurrence and co-citations.

Author Contributions: All authors have contributed equally to the development of this article. All authors have read and agreed to the published version of the manuscript.

Funding: This research received no external funding.

Institutional Review Board Statement: Not applicable.

Informed Consent Statement: Not applicable.

Conflicts of Interest: The authors declare no conflict of interest.

References

1. Diaz, C.M.G.; Sanchez, G.A.C. Analysis of the Financial Intermadiation on the stage of the XX and XXI Centuries Crisis. *Sophia-Educ.* **2011**, *7*, 106–128.
2. Mehediyev, A. The oil factor in the history of Azerbaijan. *Vopr. Istor.* **2018**, *10*, 138–144.
3. Krehebiel, T.C.; Gorman, R.F.; Erkson, O.H.; Louks, O.L.; Johnson, P.C. Advancing ecology and economics through a business-science synthesis. *Ecol. Econ.* **1999**, *14*, 183–186. [CrossRef]
4. Liebenau, J.; Alleman, J. Network resilience and its regulatory inhibitors. *Glob. Econ. Digit. Soc.* **2004**, 379–392.
5. Rose, A.; Liao, S.Y. Modeling regional economic resilience to disasters: A computable general equilibrium analysis of water service disruptions. *J. Reg. Financ.* **2004**, *45*, 75–112. [CrossRef]
6. Geng, Y.; Cote, R. Diversity in industrial ecosystems. *Int. J. Sustain. Dev. World Ecol.* **2007**, *14*, 329–335. [CrossRef]
7. Santoro, G.; Bertoldi, G.; Giachino, C.; Candelo, E. Exploring the relationship between entrepreneurial resilience and success: The moderating role of stakeholders' engagement. *J. Bus. Res.* **2020**, *119*, 142–150. [CrossRef]
8. Renko, M.; Bullough, A.; Saeed, S. How do Resilience and Self-Efficacy relate to entrepreneurial intentions in countries with varying degrees of fragility? A six-country study. *Int. Small Bus. J. Res. Entrep.* **2021**, *39*, 130–156. [CrossRef]
9. Zighan, S.; Abualqumboz, M.; Dwaikat, N.; Alkalha, Z. The role of Entrepreneurial orientation in developing SMEs resilience capabilities throught COVID-19. *Int. J. Entrep. Innov.* **2021**. [CrossRef]
10. Pittaway, L.; Cope, J. Entrepreneurship education: A systematic review of the evidence. *Int. Small Bus. J.* **2007**, *25*, 479–510. [CrossRef]
11. Mascarenhas, C.; Marques, C.; Galvão, A.R.; Santos, G. Entrepreneurial university: Towards a better understanding of past trends and future directions. *J. Enterprising Communities People Places Glob. Econ.* **2017**, *11*, 316–338. [CrossRef]
12. Wu, Y.C.J.; Wu, T. A decade of entrepreneurship education in the Asia Pacific for future directions in theory and practice. *Manag. Decis.* **2017**, *55*, 1333–1350. [CrossRef]
13. Girard, S.; Agundez, J.A.P. The effects of the oyster mortality crisis on the economics of the shellfish farming sector: Preliminary review and prospects from a case study in Marennes-Oleron Bay (France). *Mar. Policy* **2014**, *48*, 142–151. [CrossRef]
14. Park-Barjort, R.R. Samsung: An original case of knowledge transfer in economic organizations. *Enterp. Hist.* **2014**, *75*, 91–101.
15. Cerrato, D.; Alessandri, T.; Depperu, D. Economic crisis, acquisition and Firm Performance. *Long Range Plan.* **2016**, *42*, 171–185. [CrossRef]
16. Ulgen, F. Samsung: Financial development, economic crises and emerging market economies Financial Development. In *Economic Crises and Emerging Market Economies*; Routledge: London, UK, 2016; pp. 1–244.
17. Angel, K.; Menendez-Plans, C.; Orgaz Guerrero, N. Risk management: Comparative analysis of systematic risk and effect of the financial crisis on US tourism industry: Panel data research. *Int. J. Contemp. Hosp. Manag.* **2018**, *30*, 1920–1938. [CrossRef]
18. Hyz, A. *SME Finance and the Economic Crisis: The Case of Greece*; Routledge: London, UK, 2019; pp. 1–119.
19. Golubeva, O. Firms' performance during the COVID-19 outbreak: International evidence from 13 countries. *Corp. Gov. Int. J. Bus. Soc.* **2021**, *21*, 1011–1027. [CrossRef]
20. Michalski, T. Radical innovation through corporate entrepreneurship from a Competence-Based Strategic Management perspective. *Int. J. Manag. Pract.* **2006**, *2*, 22–41. [CrossRef]
21. Mahato, R.V.; Ahluwalia, S.; Walsh, S.T. The diminishing effect of VC reputation: Is it hypercompetition? *Technol. Forecast. Soc. Chang.* **2018**, *133*, 229–237. [CrossRef]
22. Vaisman, E.D.; Podshivalova, M.V. Assessment of small industry resistance to the hypercompetition threats in regions. *Econ. Reg.* **2018**, *14*, 1232–1245. [CrossRef]
23. Cooke, P. Economic globalisation and its future challenges for regional development. *Int. J. Technol. Manag.* **2003**, *26*, 401–420. [CrossRef]
24. Bellandi, M.; Carloffi, A. District internationalisation and trans-local development. *Entrep. Reg. Dev.* **2008**, *20*, 517–532. [CrossRef]
25. Smith, D.C.; James, C.D.; Griffiths, M.A. Co-brand partnerships making space for the next black girl: Backlash in social justice branding. *Psychol. Mark.* **2021**, *38*, 2314–2326. [CrossRef]
26. Surya, B.; Suriani, F.; Menne, F.; Abubakar, H.; Idris, M.; Rasyidi, E.S.; Remmang, H. Community empowerment and utilization of renewable energy: Entrepreneurial perspective for community resilience based on sustainable management of slum settlements in Makassar city, Indonesia. *Sustainability* **2021**, *13*, 3178. [CrossRef]
27. Uduji, J.I.; Okolo-Obasi, E.N.; Asongu, S.A. Oil extraction in Nigeria's Ogoniland: The role of corporate social responsibility in averting a resurgence of violence. *Resour. Policy* **2021**, *70*, 101927. [CrossRef]

28. Pless, N.M.; Maak, T.; Stahl, G.K. Promoting corporate social responsibility and sustainable development through management development: What can be learned from international service learning programs? *Hum. Resour. Manag.* **2012**, *51*, 873–903. [CrossRef]
29. Urbano, D.; Guerrero, M.; Ferreira, J.J.; Fernandes, C.I. New technology entrepreneurship initiatives: Which strategic orientations and environmental conditions matter in the new socio-economic landscape? *J. Technol. Transf.* **2019**, *44*, 1577–1602. [CrossRef]
30. Eggers, F. Masters of disasters? Challenges and opportunities for SMEs in times of crisis. *J. Bus. Res.* **2020**, *116*, 199–208. [CrossRef]
31. Sirine, H.; Andadari, R.K.; Suharti, L. Social Engagement Network and Corporate Social Entrepreneurship in Sido Muncul Company, Indonesia. *J. Asian Financ. Econ. Bus.* **2020**, *7*, 885–892. [CrossRef]
32. Argandoña, A. *El bien Común*; IESE Business School, Universidad de Navarra: Barcelona, Spain, 2011.
33. Naradda Gamage, S.K.; Ekanayake, E.M.S.; Abeyrathne, G.A.K.N.J.; Prasanna, R.P.I.R.; Jayasundara, J.S.M.B.; Rayapakshe, P.S.K. A review of global challenges and survival strategies of small and medium enterprises (SMEs). *Economies* **2020**, *8*, 79. [CrossRef]
34. Chuah, S.H.; Hoffmann, R.; Ramasamy, B.; Tan, J.H.W. Is there a spirit of Overseas Chinese Capitalism? *Small Bus. Econ.* **2016**, *47*, 1095–1118. [CrossRef]
35. Crick, J.M. Incorporating coopetition into the entrepreneurial marketing literature: Directions for future research. *J. Res. Mark. Entrep.* **2019**, *21*, 19–36. [CrossRef]
36. Carsrud, A.L.; Olm, K.W.; Thomas, J.B. Predicting entrepreneurial success: Effects of multi-dimensional achievement motivation, levels of ownership, and cooperative relationships. *Entrep. Reg. Dev.* **1989**, *1*, 237–244. [CrossRef]
37. Levine, S.S.; Preitula, M.J. Open collaboration for innovation: Principles and performance. *Organ. Sci.* **2014**, *25*, 1414–1433. [CrossRef]
38. Kimuli, S.N.L.; Orobia, L.; Sabi, H.M. Sustainability intention: Mediator of sustainability behavioral control and sustainable entrepreneurship. *World J. Entrep. Manag. Sustain. Dev.* **2020**, *2*, 81–95. [CrossRef]
39. Srikaliman, S.; Wardana, L.W.; Ambarwati, D.; Sholhin, U.; Shobrin, L.A.; Fajarah, N.; Wibowo, A. DO Creativity and Intellectual Capital matter for SME sustainability? The role of competitive advantage. *J. Asian Financ. Econ. Bus.* **2020**, *7*, 397–408. [CrossRef]
40. Goleman, D. *Ecological Intelligence: The Hidden Impacts of What We Buy*; Broadway Books: New York, NY, USA, 2010.
41. Rappaport, J. In Praise of Paradox. A Social Policy of Empowerment over Prevention. *Am. J. Community Psychol.* **1981**, *1*, 1–25. [CrossRef] [PubMed]
42. Francescato, D.; Tomai, M.; Solimeno, A. *Lavorare e Decidere Meglio in Organizzazioni Empowering ed Empowered*; Franco Angeli: Milan, Italy, 2008.
43. Mulyono, S.E.; Sutarto, J.; Malik, A.; Loretha, A.F. Community Empowerment in Entrepreneurship development based on local potential. *Int. J. Innov. Creat. Chang.* **2020**, *11*, 271–283.
44. Rocha, H.; Kunc, M.; Audretsch, D.B. Clustersm Economic Performance and social cohesion: A system dynamics approach. *Reg. Stud.* **2020**, *54*, 1098–1111. [CrossRef]
45. Rojikinnor, R.; Gani, A.J.A.; Saleh, H.; Amin, F. Empowerment throught good Governance and Public Service Quality. *J. Asian Financ. Econ. Bus.* **2020**, *7*, 491–497. [CrossRef]
46. Wempe, B. Understanding the Separation Thesis; precision after the decimal point? A response to Joakim Sandberg. *Bus. Ethics Q.* **2008**, *18*, 549–553. [CrossRef]
47. Burger-Helmchen, T. Entrepreneurial Organizations. In *Encyclopedia of Creativity, Invention, Innovation and Entrepreneurship*; Cavannis, E.G., Ed.; Springer: New York, NY, USA, 2013.
48. Mother, D.; Liberati, A.; Tetzlaff, J.; Altman, D.G. Preferred reporting items for systematic reviews and meta-analyses: The PRISMA statement. *PLoS Med.* **2009**, *6*, e1000097.
49. Matzembacher, D.E.; Raudsaar, M.; De Barcellos, M.D.; Mets, T. Sustainable entrepreneurial process: From idea generation to impact measurement. *Sustainability* **2019**, *11*, 5892. [CrossRef]
50. Matzembacher, D.E.; Raudsaar, M.; De Barcellos, M.D.; Mets, T. Business models' innovations to overcome hybridity-related tensions in sustainable entrepreneurship. *Sustainability* **2020**, *12*, 4503. [CrossRef]
51. Waltman, L.; Van Eck, V.J.; Noyons, E.C.M. A unified approach to mapping and clustering of bibliometric networks. *J. Infometrics* **2010**, *4*, 629–635. [CrossRef]
52. Badulescu, D.; Badulescu, A.; Bac, D.P.; Saveanu, T.; Florea, A.; Perticas, D. Education and Sustainable Development: From theoretical interest to specific behaviours. *J. Environ. Prot. Ecol.* **2017**, *18*, 1698–1705.
53. Moon, C.; Walmsley, A.; Apostolopoulos, N. Sustainability and Entrepreneurship Education. A Survey of 307 UN HESI signatories. In Proceedings of the 13th European Conference On Innovation And Entrepreneurship (ECIE 2018), Aveiro, Portugal, 20–21 September 2018; pp. 498–506.
54. Rios, M.M.M.; Herremans, I.M.; Wallace, J.E.; Althouse, N.; Landsale, D.; Presseur, M. Strengthening sustainability leadership competencies through university internships. *Int. J. Sustain. High. Educ.* **2018**, *19*, 739–755. [CrossRef]
55. Anderson, O. Education provides a foundation for Community Empowerment IEEE Smart Village Programs Teach Residents Job, Technical and Business, Skills. *IEE Syst. Man Cybern. Mag.* **2019**, *5*, 42–46. [CrossRef]
56. Yu, C.W.M.; Man, T.W.Y. The sustainability of enterprise education: A case study in Hong Kong. *Educ. Train.* **2007**, *49*, 138–152.
57. Urbaniec, M. Sustainability-oriented competencies in entrepreneurship education: Insights from an empirical study on Polish students. *Int. Entrep. Rev.* **2018**, *4*, 399–411.

58. Galvao, A.R.; Marques, C.S.E.; Ferreira, J.J.; Braga, V. Stakeholders' role in entrepreneurship education and training programmes with impacts on regional development. *J. Rural. Stud.* **2020**, *74*, 169–179. [CrossRef]
59. Sanchez-Hernandez, M.I.; Gallardo-Vazquez, D.; Pajuelo-Moreno, M.L. University Social Responsibility: A Modelling Framework for Student Base Analysis. In Proceedings of the 9th International Technology, Education and Development Conference, Madrid, Spain, 2–4 March 2015; pp. 2523–2530.
60. Sandoval, F.; Garcia, F. Por La Senda De Un Futuro Sustentable. Propuestas Y Acciones Con Responsabilidad Social. In *Sustainability in University Education: A Challenge for a Comprehensive Education in the Faculty of Economy and Business of the University of Chile*; Universidad de Santiago de Chile: Santiago, Chile, 2016; pp. 557–568.
61. Shier, M.L.; Van-Du, B. Framing curriculum development in social work education about social enterprises: A scoping literature review. *Soc. Work. Educ.* **2018**, *37*, 995–1014. [CrossRef]
62. Martinez-Campillo, A.; Sierra-Fernandez, M.D.; Fernandez-Santos, Y. Service-Learning for Sustainability Entrepreneurship in Rural Areas: What Is Its Global Impact on Business University Students? *Sustainability* **2019**, *11*, 5296. [CrossRef]
63. Al Mamun, A.; Ibrahim, M.D.; Bin Yusoff, M.N.H.; Fazal, S.A. Entrepreneurial Leadership, Performance, and Sustainability of Micro-Enterprises in Malaysia. *Sustainability* **2018**, *10*, 1591. [CrossRef]
64. Iqbal, Q.; Ahmad, N.H.; Halim, H.A. Insights on entrepreneurial bricolage and frugal innovation for sustainable performance. *Bus. Strategy Dev.* **2021**, *4*, 237–245. [CrossRef]
65. Özdemirci, A. Corporate entrepreneurship and strategy process: A performance based research on Istanbul market. Procedia-Social and Behavioral Sciences. In Proceedings of the 7th International Strategic Management Conference, Paris, France, 30 June–2 July 2011; Volume 24, pp. 611–626.
66. Radovic-Markovic, M.; Zizanovic, B. Fostering Green Entrepreneurship and Women's Empowerment through Education and Banks' Investments in Tourism: Evidence from Serbia. *Sustainability* **2019**, *11*, 6826. [CrossRef]
67. Nga, J.K.H.; Shamuganathan, G. The Influence of Personality Traits and Demographic Factors on Social Entrepreneurship Start Up Intentions. *J. Bus. Ethics* **2010**, *95*, 259–282.
68. Starnawska, M. Determining Critical Issues In Social Entrepreneurship Education. In Proceedings of the 6th International Conference Innovation Management, Entrepreneurship And Sustainability (IMES 2018), Prague, Czech Republic, 31 May–1 June 2018; pp. 1014–1024.
69. Tam, H.L.; Asmoah, E.; Chan, A.Y.F. Developing Social Entrepreneurship as an Intervention to Enhance Disadvantaged Young People's Sense of Self-Worth and Career Competence in Hong Kong. *Appl. Res. Qual. Life* **2021**, *19*, 1–30. [CrossRef]
70. Rus-Casas, C.; Eliche-Quesada, D.; Aguilar-Pena, J.D.; Jimenez-Castillo, G.; La Rubia, M.D. The Impact of the Entrepreneurship Promotion Programs and the Social Networks on the Sustainability Entrepreneurial Motivation of Engineering Students. *Sustainability* **2020**, *12*, 4935. [CrossRef]
71. Chatterjee, S.; Duttagupta, S.; Upadhyay, P. Sustainability of microenterprises: An empirical analysis. *Benchmarking Int. J.* **2018**, *25*, 919–931. [CrossRef]
72. Masouman, A.; Harvie, C. Regional economic modelling through an embedded econometric–inter-industry framework. *Reg. Stud.* **2018**, *52*, 1237–1249. [CrossRef]
73. Iyer, V.G. Strategic Environmental Assessment (SEA) Process towards Sustainable Construction Management Development for the Electrical, Automation and Mechanical Engineering Construction Industries Business excellence achievements. In Proceedings of the 2018 3rd International Conference On Electrical, Automation and Mechanical Engineering (EAME 2018), Xi'an, China, 24–25 June 2018.
74. Grant, J.H. Advances and Challenges in Strategic Management. *Int. J. Bus.* **2007**, *12*, 11–31.
75. Arshad, R.; Noor, A.H.R.; Yahya, A. Human Capital and Islamic-Based Social Impact Model: Small Enterprise Perspective. *Procedia Econ. Financ.* **2015**, *31*, 510–519. [CrossRef]
76. Morelli, A.; Taramelli, A.; Bozzeda, F.; Valentini, E.; Colangelo, M.A.; Cueto, Y.R. The disaster resilience assessment of coastal areas: A method for improving the stakeholders' participation. *Ocean. Coast. Manag.* **2021**, *241*, 105867. [CrossRef]
77. Beeri, I.; Gottlieb, D.; Izhaki, I.; Eshet, T.; Cohen, N. The Impact of Training on Druze Entrepreneurs' Attitudes Towards and Intended Behaviors Regarding Local Sustainability Governance: A Field Experiment at the Mount Carmel Biosphere Reserve. *Sustainability* **2020**, *12*, 4584. [CrossRef]
78. Rok, B.; Kulik, M. Circular start-up development: The case of positive impact entrepreneurship in Poland. *Corp. Gov. Int. J. Bus. Soc.* **2020**, *21*, 339–358. [CrossRef]
79. Zhu, L.; Li, M.; Metaa, N. Financial Risk Evaluation Z-Score Model for Intelligent IoT-based Enterprises. *Inf. Processing Manag.* **2021**, *58*, 102692. [CrossRef]
80. Iyer, V.G. Strategic Environmental Assessment (SEA) Process for Green Materials and Environmental Engineering Systems towards Sustainable Development-Business Excellence Achievements. In Proceedings of the 4th International Conference on Green Materials and Environmental Engineering (GMEE), Beijing, China, 28–29 October 2018.
81. Zdonek, I.; Mularczyk, A.; Polock, G. The idea of corporate social responsibility in the opinion of future managers—Comparative research between Poland and Georgia. *Sustainability* **2021**, *13*, 7045. [CrossRef]
82. Buil, M.; Aznar, J.P.; Galiana, J.; Rocafort-Marco, A. An Explanatory Study of MBA Students with Regards to Sustainability and Ethics Commitment. *Sustainability* **2016**, *8*, 280. [CrossRef]

83. Rooney, D.; McKenna, B. Should the Knowledge-Based Economy be a Savant or a Sage? Wisdom and Socially Intelligent Innovation. *Prometheus* **2005**, *23*, 307–323. [CrossRef]
84. Markopoulos, E.; Staggl, A.; Gann, E.L.; Vanharanta, H. Beyond Corporate Social Responsibility (CSR): Democratizing CSR Towards Environmental, Social and Governance Compliance. In *International Conference on Applied Human Factors and Ergonomics*; Springer: Cham, Switzerland, 2021; Volume 276, pp. 94–103.
85. Contreas-Pacheco, O.E.; Calderon, A.B. Leadership in Clusters: A Missing Link in the Academic Knowledge Corpus. *Rev. Virtual Univ. Catol. Norte* **2017**, *50*, 183–203.
86. Kolmas, M. Japan and the Kyoto Protocol: Reconstructing 'proactive' identity through environmental multilateralism. *Pac. Rev.* **2017**, *30*, 462–477. [CrossRef]
87. Sedita, S.R.; Ozeki, T. Path renewal dynamics in the Kyoto kimono cluster: How to revitalize cultural heritage through digitalization. *Eur. Panning Stud.* **2021**. [CrossRef]
88. Salameh, M.T.B.; Alraggad, M.; Harasheh, S.T. The water crisis and the conflict in the Middle East. *Sustain. Water Resour. Manag.* **2021**, *7*, 69. [CrossRef]
89. Sassenrath, G.F.; Halloram, J.M.; Archer, D.; Raper, R.L.; Hendrickson, J.; Vadas, P.; Hanson, J. Drivers Impacting the Adoption of Sustainable Agricultural Management Practices and Production Systems of the Northeast and Southeast United States. *J. Sustain. Agric.* **2010**, *34*, 680–702. [CrossRef]
90. Han, Q.; Jennings, J.E.; Liu, R.J.; Jennings, P.D. Going home and helping out? Returnees as propagators of CSR in an emerging economy. *J. Int. Bus. Stud.* **2019**, *50*, 857–872. [CrossRef]
91. Filley, T.R.; Li, M.L.; Zhuang, J.; Yu, G.R.; Sayler, G.; Ouyang, Z.Y.; Han, X.G.; Zhang, X.G.; Jiang, G.B.; Zhou, C.H.; et al. Bi-national research and education cooperation in the US-China EcoPartnership for Environmental Sustainability. *J. Renew. Sustain. Energy* **2015**, *7*, 041512. [CrossRef]
92. Rousseau, J.F. Does carbon finance make a sustainable difference? Hydropower expansion and livelihood trade-offs in the Red River valley, Yunnan Province, China. *Singap. J. Trop. Geogr.* **2017**, *38*, 90–107. [CrossRef]
93. Stroud, D.; Fairbrother, P.; Evans, C.; Blake, J. Skill development in the transition to a 'green economy': A 'varieties of capitalism' analysis. *Econ. Labour Relat. Rev.* **2014**, *25*, 10–27. [CrossRef]
94. DeLind, L.B. Where have all the houses (among other things) gone? Some critical reflections on urban agriculture. *Renew. Agric. Food Syst.* **2015**, *30*, 3–7. [CrossRef]
95. Hosseinina, G.; Ramezani, A. Factors Influencing Sustainable Entrepreneurship in Small and Medium-Sized Enterprises in Iran: A Case Study of Food Industry. *Sustainability* **2016**, *8*, 1010. [CrossRef]
96. Freeman, R.E. Business ethics at the Millenium. *Glob. Ethics Bus.* **2000**, *10*, 169–180. [CrossRef]
97. Castro, M.P.; Scheede, C.R.; Zermeno, M.G.G. The Impact of Higher Education on Entrepreneurship and the Innovation Ecosystem: A Case Study in Mexico. *Sustainability* **2019**, *11*, 5597. [CrossRef]
98. Ares, J. Estimating pesticide environmental risk scores with land use data and fugacity equilibrium models in Misiones, Argentina. *Agric. Ecosyst. Environ.* **2004**, *103*, 45–58. [CrossRef]
99. Calvo, S.; Lyon, F.; Morales, A.; Wade, J. Educating at Scale for Sustainable Development and Social Enterprise Growth: The Impact of Online Learning and a Massive Open Online Course (MOOC). *Sustainability* **2020**, *12*, 3247. [CrossRef]
100. Bernard-Hernandez, P.; Ramirez, M.; Mosquera-Montoya, M. Formal rules and its role in centralised-diffusion systems: A study of small-scale producers of oil palm in Colombia. *J. Rural. Stud.* **2021**, *83*, 215–225. [CrossRef]
101. Ajayi, C.O. The effect of Microcredit on rural household livelihood: Evidence from women micro entrepreneurs in Oyo State, Nigeria. *Sci. Pap. Ser. Manag. Econ. Eng. Agric. Rural. Dev.* **2016**, *16*, 29–36.
102. McKenna, B. Wisdom, Ethics and the Postmodern Organisation. In *Handbook on the Knowledge Economy*; Rooney, D., Hearn, G., Ninan, A., Eds.; Edward Elgar: Gloucestershire, UK, 2005; pp. 37–53.
103. Burrows, S. Precarious work, neo-liberalism and young people's experiences of employment in the Illawarra region. *Econ. Labour Relat. Rev.* **2013**, *24*, 380–396. [CrossRef]
104. Billet, A.; Dufays, S.; Friedel, S.; Staessens, M. The resilience of the cooperative model: How do cooperatives deal with the COVID-19 crisis? *Strateg. Chang.* **2021**, *30*, 99–108. [CrossRef]
105. Hayter, C.S. Toward a strategic view of higher education social responsibilities: A dynamic capabilities approach. *Strateg. Organ.* **2018**, *16*, 12–34. [CrossRef]
106. Parris, D.L.; McInnis-Bowers, C. Business Not as Usual: Developing Socially Conscious Entrepreneurs and Intrapreneurs. *J. Manag. Educ.* **2017**, *41*, 687–726. [CrossRef]
107. Yildrim, K.; Onder, M. Collaborative Role of Metropolitan Municipalities in Local Climate Protection Governance Strategies: The Case of Turkish Metropolitan Cities. *J. Environ. Assestment Policy Manag.* **2019**, *21*, 1950006. [CrossRef]
108. Yin, R.; Yin, G.; Li, L. Assessing China's ecological restoration programs: What's been done and what remains to be done? *Integr. Assess. China's Ecol. Restor. Programs* **2009**, *45*, 442–453.
109. Chocholab, P.; Tapachai, N. Organizational resilience of SMEs in the Czech Republic: An exploratory analysis. In Proceedings of the ICABR 2015: X. International Conference On Applied Business Research, Madrid, Spain, 14–18 September 2015; pp. 395–417.
110. Larsson, M.; Milestad, R.; Hahn, T.; Von Oelreich, J. The resilience of a sustainability entrepreneur in the Swedish food system. *Sustainability* **2016**, *8*, 550. [CrossRef]

111. Notman, O. The value of short-term training for Russian managers in the context of economic crisis. In Proceedings of the 10th International Days of Statistics and Economics, Prague, Czech Republic, 8–10 September 2016; pp. 1344–1352.
112. Ahmad, M. Does underconfidence matter in short-term and long-term investment decisions? Evidence from an emerging market. *Manag. Decis.* **2021**, *59*, 692–709. [CrossRef]
113. Bednar, P.; Danko, L.; Smekalova, L. Coworking spaces and creative communities: Making resilient coworking spaces through knowledge sharing and collective learning. *Eur. Plan. Stud.* **2021**. [CrossRef]
114. Deszy, S.; Miclaus, C.; Rizescu, N.; Nicu, M. Industrial sites and past pollution problems. *Environ. Eng. Manag. J.* **2004**, *3*, 861–869.
115. Žebrytė, I.; Fonseca-Vasquez, M.; Hartley, R. Emerging economy entrepreneurs and open data: Decision-making for natural disaster resilience. *J. Small Bus. Strategy* **2019**, *29*, 36–46.
116. Chlopecky, J.; Rolcikova, L.; Kratochvil, M. Decision making influence of some macroeconomic factors concerning prospective sales of silicon carbide. In Proceedings of the 15th International Multidisciplinary Scientific Geoconference SGEM 2015, Albena, Bulgaria, 18–24 June 2015; pp. 171–178.
117. Kahneman, D.; Tversky, A. *Heuristics and Biases: The Psychology of Intuitive Judgement*; Cambridge University Press: Cambridge, UK, 2002.
118. Tahler, R.H. *Misbehaving: The Making of Behavioral Economics.*; W. W. Norton & Company: New York, NY, USA, 2015.
119. Sci2 Team. *Science of Science (Sci2) Tool*; Indiana University (USA) and SciTech Strategies: Albuquerque, NM, USA, 2009.
120. Cobo, M.J.; Lòpez-Herrera, A.G.; Herrera-Viedma, E.; Herrera, F. SciMat: A new Science mapping analysis software tool. *J. Am. Soc. Inf. Sci. Technol.* **2012**, *63*, 1609–1630. [CrossRef]
121. Van Eck, N.J.; Waltman, L. CitNetExplorer: A new software tool for analyzing and visualizing citation networks. *J. Informetr.* **2014**, *8*, 802–823. [CrossRef]
122. Torres-Pruñonosa, J.; Plaza-Navas, M.A.; Dìez-Martinf, F.; Beltran-Cangròs, A. The Intellectual Structure of Social and Sustainable Public Procurement Research: A Co-Citation Analysis. *Sustainability* **2021**, *13*, 774. [CrossRef]
123. Torres-Prunonosa, J.; Plaza-Navas, M.A.; Diez-Martin, C. The source of Knowledge of the Economic and Social Value in Sport Industry Research: A co-citation Analysis. *Front. Psychol.* **2020**, *11*, 629951. [CrossRef] [PubMed]

Article

Responding to Global Challenges through Education: Entrepreneurial, Sustainable, and Pro-Environmental Education in Nordic Teacher Education Curricula

Jaana Seikkula-Leino [1,2,*], Svanborg R. Jónsdóttir [3], Marcia Håkansson-Lindqvist [2], Mats Westerberg [4] and Sofia Eriksson-Bergström [2]

[1] Renewing Learning and Collaboration, Tampere University of Applied Sciences, 333520 Tampere, Finland
[2] Department of Education, Faculty of Human Sciences, Mid Sweden University, SE-85170 Sundsvall, Sweden; marcia.hakanssonlindqvist@miun.se (M.H.-L.); sofia.eriksson-bergstrom@miun.se (S.E.-B.)
[3] School of Education, University of Iceland, 105 Reykjavík, Iceland; svanjons@hi.is
[4] Department of Social Sciences, Technology and Arts, Luleå University of Technology, SE-97187 Luleå, Sweden; mats.westerberg@ltu.se
* Correspondence: jaana.seikkula-leino@tuni.fi or jaana.seikkula-leino@miun.se

Abstract: The United Nations' Sustainable Development Goals (SDGs), and the European Union's strategies both set goals for solving environmental challenges faced by societies and communities. As part of solving these challenges, both the UN and the EU stress the development of entrepreneurial and innovative education. Teacher education plays a crucial role in these efforts, since teachers and teacher educators have a significant impact on educating citizens far into the future. In this research, we studied how Nordic (Finnish, Swedish, and Icelandic) primary teacher education curricula involve entrepreneurial, sustainable, and pro-environmental education. For this study, the authors analyzed the B.Ed. curricula of three academic teacher education institutions in Spring 2021. We used qualitative content analysis as our research method. According to the results, all three curricula incorporated both entrepreneurship education and sustainable development to some extent, although often not very explicitly. Given the urgency of problems such as global climate change, the educational goals and contents in these curricula related to entrepreneurial education and sustainable development are very limited. The idea of integrating environmental/sustainable and entrepreneurship education could be promoted in the future more explicitly, with these interdisciplinary educational themes emphasised more strongly in the curricula and education policies.

Keywords: entrepreneurship education; sustainability; sustainable education; teacher education; curriculum; Nordic education

Citation: Seikkula-Leino, J.; Jónsdóttir, S.R.; Håkansson-Lindqvist, M.; Westerberg, M.; Eriksson-Bergström, S. Responding to Global Challenges through Education: Entrepreneurial, Sustainable, and Pro-Environmental Education in Nordic Teacher Education Curricula. *Sustainability* **2021**, *13*, 12808. https://doi.org/10.3390/su132212808

Academic Editor: Gisela Cebrián

Received: 1 September 2021
Accepted: 15 November 2021
Published: 19 November 2021

Publisher's Note: MDPI stays neutral with regard to jurisdictional claims in published maps and institutional affiliations.

1. Introduction

According to the United Nations, science, technology, and innovation (STI, which includes entrepreneurial thinking) have long been recognized as one of the main drivers behind productivity increases and a key long-term lever for economic growth and prosperity [1,2]. In the European context, sustainable development and entrepreneurship are put forward as important areas for education [3–5].

Transitioning to environmentally sustainable societies also has the potential to create millions of jobs, which requires dynamic entrepreneurial competencies. However, this will require bold action to invest in people's capabilities to increase their productivity and realize their full potential [6]. Because there is great potential in entrepreneurship and entrepreneurial activity, governments have already made significant investments in innovation, entrepreneurship, and entrepreneurial education programs [7]. This is also supported by policy initiatives and economic evidence, such as those published by the Organisation for Economic Co-operation and Development (OECD) [8] and World Bank [9].

However, our existing knowledge of how entrepreneurial activities could contribute to the SDGs remains limited [2]. Many of the existing studies are conceptual and focus mainly on an individual or on organisational-level factors [10,11]. Thus, we need more research on how pro-environmental and entrepreneurial behavior can be realized in wider learning contexts [12]. In addition to this, the study of the phenomenon should be more in-depth. The recent academic debate is both too descriptive, optimistic and too indefinite integrating the research areas of entrepreneurship and sustainability. For example, according to Filsher et al. [11], despite the increasing trend towards sustainability-related entrepreneurship literature, only six of their 21 reviewed papers published after 2015 address the SDGs. In most of these papers, the SDGs are mentioned as an introductory example and not examined in depth. Furthermore, it is worth remembering that these concepts can also take on opposite meanings in the general debate. For example, the activities of companies can also pose environmental risks, and therefore entrepreneurship and sustainable education might be seen as contrary. These connotations can contribute to how "new concepts" of education, e.g., sustainable and entrepreneurship education, are understood and implemented in education and how their potential is seen more broadly.

At present, international and national strategies promote leading changes in education curricula including sustainable development and entrepreneurship. However, high-level strategy is a different matter from what is happening on the ground. As an example, for many areas of higher education, for example business, economics, and the natural sciences, the areas of sustainable development and entrepreneurship are taken for granted as part of the curriculum. Although, in the case of teacher education this may not be the case. The EU Commission Report in 2011, Entrepreneurship Education: Enabling teachers as a critical success factor states that in many EU countries, there is a large gap between the implementation of entrepreneurship education in primary schools and teaching entrepreneurship education to preservice teachers in higher education institutions. Therefore, the report stresses that the core skills linked to entrepreneurship are seldom a priority in initial teacher education. However, the report was published ten years ago. Has anything happened since then?

We also want to focus in particular on Nordic education. Nordic societies and their educational strategies naturally emphasize the responsibility and freedom of learners to develop a better society. Therefore, it could be expected that the Nordic curricula and their teacher training would proactively take these educational objectives into account. Or do we just think so? Have the Nordic countries been able to take advantage of this privilege in the planning and implementation of modern education, which may not always be possible at the global level? Some studies show that there are challenges in integrating entrepreneurship and sustainable education in curricula and teacher education. On the other hand, there are differences between universities [13–15]. However, none of these studies simultaneously look at both entrepreneurship and sustainable development in teacher education. Therefore, this study explores how teacher education in three selected cases in three Nordic countries, Finland, Sweden and Iceland, are actually designed with regard to entrepreneurship education and sustainable development. This will help us understand the future, as some of the teachers currently being trained will teach up to the 2050s. Today's curricula, therefore, have far-reaching implications for the future.

To clarify, we refer simultaneously to both mainstream concepts of *entrepreneurial education and entrepreneurship education* whilst investigating the curricula in education: Entrepreneurship education is more than content. It is also considered as a method and practice for learning [14,16]. Furthermore, in our study, we understand *pro-environmental* behaviour as "behavior that consciously seeks to minimize the negative impact of one's action on the natural and built world" [17], (p. 240). To cultivate such behaviors requires participatory teaching and learning methods that motivate and empower learners to change their behavior and become willing and capable to take action for *sustainability* which means meeting our own needs without compromising the ability of future generations to meet their own needs [18].

As in the definition of entrepreneurship education, in this study we refer simultaneously to the concepts sustainable, pro-environmental, and environmental, as if they were equivalent, even if they are not. This is due to the fact that in general education, these concepts are used freely and as corresponding to each other, even in curricula design. Therefore, if we want to study the phenomenon itself, it makes sense for us to look at how educators in general have incorporated any of those concepts related to others in their curricula writing.

The paper is structured as follows: Firstly, in the literature review section, we summarise the shift towards entrepreneurial and sustainable societies and stress teacher education and curricula in the context. Secondly, we provide an overview of the study and the methods based on content analysis. Thirdly, we present the results from teacher education curricula analysis. Finally, we discuss our key findings from the Nordic countries, and suggest some directions for the future.

2. Literature Review

2.1. Entrepreneurship Education as an Engine for Promoting Sustainable Transformation

The purpose of entrepreneurship education is to educate students to take more responsibility for themselves and their learning, to try to achieve their goals, to be creative, to discover existing opportunities, and to cope in a complicated society [19]. Another aim is for them to take an active role in the labor market and consider entrepreneurship as a natural career choice. Entrepreneurship education involves developing behavior, skills, and attributes, applied both individually and collectively, to help individuals and organizations of all kinds to create, cope with, and enjoy change and innovation [16,20–23].

Research on entrepreneurship education is based, in large measure, on a conceptual understanding of entrepreneurship and learning [4,24,25]. Some researchers have focused on identifying and analysing the core pedagogy of entrepreneurship education, characterizing it as emancipatory pedagogy, where the aim is to empower learners to become independent, creative, and active participants in society [10,19]. Prior research suggests that developing entrepreneurial mindsets is a key engine of growth and a must for sustainable development—e.g., in promoting the UN SDGs—and social cohesion, both locally and regionally, e.g., [26]. There is growing evidence of the impact of entrepreneurship education: studies indicate that entrepreneurship or entrepreneurial education can increase youths' entrepreneurial intentions and knowledge; stimulate their creativity, collaborative abilities, and self-confidence; and enhance their learning of other subjects, e.g., [27]. By developing an entrepreneurial mindset in societies, we can open an arena in which pro-environmental and sustainable solutions could be created more innovatively and co-creatively. The added value of entrepreneurship education has been understood in children's and youth education.

In addition to this, sustainable development plays a significant role in today's entrepreneurship academia and practical discussions. Shepherd and Patzelt [28] define sustainable entrepreneurship as follows: Sustainable entrepreneurship is concerned with the preservation of nature, life support, and community in the pursuit of perceived opportunities to create future products, processes, and services for profit, where profit is broadly defined to include both monetary and nonmonetary benefits to individuals, the economy, and society.

However, the link between entrepreneurship education and sustainable development has not always been recognized. According to Hermann and Bossle [29] entrepreneurial abilities such as foresight, problem-solving skills, and interdisciplinarity have been neglected in sustainability education. However, although entrepreneurship and sustainability education have separate learning objectives that are unlikely to be combined, major thematic similarities in interdisciplinary entrepreneurship and sustainability education can be identified. They are, e.g., innovation design, entrepreneurship ecosystem support, and corporate/organizational aspects. Furthermore, as Hsu and Pivec [30] argue, integrating sustainability into entrepreneurship education, including comprehensive plans in curricula, could have more potential than we are aware of in education development and promoting major goals, e.g., SDGs. As a matter of fact, Edokpolor [31] stresses the relationship between entrepreneurship education and the core values of sustainable development.

It seems that the potential and challenges of integrating entrepreneurship education and sustainable development have been recognized, and entrepreneurship education and education in sustainable development appear to be just out of reach of the school curricula in most countries, which have yet to consistently teach children to find resources to put their ideas into action. The focus on action competence as an aim of sustainability education strongly resembles the focus of entrepreneurship education on fostering entrepreneurial competence through creativity and action [32]. However, the development of societies and education takes decades. Achieving the EU's Green Transformation goal of climate neutrality by 2050 requires that the relevant competences and skills be developed by 2030. This also presupposes the integration of different competence frameworks, (e.g., EntreComp and GreenComp) and the development of corresponding educational concepts, e.g., curricula, at various levels of education. Within this context, entrepreneurship education has a central role as an engine for promoting sustainable transformation. This is also highlighted in Education 2030 by OECD [33] as environmental, economic, and social transformation and a proposed associated learning framework that encompasses disciplinary ideas, cross-cutting concepts, and social and economic practices. Thus, there is an increasing emphasis on entrepreneurship education in the field of education. The potential of entrepreneurship education has also been understood in teacher education, as teacher educators train future teachers who have a long-term impact on the future [24].

2.2. Sustainable and Pro-Environmental Education in the Nordic Context

Education for sustainable development (ESD) is meant to inculcate competencies in critical and creative thinking, imagination, and collaboration. Students need these skills to tackle the complex social, environmental, and economic issues and challenges of the modern world [18]. Instead of learning traditional discipline-focused areas, we also need to create opportunities for multidisciplinary and even phenomenon-based learning, e.g., [34], in which learners apply different perspectives to study real-world problems. To understand and solve problems related to climate change, for example, knowledge is needed from different subjects such as natural sciences, geography, psychology, economics, mathematics, and history.

Jóhannesson et al. [35] identified core characteristics that indicate sustainable development in curricula. These researchers encouraged a holistic view of sustainable development, looking at economic, environmental, and social factors as integrative entities. The characteristics were meant to reflect the goals of the United Nations Decade of Education for Sustainable Development 2005–2014 and research on environmental education and education for sustainable development.

Cars and West [36] argue that ESD can be understood as an educational ideology that came about by adding a developmental component to environmental education. Here, there are three overlapping and sometimes conflicting spheres of sustainable development—natural ecology, social issues, and economic factors. Further, ESD is meant to help people to develop attitudes, skills, and knowledge that supports them in making informed decisions that benefit themselves and others and to act upon these decisions [37]. ESD is cross-disciplinary by nature [38] and could be a catalyst for social changes and social transformations to greater equity. Thus, ESD can be seen as an application of critical pedagogy [36]. The core of sustainability education is to empower learners with the competence for action [39].

A critical distinction has been made between education for sustainable development (ESD) and *sustainability education (SE)*. ESD is defined as education that includes first and possibly second order changes, where SE is more radical and includes third order changes [40,41]. First order changes and learning take place within accepted frameworks, leaving basic values unexamined and unchanged. Second order changes involve critically reflective learning, where assumptions about the world are challenged. Third order changes are deconstructive and reconstructive, involving a deep awareness of versatile worldviews and ways of doing things, encouraging radicality and action [40,41]. ESD may be seen as a necessary journey towards SE, and *entrepreneurship education (EE)* could be a part of constructive steps towards SE. EE includes affordances that contribute to inculcating analytical and creative competences. These comptetences are needed for responsible action and provide cognitive elements (knowledge and understanding), emotions (identifying needs and problems), and action [19,40].

A comprehensive Nordic report on ESD maps education for sustainability in the Nordic countries, scrutinizing laws, regulations, national curricula, curricula of teacher education institutions, research, and reports in pre-, primary, and secondary education [40]. The report shows that the word *sustainability* is not mentioned once in the actual law on educational acts in the Nordic countries. However, the laws address issues such as democracy, human rights, equality, and respect for nature, which are all elements of sustainability education. The authors of the report also indicate that sustainability is often mentioned explicitly at the level of national curricula, special reports, and strategy papers from the Ministries of Education. One example is the Icelandic national curriculum established in 2011, which identifies sustainable development as one of six pillars of education. Sustainability is one of the core values of the Finnish national core curriculum for basic education, where it is mentioned almost 200 times. According to the Jónsson report [40], ESD has been present in the Swedish national curricula since 1994 and was written into law in the Higher Education Act of 2006. The report, however, presents a somewhat confusing picture. For example, in Iceland, educational policy seems to vacillate between strongly emphasizing sustainability, and not emphasising it at all. Iceland's law on compulsory education from 2008 has very little to say directly about sustainability. Sustainability in educational policies in Finland, Norway, and Sweden builds on a long tradition of environmental education and has been more consistent than in Iceland or Denmark. However, the Finnish, Norwegian, and Swedish educational acts fail to mention sustainability explicitly. Though 'sustainability' or variants appear almost 200 times in the Finnish national core curriculum, "the incorporation of sustainability as an educational aim or subject is often superficial" (p. 64). The authors of the report conclude that human existence has become less sustainable, and that conventional education is part of the problem and needs to be drastically redesigned [40].

2.3. *What Could Be the Opportunities of Nordic Teacher Education?*

Although the importance of teacher education has been emphasized, EE still seems to be a moderately overlooked theme in teacher education across our three countries [24,42]. Similarly, while there have clearly been attempts to include environmental education and research ideas in teacher education, these are not yet bearing much fruit – and indeed, a number of studies from around the world suggest that environmental education is not easy to fit into general teacher education programmes [13,43–48]. In summary, research emphasizes that strengthening both EE and sustainable education in teacher education would have more added value than we might think.

In the Nordic context, the development of future competencies for pro-environmental behavior may have unique potential, as the Nordic countries have a long tradition of advancing the goals of sustainable development at the national level and have been assessed as among the most SDG-ready countries. A renewed Nordic cooperation programme targeting the 2030 Agenda has the potential to help the Nordic countries become even more successful and effective, and to bring added value to the work done internationally, e.g., [49]. Furthermore, the Nordic model of education is based on national education systems that build on specific local values and practices but are influenced by international goals. Equity, participation, and welfare form the ideal Nordic model, which places value on shifting education towards more innovative, co-creative, and pro-environmental activities. The Nordic education model could be used more widely in global education development [50], as it provides an ideal "platform" to develop and test entrepreneurial and pro-environmental education initiatives. At the same time, teacher education also provides another platform for designing and implementing modern teaching, as traditionally, new trends in education first come to the fore. Teacher education institutions are even required to act as leaders in educational development in many countries. The Nordic education model and its teacher education could have more added value and impact than we think. Therefore, our research profiles how both sustainable development and EE have been taken into account in Nordic teacher education.

2.4. *Curriculum as a Tool for Educational Change?*

The definition of a curriculum assumes that: (1) a curriculum lists the courses or programs that should be offered; and (2) a curriculum is highly experiential—it demonstrates, both indirectly and directly, the abilities and skills that the individual should achieve [51]. Curricula reflect societal values and valuations. Thus, curriculum reform springs especially from a social desire for change—e.g., to realise entrepreneurial and sustainable education—and in this case, it is directed by values and ideological and political aims. Ideas concerning the 'right knowledge' and the distribution of power steer the reforms and activities [15]. General social trends and challenges such as globalisation, climate change, technological development, and the needs of labor markets direct the objectives of education and therefore also steer curricular reforms [15,52].

Traditionally, curriculum has been seen as belonging to the primary and secondary education context. Discussions and research related to higher education curriculum have sometimes been considered as questioning the autonomy of higher education institutions [53,54]. There has been little research on higher education curricula, and what there is has limited itself to specific fields [42]. Naturally, it is problematic if curricular concepts and theories coming from a primary/secondary school context are applied straightforwardly to higher education [54]. However, as Barnett and Coate [55] argue, the curriculum should be one of the core concepts used when developing higher education from research and pedagogical points of view.

It is important to examine curricula because they form the most important administrative documents that determine the content of training [42,56,57]. If entrepreneurship and sustainable education are to be systematically developed in the education system, it must be done via curricula. By looking at the three curricula documents of higher education institutions in Finland, Sweden, and Iceland, we can determine the direction of entrepreneurship and sustainable education in Nordic society. Moreover, this will indicate how these up-to-date themes of education are proceeding at some level globally. Besides our results, we will include ideas on how entrepreneurship and sustainable education could be developed in the future within teacher training, and how curriculum design can be developed more deliberately. Our study focuses especially on teacher education curricula, and more narrowly on primary teacher education on bachelor level because in our research countries, all teacher trainees receive at least bachelor level training.

Higher education institutions are independent developers of education. This element factors how they have wanted to or been able to include entrepreneurship and sustainable education in their teacher training curricula. The previously described background of the educational needs has also provoked our targets in our study, especially in teacher education, where elements such as entrepreneurship and sustainable education are integrated into curricula more or less in line with international and national strategies and documents. Therefore, we want to study how Nordic teacher education curricula have adapted entrepreneurship and sustainable education. We also want to broaden our understanding of whether "new winds" of education have been considered in this Nordic region, as might be expected.

As a summary from our literature review: Since we study how EE and sustainable development are reflected in the Nordic teacher education curricula, we emphasize curriculum research as the central administrative documents that guide the development and implementation of instruction. We also emphasize teacher education because it has far-reaching implications for the future. Furthermore, we also focus on research in Nordic education because we believe that Nordic education can be a pioneer in the field. Therefore, our findings can give preliminary indicators into where education is moving onto. Based on this, the main themes of our research are entrepreneurship education (EE), sustainable development, environmental education, teacher education, curriculum, and Nordic education. Furthermore, other related concepts, such as entrepreneurial learning, and SDGs have been presented in the literature review to broaden the understanding of our study and its methodological choices. Next, we will describe our research question and the chosen methodology of our study.

3. Research Question and Methodology

To investigate how and to what extent the Nordic countries incorporate entrepreneurial and sustainable education into their teacher education curricula, we pose the following research question:

- How does entrepreneurial, sustainable, and pro-environmental education emerge in Nordic (Finnish, Swedish, and Icelandic) primary teacher education curricula?

For this study, the authors analysed the B.Ed. curricula of the three academic teacher education providers in Finland, Sweden and Iceland, in April–May 2021. In the analysis, we looked for specific types of curricular topics or subjects referring to sustainability and entrepreneurship education (EE), (and concepts related), since we consider these issues are essential in two respects: First, we need to have some evidence to guide understanding of where these topics are situated in the framework documents (e.g., indicate the extent to which the learning framework supports "environmental, economic and social transformation" (28); and, second, to draw attention more fully—especially in conclusion—to strategies and future curricula design to address the observed gaps.

The curricula were obtained online. Qualitative content analysis was the research method used to interpret the content of text data through the systematic classification process of coding and identifying themes or patterns. Content analysis is usually used with a study design whose aim is to describe a phenomenon [58]. The curricula were read first generally, and then reflectively, with the aim of finding explicit and inexplicit references to the following concepts: entrepreneurship education; entrepreneurial learning/skills/competencies; innovative/creative learning/education; sustainable education, environmental education; SDG. The concepts were then identified and analysed separately. A somewhat similar study was found in this area. For example, Seikkula-Leino et al. [15] studied entrepreneurship education in teacher education curricula. As in this study, we also utilized a similar content analysis technique found to work in that study. At this point, we felt it was essential to analyze the curricula first because they are the primary documents that guide the implementation of teaching. Therefore, we did not yet proceed with qualitative interviews, for example, or quantitative surveys.

In analyzing the material, we thoroughly reviewed the descriptions of the degree programs and their courses. We analyzed the objectives, contents, expected results, and course evaluation criteria. In addition, we reviewed the course learning material. A broader conceptual bank, as previously described, was to support the evaluation, e.g., the concept of EE is often not explicitly used but is referred to by other concepts (such as creativity and innovation). The concept bank made it possible to evaluate the material comprehensively. Suppose the main concepts used in this study are directly recorded in the title and objectives of the course. In that case, we consider it more important than the fact that, for example, subject teaching uses primary education school material, such as the national curriculum, which includes entrepreneurship and sustainable development. So we focused primarily on what the goals and contents are in the bachelors' programs of our study. However, the references were not always so easy to find. Therefore, as described, we also extended looking at the learning material used to get inside the phenomenon somehow.

- At a minimum, we assessed each curriculum for the following characteristics: expediency, authenticity, relevance, administrative approval within the organisation, accuracy, and objectivity [59]. When it was clear that a curriculum steered the organisation's actions, all these characteristics were present. However, when curricula were obtained online, authenticity was an important factor to consider.
- Overall, this type of even minor pilot study with three universities is a good place to start looking at how these global aims are manifested in central educational documents. As Cohen et al. argue [59], the generalizability of single experiments (e.g., case and pilot studies) can be further extended through wider replication or multiple experiment strategies, allowing single pilot studies to contribute to the development of a growing pool of data, and allowing the key findings to be more broadly generalized in the future.

Below we describe in detail of the case examples of teacher education in Finland, Sweden, and Iceland. This is followed by analysis of their curricula, discussion, and conclusions.

4. Case Study Overview

Next, we briefly present the case of educational organizations, and their country context in teacher education, involved in the study from which the three curricula cases were collected and analyzed to understand how entrepreneurship education and sustainable/pro-environmental education are involved in teacher education. We chose only one teacher training organization from each country because mainly one university provides the most teacher training in Iceland.

4.1. Finnish Teacher Education: University of Turku, Faculty of Education, Teacher Education, Rauma Campus

In Finland, teacher training is arranged by universities and vocational institutes of higher education. They train pre-school, classroom, subject, special education, and vocational teachers. Academic teacher education is offered by 12 higher education units and their 13 teacher training schools [42]. Higher education institutions decide independently on the contents of teacher education, and emphasise the link between teaching and research. All teacher education also involves pedagogical studies and guided teaching practice. These are realized in the universities' own schools for teaching practice or other schools nominated for the purpose [60].

The teacher education curriculum at the University of Turku, Rauma Campus program includes studies in educational science, teaching internships, and multidisciplinary studies in the subjects and subject areas taught in basic education. The graduating classroom (primary school) teacher is prepared for both independent work and interprofessional cooperation as a teacher and educator. The aim of the degree program is the ability to meet and teach students from different cultural backgrounds and abilities. Most of those who graduate as class teachers work in teaching positions. However, the training also equips for administrative, planning, research and development tasks in the field of education. In Rauma Campus is also a teacher training school educating pupils, students, and student teachers [61].

4.2. Swedish Teacher Education: Mid Sweden University, Department of Education, Campus Sundsvall

The current teacher education system was introduced in 2011 as an outcome of the many official reports by the Swedish government that examined Swedish teacher education. Swedish teacher education has been fundamentally reformed several times since the Second World War. The teacher education that exists in Sweden today emphasizes subject knowledge, and thus gives the academic education ideology more space than before. All teacher education is run by a college or university and is nationally established and governed in accordance with the Higher Education Act (SFS 1992:1434) and the Higher

Education Ordinance (SFS 1993:100), as is all higher education. Twenty-eight different universities offer teacher education programs.

In Sweden, teacher education is organized through different programs that correspond to different ages of pupils in school. The Swedish education system includes preschool to high school, for children from 1 to 16 years old. The current teacher education system has reintroduced a clear division with different programmes of education for different teacher categories: grades 1–3, grades 4–6, grades 7–9, and upper secondary school. The argument is that pupils of different ages require different kinds of knowledge and skills. The programs include different subject areas such as educational science, didactics, studies in specific subjects, and internship education. The internship education requires the student, under supervision, to plan and carry out activities in the school. These 30 European Credit Transfer System (ECTS) credits, in one term, are distributed over various shorter periods throughout the program. Teacher education is vocationally oriented and aims for the student to develop a scientific and pedagogical approach, theoretical understanding, practical knowledge, as well as to develop as a person.

At Mid Sweden University, the teacher education is organized with a campus/distance learning model in which students conduct their studies in their hometown and spend only four to five weeks per semester on campus with three to five days of scheduled activities per week.

4.3. Icelandic Teacher Education: University of Iceland, School of Education

In 2008, teacher education in Iceland changed from a three-year B.Ed. degree to a five-year Master's degree (Act nr. 87/2008). To become licensed teachers before 2008, most students enrolled in three-year programmes at the Iceland University of Education (later the School of Education (SoI) at the University of Iceland), or the University of Akureyri. Those who already had a B.Ed. degree kept their license, but many have chosen to add a master's degree. Vocational education teachers require 60 ECTS in teacher certification studies in addition to a final diploma in their vocation (e.g., master craftsperson). The premise behind adding the master's level in 2008 was that teachers needed to be involved in research and the development of knowledge and thus strengthen their professional self-image [62]. According to the current law governing teacher education in Iceland (Act nr. 95/2019), student teachers must complete a master's degree and have both general competence as well as a specific competence such as completing at least 90 ECTS in a specific school subject.

The fundamental B.Ed. degree at the SoI is a 180 ECTS programme of academic and practical studies for those who intend to teach grades 1–5 in compulsory schools. The goal is for students to have knowledge of children's development, how they learn and communicate, literacy and teaching reading, and use of language. Emphasis is placed on the main learning areas and subjects at the primary level [63]. Theoretical content and field practice are woven into courses; this includes the interaction of theory and practice. Since the law requiring master's level education took effect in 2011, there have been contradicting pulls and conflicts in the development of teacher education in Iceland [64]. Conflicts have emerged between teacher education programs and the State about who is responsible for teacher education and what it should contain. Within the SoI itself, the development of the programme has involved arguments and conflicts between a focus on specialisation versus a focus on breadth of knowledge [64].

After the three case presentations described above, we look at the study results: an analysis of curricula from these three teacher training units.

5. Results

The following Table 1 describes the outcomes of curricula analysis of three universities step by step. Our data show that EE and sustainable development are taken into account to some extent in teacher education curricula in the Nordic countries. All in all, the teaching units related to entrepreneurship and sustainable education are part of, for example, subject studies or optional studies. The goals and contents of biology and other science, for example, include the starting points for environmental education, thus including sustainable development. Moreover, our results indicate that teaching these themes is not stressed at any particular year level.

Considering how much these themes are discussed today, these educational goals and contents are scarce. In Finland, entrepreneurship/entrepreneurial and sustainable education are widely approached, primarily through the national basic education core curriculum. These studies are available in both compulsory and optional studies. Education is also provided in the preparation and implementation of the curriculum. Thus, the curriculum itself seems to assume that future teachers will be somehow trained to implement both entrepreneurial learning and sustainable development. On the other hand, the aims and contents of the teacher education curriculum do not explicitly mention this elsewhere.

In Iceland, entrepreneurship and entrepreneurial education are not visible as distinctive elements in the University of Iceland B.Ed. program. This element "entrepreneurial learning" is mentioned once in the compulsory school core curriculum, however initiative is often mentioned there, often in relation to creativity and or independence. Neither were found in the B.Ed program for primary education teachers. The sustainability concept (34 times) is also clearly visible in the Icelandic compulsory school general curriculum and so is creativity (38 times in different compositions). Similarly to the Finnish case, the conclusion is that the aims and contents of the B.Ed. teacher education curriculum for primary school level at the School of Education (SoI) only mention these concepts in optional studies.

In Sweden, entrepreneurship/entrepreneurial education is not an explicit element in primary education for teacher students. Although entrepreneurship is mentioned twice in the general curriculum, one instance refers to entrepreneurship as a fundamental goal and task of the schools. The other instance refers to the 7–9 school level, where this concept is not included in the program plan for student teachers. Neither is the concept of sustainable development found explicitly in the primary teacher education plan. However, the concept of sustainable development is explicit (38 times) in the Swedish curriculum for the national core curriculum for primary education that teacher education utilizes in their education. Here, this concept is general for the fundamental goals and tasks of the schools and for the primary level of school. The concept of creativity/creative ability is seen in the curriculum, but it is not included in the goals for the primary education program plan. Thus, the Swedish case appears to be in line with the Icelandic and Finnish cases.

Table 1. The Content Analysis of Nordic Teacher Education Curricula with Three Case Examples.

Curriculum Case Studies: Country Examples from Primary Teacher Education, Bachelor's Degree Programme/Steps and Outcomes by Content Analysis	(1) **Curriculum Case Finland:** ***Class Teacher Education (BA), University of Turku, Faculty of Education, Class Teacher Education, Rauma***	(2) **Curriculum Case Sweden: Mid Sweden University**	(3) **Curriculum Case Iceland University of Iceland, School of Education, Primary Education B.Ed.**
The content was read several times in order to build an overall picture of the curriculum	General notes: The programme (180 ECTS) consists of studies in educational science, practical training and studies that provide pedagogical skills required for positions in primary education. The aim of the degree programme is to educate well-qualified academic professionals, active future makers, critical and ethically responsible experts and researchers in the educational field. At the core of a teacher's work lies the understanding and supporting of a child's and group's development. The dialogue between theory and practice takes place particularly during teaching practice periods, which offer a holistic view of a teacher's work. *The objective of the programme is to develop the following areas of competence:* 1. Communicative competence. Student is able to act collaboratively and is capable of communicating in different interactional situations. 2. Pedagogical competence. Student knows the basics of curriculum planning and is able to plan, implement, evaluate and develop learning processes and learning environments. 3. Intellectual competence. Student has basic knowledge of education as a science and contents of the multidisciplinary subjects taught in basic education. Student base his/her actions and professional development on scientific thinking. Student understands the principles of the societal and cultural basics of childhood. 4. Ethical competence. Student is able to identify and analyse his/her actions from an ethical viewpoint and act in accordance with ethical principles.	General notes: The programme (240 ECTS) consists of studies in educational science, practical training and studies that provide pedagogical skills required for positions in primary education, year 1–3 and year 4–6: *Primary School Teacher Education Programme in Pre-school Class and School Years 1–3 (Lärarutbildning-Grundlärare med inriktning mot arbete I förskoleklass och grundskolans årskurs 1–3, 240 hp)* *Primary School Teacher Education Programme, Years 4–6 (Lärarutbildning-Grundlärare med inriktning mot arbete igrundskolans årskurs 4–6, 240 hp)* *Aim* *The aim of the program is to for the student through theoretical, scientific and practice-based studies support students with the knowledge and skills needed to be able to work independently as teachers in year 0. 1–3.* Subject and subject-didactic studies totalling 165 ECTS in Swedish, Mathematics, English, Civics, Natural science and Technology. For Swedish and Mathematics, 30 ECTS are required and for English 15 ECTS. Further, *General Education*, 60 ECTS and 60 ECTS Practice-based training. Of the subject and subject-didactic studies 15 ECTS must be subject-related practice-based training	General notes: The programme is a three year 180 ECTS BEd studies (of five obligatory with two years on master's level to get a teaching permit). It consists of academic and practical studies emphasising teaching and learning of grades 1–5. The programme focuses on: • Childrens' development, communicative competence and use of language, learning to read, how to teach reading as well as first language learning. • Rich emphasis is on the importance of collaboration with parents and the main learning subjects in compulsory school. • The studies are conducted in close collaboration with the field (schools etc.) and are integrated with theoretical preparation for further studies and work in compulsory school. The studies consist of the following elements and subjects: • Icelandic • Literacy and teaching reading • Developmental and learning psychology • Methodology and educational research • Mathematics • Teaching primary school students • Curriculum and assessment • Childrens' literature • Speech and written text Responses/solutions to challenges in play and learning By reading the obligatory courses and bound electives, the following objectives seem to guide the studies. The competences thus extracted are (such analysed – not presented directly): 1. Knowledge of curricula guiding primary school education in Iceland. 2. Pedagogical competence. The teacher student knows the basics of curriculum planning and can plan, implement, evaluate and develop versatile learning processes and learning environments for different learners. 3. Intellectual competence. Student has basic knowledge of developmental psychology and learning challenges. Has knowledge of basic contents and teaching strategies of fundamental subjects (Icelandic, reading, writing, mathematics) as well as of integration of subjects. 4. The student can identify his/ her professional working theory and can understand his/her actions in relation to personal and academic theories.

Table 1. *Cont.*

Curriculum Case Studies: Country Examples from Primary Teacher Education, Bachelor's Degree Programme/Steps and Outcomes by Content Analysis	(1) Curriculum Case Finland: *Class Teacher Education (BA), University of Turku, Faculty of Education, Class Teacher Education, Rauma*	(2) Curriculum Case Sweden: Mid Sweden University	(3) Curriculum Case Iceland University of Iceland, School of Education, Primary Education B.Ed.
The curriculum was read reflectively, the aim being to find explicit refers to concepts: entrepreneurship education; entrepreneurial: learning/skill/competencies; innovative/creative learning/education; sustainable education, environmental education, SDG	Explicit refers: 1. Elective study: Entrepreneurship education and entrepreneurial pedagogy in early learning, 5 ECT 2. Elective study: Advanced course in nature and environmental education, 3 ECT, (refers also to sustainable education and SDGs	Explicit refers: Primary School Teacher Education Programme, and Years 1–3; Years 4–6 (Utbildningsplan för: Lärarutbildning–Grundlärare med inriktning mot arbete I grundskolans årskurs 4–6, 240 hp) *Reflective reading–no references to environment; entrepreneurship; innovative learning. Same for years 1–3 and years 4–6.* Values–the history of school, organization and conditions, values, including basic democratic values and human rights. (p. 2)	Explicit refers: No explicit references are made to entrepreneurship (or other forms of entre- enter- entrepreneurial) or action competence. Hardly any direct references to creativity except in two titles of obligatory courses and ways of working but not as a competence. One bound elective course (visual arts) has creativity as a competence aim. Innovative or innovation is not to be found. Sustainability is only mentioned once as an element within one course (5 ECTS Integrative and creative work) in the third year. In the same course critical and creative thinking is presented as a core thread.
The curriculum was read more reflectively and analytically, the aim being to find inexplicit refers to concepts/themes: entrepreneurship education; entrepreneurial learning/skills competencies; innovative/creative learning/education; sustainable education, environmental education, SDG	(1) Orientation to Teaching Practice in Elementary School, 4 ECTS: National Core Curriculum for Basic Education as a learning material which includes several refers to Multidisciplinary Learning Modules (e.g., Working life and entrepreneurial competence; Participation, influence, and building a sustainable future) for integrating learning and for increasing the dialogue between different subjects. Furthermore, the subject studies in core curricula involve multidisciplinary learning themes. (2) Didactic Teaching Practice in Elementary School, 8 ECTS: National Core Curriculum for Basic Education (see the description above) (3) Multidisciplinary Stud. in the Subjects and Cross-Curricular Themes Taught in Basic Educ. National Core Curriculum for Basic Education as a learning material (see the description above). Other learning material to support local/school based curriculum design and its implementation which have aims and contents for entrepreneurship education and sustainability. (4) Biology and Health Education, 4 ECTS: Some of the learning material involve topics such as environmental and nature protection. (5) Introduction to Craft, Design and Technology in Primary Education, 6 ECTS: National Core Curriculum for Basic Education as a learning material (see the description above). Other learning material of entrepreneurship education & arts and crafts education. (6) Maker culture in arts and crafts, 3–4 ECTS, (inexplicit refers to entrepreneurial way of working)	In the national core curriculum: *Curriculum for the compulsory school, preschool class and school-age educare, revised in 2018.* (1) There are two references to entrepreneurship, entrepreneurial learning Entrepreneurship is referred to in the general goals, and as a specific goal in the subject Civics. (2) Sustainable development (38 references) to all school forms Here the concept sustainable development is seen in the general goals as well as specific course goals. Home and consumer education, Biology, Chemistry, History, Crafts and Technology. Examples: Chemistry–sustainable development is presented as a general goal, is stated implicitly in year 1–3 and explicitly in year 4–6 regarding grade levels. Biology–implicit goal for 1–3, explicit for 4–6 Creativity The concept creativity appears 15 times. Here, the concept is seen in the fundamental values an tasks, as well as subject goals for Art, Music and Crafts. The concept creative ability occurs once. This is stated as an ability which pupils should acquire.	In the curriculum for Primary Teacher Education, Bachelor's Degree the signs of these issues or emphasis were either vague or only in some courses or not at all to be seen. Direct references to learning about or using the national core curriculum were to be seen in courses about: Reading and writing, information technology, mathematics, Icelandic, general introductory course, one course about visual arts, drama and music and in a course about curricular studies. However, none of these references were linked to or about the concepts and ideas we looked for. Indicators of emphasis in the spirit of entrepreneurship education and sustainability education can be seen in the description of different courses. For example, in the only course that mentions sustainability there is an indication of ways of working that consider student experience and context: "An emphasis will be put on students' different experience and premises built on individualised learning and inclusion in a multicultural society where critical and creative thinking will be a core thread throughout the course". Similarly in the same course critical and analytical thinking and independent work is encouraged in the aim: The student will be able to discuss critically methods, ideas and issues in assessment of learning. In the bound choice obligatory course about *philosophy and ideas in education*, main ideas and ideologies in education are presented and might therefore cover ideas such as innovation and or entrepreneurship education and sustainability education although no direct mention is in the course's description. These inexplicit indicators are all dependent on interpretation and guesswork and not clearly visible aims that can be associated with education for sustainable development or entrepreneurship education. It must be mentioned that teacher students can over the three years choose all in all ECTS from 59 courses (5 or 10 ECTS). In these courses some of them have explicit mentions of sustainability, initiative and creativity but they are not courses that all or most students choose and is up to chance which ones students choose.

Table 1. *Cont.*

Curriculum Case Studies: Country Examples from Primary Teacher Education, Bachelor's Degree Programme/Steps and Outcomes by Content Analysis	(1) Curriculum Case Finland: *Class Teacher Education (BA), University of Turku, Faculty of Education, Class Teacher Education, Rauma*	(2) Curriculum Case Sweden: Mid Sweden University	(3) Curriculum Case Iceland University of Iceland, School of Education, Primary Education B.Ed.
Case Conclusions	Entrepreneurship/entrepreneurial and sustainable education are widely approached, primarily through the basic education core curriculum. These studies are available in both compulsory and optional studies. Education is also provided in the preparation and implementation of the curriculum. Thus, it could be assumed that future teachers will be somehow trained to implement both entrepreneurial learning and sustainable development. On the other hand, the aims and contents of the teacher education curriculum hardly explicitly mention this elsewhere than in optional studies.	Entrepreneurship/entrepreneurial learning and sustainable development are approached in the general core curricula. However, in regard to the many goals and values, these concepts are few. The same is true for concepts referring to creativity. In the education program plans for primary education student teachers the concepts of entrepreneurship, sustainable development and creativity are not mention explicitly. This may mean that primary teacher student meet these concepts and content in specific courses. However, as prioritized elements of education, it is most likely that these concepts should be made more explicit to support teacher students in their work.	Entrepreneurship/entrepreneurial education is not visible as a distinctive element in the University of Iceland, School of Education, primary education B.Ed. program. This element "entrepreneurial learning" is once mentioned in the compulsory school core curriculum–however initiative is often mentioned there, often in relation to creativity and or independence. Neither were found in the B.Ed program. The sustainability concept (34 times) is also clearly visible in the compulsory school general curriculum and so is creativity (38 times in different compositions).

Note. Adapted from University of Turku, Studying at the Faculty of Education [65].

6. Discussion

To investigate how and to what extent the Nordic countries incorporate entrepreneurial and sustainable education into their teacher education curricula, we aimed to study how entrepreneurial, sustainable, and pro-environmental education emerge in Nordic (Finnish, Swedish, and Icelandic) primary teacher education curricula. As this is a study of primary teacher education, which also aims to teach primary level education goals and contents, our results highlight this interdisciplinarity. For example, the contents of a teacher education curriculum often include references to primary-level materials, such as the national core curricula for primary education.

In our study results indicate that teacher education curricula have somehow inexplicitly refers to EE, entrepreneurial learning, sustainable education, sustainability, environmental, and pro-environmental education. However, none of the curricula clearly and explicitly address these themes in three Nordic countries in our study. Our data also demonstrate that these primary level teacher education objectives focus on foundational learning and pedagogical activities. Might it not be surprising to see the slight emphasis on higher-order phenomena like sustainable development or EE? However, it is also notable that these academic studies in teacher education always involve, e.g., traditional subject studies that have ancestral roots in disciplines created during hundreds of years throughout academia. Therefore, we could also question if it is enough today that, e.g., studies of math, literature, history, and sciences involve randomly and inexplicitly sustainable and entrepreneurial education? Therefore, we would challenge curricula design in education: Do these crucial and transversal areas of education need to be more explicit in modern teacher education curricula?

Strategies guiding education and teacher education have led to the integration of cross-cutting themes such as sustainable development and EE into teaching, e.g., [2,8,9]. However, there is a difference between what the steering documents say and what happens in practice. Higher education institutions are also autonomous in deciding the content of their teaching. For example, if sustainable development or EE are not seen as essential themes in teaching, they do not necessarily have to be implemented [13,43–48]. Therefore, the activities of education organizations must also be viewed critically in the light of entrepreneurship and sustainable development.

Based on our results and their analysis, we propose the following practices for teacher education and curricula. First, EE and sustainable development would be explicitly addressed in teacher education curricula, its course descriptions, objectives, contents, and pedagogy, thus promoting entrepreneurial, e.g., critical pedagogy [36], and emancipatory pedagogy [10,19]. Second, it would be essential for students to have their own experiences of utilizing this type of education and pedagogy. These experiences affect what they teach, for example, in primary education and how.

While we would expect the Nordic model of education to promote e.g., entrepreneurship and sustainable education, e.g., [49,50], it seems that the importance of these themes has not been understood in-depth in the Nordic countries, and they do not emerge as clearly expressed in these three teacher education curricula in this case study. It seems that the Nordic countries are not significant forerunners in these goals, even if it could have been assumed.

Is it possible that the default is that educational institutions and teachers are sufficiently vigilant and therefore not sufficiently aware of the issue? Or maybe the teaching is done despite the curricula? Although the importance of cross-cutting educational themes is recognized, EE, for example, has also given rise to a wide range of debates. There have been discussions, e.g., of how EE is only related to business activities, even though EE promotes the skills needed in working life on a large scale [66]. Could this have caused some confusion in the Nordic teacher education? However, one should note that the corresponding education debate is not relevant to sustainable education. Therefore, at least not across the board, this conclusion about the background factors influencing the results of this study is not entirely relevant.

We could find out about this in the future studies, e.g., with surveys and interviews to develop our understanding of the phenomenon. The idea on integrating pro-environmental/sustainable and EE could be promoted in the future more explicitly, in which these interdisciplinary educational themes are taken into account more strongly in the curricula. In addition, it would be interesting to share experiences from these practices and at the same time seek models for so-called good practices, and developing communicative networks for teacher educators in the Nordic and global context which may help to push teacher education forward in terms of sustainable and entrepreneurial education. In the future, we could also include other Nordic countries, such as Norway and Denmark. On the other hand, we could deliver a broader international study. Undoubtedly, case and action studies would gather interesting information from these activities; these could be used to further accumulate interesting teaching practices, for example in teacher education. These processes could provide a meaningful basis for further curricula development. Furthermore, an analysis of the relevant global and national policy documents could provide insights into how these competences have transformed, and how these policies may be further improved to support sustainable and entrepreneurial education development.

7. Conclusions

In a changing modern era, a developed ability for creativity is an important attribute [67]. Today creative and innovative competencies are often put in relation to the climate crisis and to sustainability. Caiman and Lundegard [68] argue for the need of an education that supports and stimulates creative processes that can serve as a tool in the creation of a more sustainable world. Maybe it is not enough mainly to teach children many facts about sustainability. According to Hedefalk [69], starting from a problem and instead engaging the children in finding solutions can stimulate children and students to create an understanding, for example, of the underlying conflicts and interests that may have caused an environmental impact. Processes of innovative thinking, such as creating new and imagining things that do not exist, stimulate courage and belief that the future can be influenced and changed. Thinking about the potential of entrepreneurship and sustainable education, we wanted to find out how these types of issues have been taken into account in teacher education. We wanted to focus in this way initially on curricula, because by default, the content and activities of teaching are based on them.

In this research, we have studied how Nordic (Finnish, Swedish, and Icelandic) primary teacher education curricula involve entrepreneurial, sustainable, and pro-environmental education. According to our results, EE and sustainable development are taken into account to some extent in curricula. Deliberating how much of these themes are discussed today, we conclude that these educational goals and contents are limited. However, considering the scale of the phenomenon, this is a very small opening to explore the theme only from a few curricula point of view.

We have shown that multidisciplinary research on a theme in one study has its challenges. Nevertheless, on the other hand, adopting this approach has its advantages: EE and sustainable development education have much in common. For example, education for sustainable development (ESD) may be seen as a necessary journey towards sustainable education (SE), and EE could be a part of constructive steps towards sustainable education. EE includes affordances that contribute to inculcating analytical and creative competences. These comptetences are needed for responsible actions and provide cognitive elements (knowledge and understanding), emotions (identifying needs and problems) and action [19,40]. Taken together, we need entrepreneurial competencies to promote sustainable development. Our research, combining different entry angles, is a significant case opening in studying how to solve global problems by human thinking and behavior.

While our approach, which is built on the broader framework documents, emphasizes the complementarity of sustainable and entrepreneurial development, there may be tensions and even contradictions (e.g., development of enterprises built on wasteful consumption). A more solid justification for how these concerns might be addressed, as broadly as possible, and how more complementary models might be fostered would be beneficial. Critical thinking, creativity, teamwork, and the study of "real-world problems," as in phenomenon-based learning (29), could also be effective in addressing sustainability and entrepreneurship. In fact, they might be already occurring, as we previously discussed, or, conversely, are they somehow squeezed out or distorted by an emphasis on other priority areas within the respective curricula?

On the other hand, it is good to remember that EE as a concept is also good to transport in the development of the education system consciously. EE is intended to develop society in the direction of both developing and continuing entrepreneurship. If other concepts replace this concept, there is a risk that the aims and dimensions of entrepreneurship education will be blurred.

Our pilot study gives preliminary indicators into where education is moving onto. The results show that education does not respond adequately to societal hashes and even crises such as climate change. Therefore, it is also clear that we need more systematic policy guidance on the integration of interdisciplinary themes in education, such as EE and sustainable development. Moreover, we could consider who should take responsibility for developing these essential issues in education. How should this be handled? How is curricula development guided? Guidelines and actions for the future need to be more concise and explicit goals are needed to support these important areas of knowledge and skills. Furthermore, if universities have autonomy reflecting the needs for education from society, how could universities be motivated to such educational issues, if they are so crucial?

Finally summed up, we could also emphasize the importance of teacher education in the development of societies as a whole. Teacher education has a significant impact on educating citizens far into the future. Thinking about the changing world, we argue that student teachers should be prepared with tools and assignments that stimulate their reflections, creativity, and courage, which are important entrepreneurial attitudes to acquire in order to meet the demands of creating education for the future entrepreneurial citizens, where sustainable development is commonplace in their lives. Thinking, for example, of critical issues such as global warming, we should act now. We are currently training teachers who will soon be entering the workforce. These teachers train the whole nation, and their activities have a long-term impact on the future. If we want to influence the entrepreneurial and sustainable thinking of both children and young people, we should take better account of teacher education. This is also a road to implementing global strategies to enhance life-quality and well-being on a large scale.

Although our research opening is small at this stage, we see its added value, especially in how this opens the door to more advanced curricula development, educational activities, the creation of clearer guiding policy documents, and the research in the field [59] to promote entrepreneurial and sustainable education development in societies.

Author Contributions: Conceptualization, J.S.-L., S.R.J., M.H.-L. and M.W.; methodology, J.S.-L., S.R.J. and M.W.; validation, J.S.-L., S.R.J. and M.H.-L.; formal analysis, J.S.-L., S.R.J., M.H.-L. and S.E.-B.; writing—original draft preparation, J.S.-L., S.R.J., M.H.-L., S.E.-B.; writing—review and editing, J.S.-L., S.R.J. and M.H.-L. All authors have read and agreed to the published version of the manuscript.

Funding: This research received no external funding.

Institutional Review Board Statement: Not applicable.

Informed Consent Statement: Not applicable.

Data Availability Statement: There is no data in this study.

Conflicts of Interest: The authors declare no conflict of interest.

References

1. Giovannini, E.; Roure, F. The inclusion of Science, Technology and Innovation (STI) in the Financing of the 17 Sustainable Development Goals (SDGs). *Ann. Mines Responsab. Environ.* **2017**, *88*, 40–44. [CrossRef]
2. Apostolopoulos, N.; Al-Dajani, H.; Holt, D.; Jones, P.; Newbery, R. Entrepreneurship and the sustainable development goals. In *Entrepreneurship and the Sustainable Development Goals, Contemporary Issues in Entrepreneurship Research*; Emerald Publishing Limited: Bradford, UK, 2018; Volume 8, pp. 1–7.
3. European Commission Joint Research Centre. *EntreComp into Action: Get Inspired, Make It Happen*; EU, Publications Office: Brussels, Belgium, 2018. [CrossRef]
4. Seikkula-Leino, J.; Salomaa, M. Entrepreneurial Competencies and Organisational Change—Assessing Entrepreneurial Staff Competencies within Higher Education Institutions. *Sustainability* **2020**, *12*, 7323. [CrossRef]
5. Seikkula-Leino, J.; Salomaa, M.; Jónsdóttir, S.; McCallum, E.; Israel, H. EU Policies Driving Entrepreneurial Competences—Reflections from the Case of EntreComp. *Sustainability* **2021**, *13*, 8178. [CrossRef]
6. International Labour Organization Skills for a Greener Future. Key Findings. 2019. Available online: https://www.ilo.org/skills/projects/WCMS_709121/lang--en/index.htm (accessed on 14 November 2021).
7. Ruskovaara, E.; Pihkala, T.; Seikkula-Leino, J.; Järvinen, M.R. Broadening the resource base for entrepreneurship education through teachers' networking activities. *Teach. Teach. Educ.* **2015**, *47*, 62–70. [CrossRef]
8. OECD Publishing. *OECD Measuring Innovation in Education: A New Perspective, Educational Research and Innovation*; OECD Publishing: Pairs, France, 2014.
9. Valerio, A.; Parton, B.; Robb, A. *Entrepreneurship Education and Training Programs around the World: Dimensions for Success*; World Bank Publications: Washington, DC, USA, 2014.
10. Jónsdóttir, S.R.; Macdonald, M.A. The feasibility of innovation and entrepreneurial education in middle schools. *J. Small Bus. Enterp. Dev.* **2019**, *26*, 255–272. [CrossRef]
11. Filsher, M.; Kraus, S.; Roig-Tiemo, N.; Kailer, N.; Fischer, U. Entrepreneurship as a Catalyst for Sustainable Development. Opening the Black Box. *Sustainability* **2019**, *11*, 4503. [CrossRef]
12. Apostolopoulos, N.; Liargovas, P. Unlock Local Forces and Improve Legitimacy: A Decision Making Scheme in the European Union Towards Environmental Change. *Eur. Policy Anal.* **2018**, *4*, 146–165. [CrossRef]
13. Juntunen, M.K.; Aksela, M.K. Education for sustainable development in chemistry—challenges, possibilities and pedagogical models in Finland and elsewhere. *Chem. Educ. Res. Pract.* **2014**, *15*, 488–500. [CrossRef]
14. Seikkula-Leino, J. *Opetussuunnitelmauudistus Ja Yrittäjyyskasvatuksen Toteuttaminen [Curriculum Reform and the Implementation of Entrepreneurship Education]. Opetusministeriön Julkaisuja [Publications of Finnish Ministry of Education] 2007: Koulutus- Ja Tiedepolitiikan Osasto [The Department of Education and Science Policy]*; Yliopistopainos: Helsinki, Finland, 2007.
15. Seikkula-Leino, J.; Ruskovaara, E.; Hannula, H.; Saarivirta, T. Facing the Changing Demands of Europe: Integrating Entrepreneurship Education in Finnish Teacher Training Curricula. *Eur. Educ. Res. J.* **2012**, *11*, 382–399. [CrossRef]
16. Gibb, A. *The Future of Entrepreneurship Education—Determining the Basis for Coherent Policy and Practice?* Kyrö, P., Carrier, C., Eds.; The Dynamics of Learning Entrepreneurship in a Cross-Cultural University Context; Entrepreneurship Education; University of Tampere, Research Centre for Vocational and Professional Education: Hämeenlinna, Finland, 2005.
17. Kollmuss, A.; Agyeman, J. Mind the Gap: Why People Act Environmentally and What Are the Barriers to pro-Environmental Behaviour. *Environ. Educ. Res.* **2002**, *8*, 239–260. [CrossRef]
18. United Nations Transforming Our World: The 2030 Agenda for Sustainable Development. 2015. Available online: https://sustainabledevelopment.un.org/content/documents/21252030%20Agenda%20for%20Sustainable%20Development%20web.pdf (accessed on 14 November 2021).
19. Jónsdóttir, S.R.; Gunnarsdóttir, R. *The Road to Independence: Emancipatory Pedagogy*; Sense: Rotterdam, The Netherlands, 2017.
20. Gibb, A. Entrepreneurship/Enterprise Education in Schools and Colleges: Are We Really Building the Onion or Peeling It Away? *Paper Presented in National Council for Graduate Entrepreneurship, Working Paper 039/2006.* 2006. Available online: http://ncge.com/communities/research/reference/detail/880/7 (accessed on 14 November 2021).
21. Pittaway, L.; Cope, J. Entrepreneurship Education: A Systematic Review of the Evidence. *Int. Small Bus. J.* **2007**, *25*, 479–510. [CrossRef]
22. Pittaway, L.; Cope, J. Simulating Entrepreneurial Learning: Integrating Experiential and Collaborative Approaches to Learning. *Manag. Learn.* **2007**, *38*, 211–231. [CrossRef]
23. Seikkula-Leino, J.; Ruskovaara, E.; Pihkala, T.; Rodríguez, I.D.; Delfino, J. Developing entrepreneurship education in Europe: Teachers' commitment to entrepreneurship education in the UK, Finland and Spain. In *The Role and Impact of Entrepreneurship Education*; Edward Elgar Publishing: Cheltenham, UK, 2019; pp. 130–145.
24. Deveci, I.; Seikkula-Leino, J. A Review of Entrepreneurship Education in Teacher Education. *Malays. J. Learn. Instr.* **2018**, *15*, 105–148. [CrossRef]
25. Kyro, P.; Seikkula-Leino, J.; Mylläri, J.; Borch, O.; Fayolle, A.; Ljunggren, E. Meta Processes of Entrepreneurial and Enterprising Learning: The Dialogue between Cognitive, Conative and Affective Constructs. *Entrep. Res. Eur.* **2011**. [CrossRef]
26. Hynes, B.; Richardson, I. Entrepreneurship Education: A Mechanism for Engaging and Exchanging with the Small Business Sector. *Educ. + Train.* **2007**, *49*, 732–744. [CrossRef]

27. European Commission Entrepreneurship 2020 Action Plan: Reigniting the Entrepreneurial Spirit in Europe. 2013. Available online: https://www.smefinanceforum.org/post/the-entrepreneurship-2020-action-plan-reigniting-the-entrepreneurial-spirit-in-europe (accessed on 14 November 2021).
28. Shepherd, D.A.; Patzelt, H. The new field of sustainable entrepreneurship: Studying entrepreneurial action linking "what is to be sustained" with "what is to be developed". *Entrep. Theory Pract.* **2011**, *35*, 137–163. [CrossRef]
29. Hermann, R.R.; Bossle, M.B. Bringing an entrepreneurial focus to sustainability education: A teaching framework based on content analysis. *J. Clean. Prod.* **2020**, *246*, 119038. [CrossRef]
30. Hsu, J.; Pivec, M. Integration of Sustainability Awareness in Entrepreneurship Education. *Sustainability* **2021**, *13*, 4934. [CrossRef]
31. Edokpolor, J.E. Entrepreneurship education and sustainable development: Mediating role of entrepreneurial skills. *Asia Pac. J. Innov. Entrep.* **2020**, *14*, 329–339. [CrossRef]
32. Bacigalupo, M.; Kampylis, P.; Punie, Y.; Van den Brande, G. *EntreComp: The Entrepreneurship Competence Framework*; Publication Office of the European Union: Luxembourg, 2016.
33. OECD. *OECD Learning Framework 2030*; OECD Publishing: Pairs, France, 2018.
34. Lonka, K.; Makkonen, J.; Berg, M.; Talvio, M.; Maksniemi, E.; Kruskopf, M.; Lammassaari, H.; Hietajärvi, L.; Westling, S.K. Phenomenal Learning from Finland. *Edita* **2018**. Available online: https://researchportal.helsinki.fi/en/publications/phenomenal-learning-from-finland (accessed on 14 November 2021).
35. Jóhannesson, I.Á.; Norðdahl, K.; Óskarsdóttir, G.; Pálsdóttir, A.; Pétursdóttir, B. Curriculum analysis and education for sustainable development in Iceland. *Environ. Educ. Res.* **2011**, *17*, 375–391. [CrossRef]
36. Cars, M.; West, E. Education for sustainable society: Attainments and good practices in Sweden during the United Nations Decade for Education for Sustainable Development (UNDESD). *Environ. Dev. Sustain.* **2014**, *17*, 1–21. [CrossRef]
37. *UNESCO Report on the UN Decade of Education for Sustainable Development, Shaping the Education of Tomorrow, Abridged*; UNESCO: Paris, France, 2012. Available online: https://sustainabledevelopment.un.org/content/documents/919unesco1.pdf (accessed on 14 November 2021).
38. Rosen, M.A. Sustainability: A Crucial Quest for Humanity—Welcome to a New Open Access Journal for a Growing Multidisciplinary Community. *Sustainability* **2009**, *1*, 1. [CrossRef]
39. Mogensen, F.; Schnack, K. The action competence approach and the 'new' discourses of education for sustainable development, competence and quality criteria. *Environ. Educ. Res.* **2010**, *16*, 59–74. [CrossRef]
40. Jónsson, Ó.P.; Guðmundsson, B.; Øyehaug, A.B.; Didham, R.J.; Wolff, L.-A.; Bengtsson, S.; Lysgaard, J.A.; Gunnarsdóttir, B.S.; Árnadóttir, S.M.; Rømoen, J.; et al. Mapping Education for Sustainability in the Nordic Countries. In *Policies for the promotion of BECCS in the Nordic Countries*; Nordic Council of Ministers: Copenhagen, Denmark, 2021.
41. Sterling, S. *Sustainable Education: Re-Visioning Learning and Change*; Green Books: Cambridge, UK, 2001.
42. Seikkula-Leino, J.; Satuvuori, T.; Ruskovaara, E.; Hannula, H. How Do Finnish Teacher Educators Implement Entrepreneur- Ship Education? *Educ. + Train.* **2015**, *57*, 392–404. [CrossRef]
43. Franzen, R.L. Environmental Education in Teacher Education Programs: Incorporation and Use of Professional Guideline. *J. Sustain. Educ.* **2017**, *16*, 1–18.
44. McConnell, B. Teacher Education in Environmental Education—Does it Work? *Aust. J. Environ. Educ.* **2001**, *17*, 35–39. [CrossRef]
45. Moore, J. Barriers and pathways to creating sustainability education programs: Policy, rhetoric and reality. *Environ. Educ. Res.* **2005**, *11*, 537–555. [CrossRef]
46. Reddy, C. Environmental education in teacher education: A viewpoint exploring options in South Africa. *South. Afr. J. Environ. Educ.* **2017**, *33*, 117. [CrossRef]
47. Ormond, C.; Zandvliet, D.; McLaren, M.; Robertson, P.; Leddy, S.; Metcalfe, S. Environmental Education as Teacher Education: Melancholic Reflections from an Emerging Community of Practice. *Can. J. Environ. Educ.* **2014**, *19*, 160–179.
48. Van Petegem, P.; Blieck, A.; Pauw, J.B.-D. Evaluating the Implementation Process of Environmental Education in Preservice Teacher Education: Two Case Studies. *J. Environ. Educ.* **2007**, *38*, 47–54. [CrossRef]
49. Halonen, M.; Persson, Å.; Sepponen, S.; Siebert, C.K.; Bröckl, M.; Vaahtera, A.; Quinn, S.; Trimmer, C.; Isokangas, A. Sustainable Development Action—The Nordic Way. In *Policies for the promotion of BECCS in the Nordic Countries*; Nordic Council of Ministers: Copenhagen, Denmark, 2017.
50. Blossing, U.; Imsen, G.L.; Moos, L. *"A School for All". Encounters Neo-Liberal Policy. Policy Implications of Research in Education*; Springer: Dordrecht, The Netherland, 2014.
51. Bobbitt, F. Scientific Method in Curriculum Making. In *The Curriculum Studies Reader*; Flinders, D.J., Thornton, S.J., Eds.; Routledge Falmer: New York, NY, USA, 2004.
52. Letschert, J.; Kessels, J.; Akker, J.V.D.; Kuiper, W.; Hameyer, U. Social and Political Factors in the Process of Curriculum Change. In *Curriculum Landscapes and Trends*; Springer: Berlin/Heidelberg, Germany, 2004; pp. 157–176.
53. Leathwood, C.; Phillips, D. Developing curriculum evaluation research in higher education: Process, politics and practicalities. *High. Educ.* **2000**, *40*, 313–330. [CrossRef]
54. Annala, J.; Mäkinen, M. Korkeakoulutuksen Opetussuunnitelma—Kohti Tietämisen Ja Taitamisen Päämääriä [Curriculum in Higher Education—towards the Goals of Knowing and Mastering]. *Kasvatus* **2011**, *42*, 6–18.
55. Barnett, R.; Coate, K. *Engaging the Curriculum in Higher Education*; Open University Press/McGraw-Hill: Buckingham, UK, 2005.
56. Carl, A.E. *Teacher Empowerment through Curriculum Development: Theory into Practice*, 3rd ed.; Juta: Cape Town, South Africa, 2009.

57. Zhong, Q.-Q. Curriculum reform in China: Challenges and reflections. *Front. Educ. China* **2006**, *1*, 370–382. [CrossRef]
58. Hsieh, H.-F.; Shannon, S.E. Three Approaches to Qualitative Content Analysis. *Qual. Health Res.* **2005**, *15*, 1277–1288. [CrossRef]
59. Cohen, L.; Manion, L.; Morrison, K. *Research Methods in Education*; Routledge: Abington, UK, 2017.
60. The Finnish Ministry of Education and Culture, Teacher Education in Finland. 2016. Available online: https://okm.fi/documents/1410845/4150027/Teacher+education+in+Finland/57c88304-216b-41a7-ab36-7ddd4597b925 (accessed on 14 November 2021).
61. University of Turku International Bachelor's and Master's Degree Programmes at the University of Turku. 2021. Available online: https://www.utu.fi/en/study-at-utu/bachelors-and-masters-degree-programmes (accessed on 14 November 2021).
62. Aðalsteinsdóttir, K. Að Styrkja "Haldreipi Skólastarfsins". Menntun Grunnskólakennara á Íslandi í 100 Ár [Strengthening the "Mainstay of Schoolwork"-Teacher Training in Iceland for One Hundred Years]. Tímarit um menntarannsóknir. *J. Educ. Res.* **2007**, 57–82. Available online: https://skemman.is/handle/1946/15099?locale=en (accessed on 14 November 2021).
63. Menntavísindasvið Grunnskólakennsla Yngri Barna [Teaching Primary Level in Compulsory School]. Available online: https://www.hi.is/grunnskolakennsla_yngri_barna (accessed on 14 November 2021).
64. Sigurðardóttir, A.K.; Jóhannesson, I.Á.; Óskarsdóttir, G. Challenges, Contradictions and Continuity in Creating a Five-Year Teacher Education Programme in Iceland. *Educ. North* **2018**, *25*, 135–154. [CrossRef]
65. University of Turku, Studying at the Faculty of Education. Available online: https://www.utu.fi/en/university/faculty-of-education/studying. (accessed on 14 November 2021).
66. Jónsdóttir, S.R.; Þórólfsson, M.; Karlsdóttir, J.; Finnbogason, G.E. Að Uppfæra Ísland: Sýn Stjórnenda Íslenskra Framhaldsskóla á Nýsköpunar-Og Frumkvöðlamennt Og Framkvæmd Námssviðsins í Námskrárfræðilegu Ljósi. [Upgrading Iceland: Icelandic Secondary School Administrators' Views about Innovation and Entrepreneurial Education (IEE) within the Context of Curriculum Studies. In Icelandic, Long Abstract in English]. Netla—Veftímarit Um Uppeldi Og Menntun. *Netla* **2014**. Available online: https://ojs.hi.is/netla/article/view/1944 (accessed on 14 November 2021).
67. European Commission Rethinking Education Strategy; 2012. Available online: https://ec.europa.eu/commission/presscorner/detail/en/IP_12_1233 (accessed on 14 November 2021).
68. Caiman, C.; Lundegard, I. *Kreativitet För Hållbar Utveckling*; Swedish National Agency for Education: Stockholm, Sweden, 2018.
69. Hedefalk, M. *Förskola För Hållbar Utveckling: Förutsättningar För Barns Utveckling Av Handlingskompetens För Hållbar Utveckling*; Uppsala University: Diss, UK, 2014.

 *sustainability*

Article

Education for Social Change: The Case of Teacher Education in Wales

Rebecca Weicht [1] and Svanborg R. Jónsdóttir [2,*]

[1] Department of Strategy, Enterprise and Sustainability, Manchester Metropolitan University, All Saints, Manchester M15 6BH, UK; r.weicht@mmu.ac.uk
[2] School of Education, University of Iceland, Stakkahlíð 1, Reykjavík 105, Iceland
* Correspondence: svanjons@hi.is

Abstract: Entrepreneurial education offers valuable opportunities for teachers to foster and enhance creativity and action competence, which are also important for sustainability education. The University of Wales Trinity Saint David (UWTSD) is a leader in the development of entrepreneurial education in teacher education both in Wales and internationally. The objective of this article is to shed light on how an entrepreneurial education approach can help foster social change. The aim of this study is to learn from teacher educators at UWTSD about how they support creativity, innovation, and an enterprising mindset in their learners. A case study approach is applied. By analysing documentary evidence such as module and assignment handbooks, we explore how teacher educators at UWTSD deliver entrepreneurial education for social change. Our findings indicate that UWTSD's development of entrepreneurial education in teacher training has enabled constructive learning, cultivating creativity and action competence. We provide examples that display how the intentions of the Curriculum for Wales and entrepreneurial education approaches of the UWTSD emerge in practice. These examples show outcomes of the entrepreneurial projects that evince the enactment of social change. The findings also show that the educational policy of Wales supports entrepreneurial education throughout all levels of the educational system.

Keywords: entrepreneurial education; sustainability education; social change; creativity; innovation; action competence

Citation: Weicht, R.; Jónsdóttir, S.R.. Education for Social Change: The Case of Teacher Education in Wales. *Sustainability* **2021**, *14*, 8574. https://doi.org/10.3390/su13158574

Academic Editors: Jaana Seikkula-Leino, Mats Westerberg, Priti Verma and Maria Salomaa

Received: 16 June 2021
Accepted: 26 July 2021
Published: 31 July 2021

Publisher's Note: MDPI stays neutral with regard to jurisdictional claims in published maps and institutional affiliations.

1. Introduction

Entrepreneurial education has for some time been seen primarily as education about business, with its importance in its potential contribution to the economic progress of societies. Here, we follow the definition of entrepreneurial education as a "catch-all" term that comprises both enterprise and entrepreneurship as outlined in the guidance document "Enterprise and Entrepreneurship Education: Guidance for UK Higher Education Providers" by the UK's quality body for higher education, The Quality Assurance Agency (QAA) [1]. Enterprise education denotes the development of students' capabilities as critical and future-orientated thinkers who are civic-minded and socially responsible. In contrast, entrepreneurship education focuses on fostering the competencies outlined in enterprise education, but within the specific context of creating a new venture [1]. In recent decades, the understanding of entrepreneurial education has widened to encompass an area of learning that cultivates creativity, action, and critical thinking [2–6]. Acknowledging this view, entrepreneurial education can be seen as enhancing personal and cultural growth, economic and technological development, and scientific discovery [7,8]. Some researchers describe the core pedagogy of entrepreneurial education as *emancipatory pedagogy*, where the learners have ample agency and the teacher gradually moves from strong framing towards giving learners total control of their projects [9,10]. To respond to the uncertainties of the future and to imagine possible futures, people in the modern world need the ability to be creative and innovative, as this is important for dealing with the intertwined challenges of economic, social and environmental issues [11]. The development of *action competence* is also a key to *sustainability education* [12]. Learners of the 21st century need a broad skillset to function in a sustainable world, including collaboration, problem framing, critical thinking, innovation, and creativity [13,14]. Sustainability and education for sustainability are complex endeavours that must build on an understanding of the interconnectedness and multidisciplinarity of the economic, social, institutional, and environmental aspects of society [15–17]. Entrepreneurial education can also drive the changes in education needed to support and inculcate competences for sustainability [5,18].

Yet, while across EU member states (and beyond) there is a consensus for the need for entrepreneurial skills—which are acknowledged to be key to learners' personal and professional lives [19,20]—the teaching of entrepreneurial skills in Europe's schools is patchy. Eurydice [21] found entrepreneurship education in schools to be fragmented and not yet prioritised. Specifically, the researchers found that over half of the researched countries had few or no guidelines on entrepreneurship education teaching methods, and that it was rarely addressed in initial teacher education (but more common in teacher continuous professional development [21]). Equally, no country had fully mainstreamed entrepreneurship education [21].

It is against this background that we can learn from Wales, which has been at the frontier of change in advancing entrepreneurial education both in policy and in practice [22]. The Welsh education sector benefits from over a decade of experience in entrepreneurial teacher training, and Welsh education policy has influenced European-level education policy in relation to entrepreneurial education development [23]. Among other activities, Welsh researchers and educationalists helped develop the European Commission's "Entrepreneurship Competence Framework" (EntreComp) [23], which is the foundation for the EU's Entrepreneurship key competence for lifelong learning [15,20,22,24].

One reason that Wales is advanced in delivering entrepreneurial education is that the Welsh education system has responded swiftly and strategically to repeated criticism of underperformance [25]. In the 2014 Programme for International Assessment (PISA) study, the Organisation for Economic Co-operation and Development (OECD) found that the Welsh education system was producing a high number of "low performers" and that schools were unable to meet learners' needs [26] (p. 7). They also found that inequality persisted because educational results still closely correlated with socio-economic status [25,27]. The OECD thus concluded that the Welsh education system needed a "radical restructuring" [25] (p. 318). Today, the Welsh Government is in the process of rolling out a new curriculum ("Curriculum for Wales", CFW) specifically focussed on skills and on teaching learners to become "ambitious, capable learners, ready to learn throughout their lives" and "enterprising, creative contributors, ready to play a full part in life and work" [28] (p. 11). One of the interesting core concepts in the CFW is *cynefin*, a sense of belonging and identity in a historic, cultural and social context, providing a foundation for a local and international citizenship [28].

However, delivering on such ambitious goals requires enterprising educators [29] such as those at the University of Wales Trinity Saint David (UWTSD). Here, we explore this university as a case study in learning how to help teachers become entrepreneurial educators and successfully deliver the new CFW. Educators and researchers from UWTSD have left a footprint at the international, European, and national levels [30,31]. Among others, they lent their expertise to develop a framework and national teacher training course in North Macedonia in a World Bank-funded programme [21]. Equally, UWTSD's newly developed Doctorate in Education (EdD), which was informed by and integrates the EU's EntreComp framework, has been featured in the "EntreComp into Action" user guide to the framework [32]. At national level, UWTSD professor Andy Penaluna chaired the development of the 2018 updated version of the QAA Guidance for UK Higher Education Providers on Enterprise and Entrepreneurship Education [33].

The purpose of this study is to present an example of educational policy about entrepreneurial education that can be a model for other policy makers in education interested in sustainability thinking and actions for social change. The aim is to shed light on how Wales has developed entrepreneurial education and how UWTSD has put this policy in practice.

In the following sections, we outline our theoretical framework as derived from Wales' proposal for a "Curriculum for Wales" that is currently being rolled out and seeks to create an entrepreneurial culture. We then present our methods section to finally compare how the core demands of the new curriculum have already been implemented in UWTSD's education, and teacher training specifically, for some time.

2. Background

The new Curriculum for Wales (CFW) is a strategic response to criticism on the educational system in Wales. In this section, we first introduce the core elements in the CFW to highlight the innovative aspects in its pedagogy, approaches, assessment, and ideologies. The second part of this section will then outline the theoretical framework that derives from our analysis of the CFW.

2.1. Developing the Curriculum for Wales

Wales' new curriculum is the culmination of review and change processes that started over a decade ago. Following a series of bad results in international comparisons, the Welsh government sought to create an entrepreneurial culture both through its curriculum and assessment approaches [22,25]. As early as 2006, a review of initial teacher education in Wales was undertaken and recommendations made on how to improve it [34]. In 2013, as part of a multi-step plan to better standards in Welsh schools, Professor Ralph Tabberer, a former Director-General of Schools in England, also reviewed initial teacher training [35]. Tabberer found the quality of initial teacher training in Wales to be no better than "adequate" and pointed to problems in recruitment, quality, and consistency, as well as a lack of competition among initial teacher training providers in Wales [35] (p. 14). He made 15 recommendations to inform future policy decisions and to raise the quality and consistency of teaching and assessment in initial teacher training, including to improve leadership in the sector and the status of teachers in Welsh society to attract the best candidates to the profession [35].

In 2015, the "Successful Futures; Independent Review of Curriculum and Assessment Arrangements in Wales" report (or Donaldson review) sought to determine how Welsh schools can prepare learners for an uncertain future [33]. Donaldson outlined that education should help learners become enterprising and creative contributors who are ready to play their full part in life and work, as well as ethical and informed citizens who are ready to be citizens of Wales and the world [33]. He focussed his proposal on how to help learners become active citizens and lifelong learners [33]. This played out in a pedagogy focussed on methods to develop skills with an emphasis on progression [33]. In the same manner, assessment should be formative and become an "essential" part of teaching [33] (p. 76). Donaldson's recommendations for how to adapt education in Wales for the future were accepted in full by the Welsh government and became the blueprint for the new CFW, which will be rolled out as of the 2022/23 school year and aims for every child and young person (aged 3 to 16) to become:

1. Ambitious, capable learners, ready to learn throughout their lives;
2. Enterprising, creative contributors, ready to play a full part in life and work;
3. Ethical, informed citizens of Wales and the world;
4. Healthy, confident individuals, ready to lead fulfilling lives as valued members of society [28] (p. 11).

The CFW has been described as a "bold new vision for curriculum, teaching and learning" and a "radical departure from the top-down, teacher proof policy of the previous National Curriculum" [36] (p. 181–182). Others noted that "pupils will encounter knowledge very differently from previous generations" because of its move away from subjects and the autonomy it offers schools and teachers on how to deliver content, as well as its learner-centred focus and focus on transversal skills [37] (p. 7). Power et al. report that teachers are both excited and nervous about the new curriculum [24]. They are excited because they see the new curriculum as less "prescriptive" and thus suffering less from burdensome administrative work [38] (p. 5), but also nervous in the face of the unknown.

The CFW outlines the details of the new curriculum's aims and objectives [28]. It explains that the aim for learners to be ambitious and capable means that they are both capable of solving problems, and also enjoy the challenge of solving them. To be enterprising, creative contributors requires learners to be creative in their approach to solving problems and to "give of their energy and skills so that other people will benefit" [28] (p. 24). To be an ethical, informed citizen means to consider others, the environment, and one's actions, and to be ready to engage with contemporary issues based on one's values [28].

Specific skills are considered "integral" to enabling the four purposes of the CFW [28] (p. 25). They are 1. Creativity and innovation, 2. Critical thinking and problem-solving, 3. Personal effectiveness, and 4. Planning and organising. Creativity and innovation mean for learners to be curious and inquisitive. Personal effectiveness means that learners are able to evaluate their thinking and mistakes and to be able to identify and recognise different types of value [28]. Planning and organising is the ability to put ideas into action. The guidance specifically states:

When developing these skills, learners should:

- Develop an appreciation of sustainable development and the challenges facing humanity;
- Be afforded the space to generate creative ideas and to critically evaluate alternatives—in an ever-changing world, flexibility and the ability to develop more ideas will enable learners to consider a wider range of alternative solutions when things change [28] (p. 26).

The curriculum clearly calls for empowering learners to become active agents of building a socially just and sustainable society. The guidance asks learners to "appreciate the contribution they and others can make within their different communities and to develop and explore their responses to local, national and global matters" [28] (p. 30).

Until the CFW is rolled out across Wales, so-called "Pioneer Schools" are tasked with the development of the curriculum in more detail. The CFW guidance document so far only outlines six "areas": Expressive Arts, Health and Well-being, Humanities; Language, Literacy and Communication; Mathematics and Numeracy; and Science and Technology. It does not break down knowledge into subjects that should be taught at different levels [28]. Instead, the focus is on interdisciplinary, and student-centred, active learning with real-life relevance [24]. The Welsh government believes that schools are best placed to make decisions about learners' needs and has thus tasked schools in their detailed guidance to develop the curriculum based on "What matters" statements [24]. Upon implementation in schools, it is then up to the schools to develop a vision for themselves and design a curriculum to implement that vision in their school [28].

2.2. What the CFW Means for Teacher Agency and Expectations

Such an innovative approach to curriculum and teaching demands a lot of autonomy and agency from teachers [28]. It is here that teacher education that focuses on developing innovation and action skills of teachers becomes crucial. Newton et al., in surveying teachers and headteachers, found that teacher perception of the new curriculum depends on their schools [38]. They found that, while the perception of the new curriculum is shaded by the "bad" perception of the current curriculum, respondents felt that the Pioneer Schools had a more positive outlook than schools outside the Pioneer School network [38] (p. 9). This may be because the Pioneer Schools are better prepared and have access to more and better resources. Their survey respondents also described the Pioneer Schools as "innovative" and "progressive" places where teachers already used the teaching methods as laid out in the Successful Futures report, the foundation for the CFW [38]. They quote one teacher as stating:

> There's nothing really new in Donaldson it's just good teaching, and the good teachers have been teaching in an Donaldson-esque way for a considerable length of time, it's just they didn't know what it was. It's just good teaching—making sure that it's relevant to the pupils [38] (p. 39).

Teachers who positively anticipate the new curriculum expect to see benefits from the focus on progression over attainment [38]. They expect that this approach allows for greater recognition of the different ways of learning achievements. For example, the focus on formative assessment is expected to be "more likely to support multiple pathways to learning" [38] (p. 13). Equally, the autonomy to move away from the previous model of tests and exams is viewed positively [25].

Therefore, the importance of teacher training in (successfully) delivering the new curriculum lies in giving teachers the tools and fostering the mindset to try new ideas and use the space they are given. It is here that the importance of innovative teacher training comes in. Teacher trainers and educators alike need to share good practices to help teachers adapt to the mindset shift that the CFW demands. In the words of Welsh educators:

> "It's a difficult one because it's 'change your mindset' more than resource."
>
> "I think a lot of heads will need to become far more creative and change their mindsets, look at the curriculum design issue. It's not going to be a box ticking exercise thank god, we've had that. This has got to be a lot more evolved and it's got to be a change of mindset."
>
> "Have they got the skills to do those things because we've never taught in that particular way and you can't just suddenly change the mindset of a profession that's almost going to take a generation to re-educate that profession to do things differently" [38] (p. 47).

2.3. *Curriculum for Wales Delivery and Assessment*

The following section outlines in detail how education delivery and assessment are described in the CFW, as well as where there is an overlap with sustainability education and education for social change. The understanding of how the CFW is meant to be delivered, how assessment is meant to take place, and how the CFW views active and responsible citizens will then guide the empirical analysis in the following sections [39].

2.3.1. CFW Delivery

The CFW specifies that the delivery of the curriculum should make use of external practitioners and their expertise. For example, in the Expressive Arts Area of Learning and Experience, this may include visits to theatres and galleries as well as bringing the expertise of external practitioners into the classroom [28]. Relatedly, to enhance learners' skills, learning and teaching should be delivered using a range of teaching and learning approaches, including digital ones. For instance, in the Mathematics and Numeracy Area of Learning and Experience, learners may work together using digital skills and to solve a problem and develop an algorithm that supports the understanding of patterns. They may also use digital tools to create graphs from spreadsheets, for example [28].

Similarly, learning and teaching should take place in a range of contexts and be cross-curricular. The CFW is cross-disciplinary within and across Areas. It also sets out three mandatory cross-curricular skills: literacy, numeracy, and digital competence [28] (p. 13). The mandatory cross-curricular skills "are essential to all learning and the ability to unlock knowledge. They enable learners to access the breadth of a school's curriculum and the wealth of opportunities it offers, equipping them with the lifelong skills to realise the four purposes" [28] (p. 27). For each of the six thematic areas, the mandatory cross-curricular skills are mapped out. For example, in the Well-being Area of Learning and Experience, numeracy "is a key enabler in making a number of informed decisions, in particular managing money and supporting good financial decision-making and critically engaging with social norms around money. Numeracy also plays a role in purchasing and preparing food to support nutrition" [28] (p. 85).

Learning and teaching should also ensure exposure to local, national, and international contexts at different stages of development. In addition, learning and teaching should take place in authentic contexts. For example, collaboration with individuals and employers provides learners with opportunities to learn about work, employment, and the skills valued in the workplace [28] (p. 44). This may also lead them to develop enterprise activities, which can provide authentic learning experiences that contribute to their development as enterprising, creative contributors to society. Equally, engaging parents and caregivers, school partners, and the local community can create authentic, contextualised learning opportunities. For instance, the overlap between the Humanities Area and the Mathematics and Numeracy Area might include the collection of a range of qualitative and quantitative primary data [28].

Furthermore, learning and teaching should allow for learners to develop their skills (e.g., critical thinking, problem solving, and decision-making) and for them to generate different types of value (financial, cultural, social, and learning) [28]. The integral skill "creativity and innovation" should support learners in creating different types of value.

Lastly, learning and teaching may be based on a whole-school approach. For example, in the Languages, Literacy and Communication Area, the CFW refers to the Content and Language Integrated Learning (CLIL) approach. It is stated that effective language learning requires a "systematic whole-school approach" that requires that schools be "aware how best to ensure progression for all learners in all their languages, for example through immersion, Content and Language Integrated Learning (CLIL) or plurilingual activities" [28] (p. 160).

All in all, the delivery of the CFW appears to be tailored as a holistic design that emphasises the interconnectedness of the elements within the whole, and where each area of learning, skill, or competence engages learners in a meaningful way.

2.3.2. Assessment in the CFW

Assessment in the CFW is formative and progression-focussed [28]. It should be ongoing to help the learner identify their strengths and improve their weaker areas. It should guide the learner to the steps needed to progress. The "overarching purpose is to support" and move learning forward [28] (p. 9). Progress is measured based on the statements of what matters [28]. The additional principles of progression aim to give educators a better understanding of progression [28]. They apply across the curriculum and explain what progress may look like and which principles underpin progress. These principles are: increasing breadth and depth of knowledge, deepening understanding of the ideas and disciplines within the Areas, refinement and growing sophistication in the use and application of skills, making connections and transferring learning into new contexts, and increasing effectiveness [28] (p. 129–131). In short, evaluation and improvement through reflection are core to the new curriculum and make "a vital contribution to raising the quality of education and standards of achievement" [28] (p. 229).

Assessment should also be holistic in providing insights into the learner's learning needs. It should include a wide range of assessment approaches to provide a full picture of the learner's development. It should include assessments by educators as well as the learner themselves. This may take place, for example, via portfolios that allow the learner to visualise their progress over time. Assessment should occur not just in the school but in exchange with the wider world as well. There should be engagement between the learner and the world around them, including parents or caregivers and practitioners. For instance, in relation to learning about careers and work-related experiences, the CFW indicates that experiences should stimulate an interest in different careers and that learning should take place in practical ways. Entrepreneurial activities, for example, necessitate reflection as a learning skill. They relate to practical activities such as business start-ups or venture-creation programmes [28] (p. 43). Similarly, in the arts, self-evaluation and reflection are part of the integral skills of critical thinking and problem-solving. The CFW explains:

> Refining work is encouraged throughout one of the statements of what matters in this Area, with the aim of building skills in self-evaluation and reflection. The evaluation involved in the creative process enables learners to develop reflective, questioning and problem-solving skills, as well as to challenge perceptions and identify solutions. Learners may demonstrate resilience in applying critical appraisal of their work and be expected to respond positively to critical feedback. Learners can develop problem-solving skills by experimenting with a variety of arts and artistic techniques [28] (p. 65).

2.3.3. Active, Informed, and Responsible Citizens

Many of the elements of the CFW indicate a will to inculcate in learners a competence for action in developing a socially just and sustainable society. Learners are meant to engage with important issues facing humanity, such as sustainability and social change, and to develop the skills necessary to do so. They are expected to learn to become active, informed, and responsible citizens and consumers who can identify with and contribute to their communities and reflect on the impacts of their actions. As in entrepreneurial education, there is an emphasis on a connection with society, authentic contexts, and cross-curricular areas through real life experiences. Learners are meant to learn to exercise their democratic rights, imagine possible futures, and take social action. They are expected to know or participate in enterprise and entrepreneurial activities and social action projects [28] (p. 123). Learners' creativity is meant to be stimulated and their capacity to produce solutions should grow as they engage with ethical issues of sustainability and business [28]. They must be able to make responsible decisions when acting socially, politically, economically, and entrepreneurially [28] (p. 102).

The CFW suggests that learners can get to know and explore the multiple and connected issues of sustainability through entrepreneurial education. Entrepreneurial education affords them opportunities to "understand the interconnected nature of economic, environmental and social sustainability; justice and authority; and the need to live in and contribute to a fair and inclusive society" [28] (p. 102) as learners get to experience real-life enterprises applying their own creativity and action competence in collaboration with others. A fascinating concept is presented in the CFW, *cynefin*, taken from Welsh:

> The place where we feel we belong, where the people and landscape around us are familiar, and the sights and sounds are reassuringly recognisable. Though often translated as 'habitat', cynefin is not just a place in a physical or geographical sense: it is the historic, cultural and social place which has shaped and continues to shape the community which inhabits it" [28] (p. 241).

Cynefin seems to embrace the individual, the local and global in understanding oneself as a part of a community and culture and realising how choices we all make can have impacts on society.

Tables 1–3 summarise the major themes of how the CFW should be delivered, assessed, and where it overlaps with education for social change and sustainability through entrepreneurial education. We will use these tables to identify how and where the characteristics from the CFW emerge in UWTSD documents.

Table 1. Learning and teaching delivery as outlined in the CFW.

Learning and Teaching in the CFW
• Learning and teaching should be collaborative and cross-disciplinary (pp. 6, 47, 50)
• Learning and teaching should make use of external practitioners (e.g., pp. 53, 88, 227–228)
• Learning and teaching should be delivered using a range of teaching and learning approaches (e.g., pp. 35, 57, 89, 116, 119)
• Learning and teaching should take place in a range of contexts and be cross-curricular (e.g., pp. 8, 24, 26, 34, 44)
• Learning and teaching should allow for learners to develop their skills (e.g., critical thinking, problem solving, decision-making) and for them to generate different types of value (financial, cultural, social, learning value; e.g., pp. 6, 23–26)
• Learning and teaching should ensure exposure to local, national and international contexts at different stages of development (e.g., pp. 30, 102)
• Learning and teaching should take place in authentic context(s) (e.g., pp. 66, 96)

Table 2. Assessment as outlined in the CFW.

Assessment in the CFW
• Assessment should enable reflection on learner progress over time (e.g., it should inform a learner on their strengths and achievements, as well as areas for improvement and, if relevant, barriers to learning; e.g., p. 8)
• Assessment should also enable reflection on group progress over time (e.g., at school level)
• A wide range of assessment approaches should be used to provide a holistic picture of learners' development (pp. 6, 31)
• There should be engagement between the learner and the world outside of school, incl. parents or carers, and practitioners (p. 226)
• Learners should participate in the assessment process (e.g., reflect on their learning journey; e.g., pp. 51, 92, 157, 185)
• As learners progress, they should become increasingly effective. This includes increasingly successful approaches to self-evaluation, the identification of their next steps in learning and more effective means of self-regulation (p. 30)

Table 3. Education for social change as outlined in the CFW.

Education for Social Change in the CFW
• Learners should be empowered to become active agents of building a socially just and sustainable society (e.g., pp. 12, 19, 30, 41, 42, 76, 97, 98, 102)
• Learner engagement is emphasised. Learning should take place in authentic contexts across curricular areas (pp. 48–50)
• Learners should adopt an enterprising spirit and action competence. Learners should be able to create value of different kinds—financial, cultural, and social (p. 25)
• Learners should become enterprising in managing their own and others' resources, valuing failure as a part of the creative process, and relatedly strengthening their employability skills (e.g., pp. 65, 73, 85, 98, 123)
• Learners should become sustainable citizens through a sustainable education, and should be able to respond to challenges of a social, economic and environmental nature (e.g., pp. 45, 70, 98, 102, 120)
• Learners should be able to make responsible decisions, to act as caring, participative citizens of their local, national, and global communities, committed to justice, diversity and the protection of the environment (e.g., pp. 70, 98, 102)

3. Methods

We apply a case study methodology to achieve our research objective. With its focus on understanding the how and why of a social phenomenon in context, a case study lends itself to learning from Wales, and UWTSD in particular, to understand how educators are supported to cultivate competences for social change in learners [40].

To address our research aim, we drew on more than a dozen data sources. We drew on peer-reviewed papers, internal and external UWTSD documents such as Annual Reports as well as programme handbooks, and websites or blogs (see Table 4). For instance, where they were available, we sought to primarily rely on peer-reviewed papers to highlight the entrepreneurial education approach used by UWTSD (e.g., [37,41]). This means that our data sources had already passed a level of quality control through peer-review. We considered it appropriate to draw on websites such as project websites and, for example, their blogs (e.g., [42,43]) to illustrate how some of these concepts are applied in practice. All data are listed in the reference list. Most documents are available online and can be accessed freely. All accessed materials were in English. By triangulating these different forms of data (academic papers, blog posts, project and annual reports, websites), we seek for our case study to become a rich and robust account that is comprehensive and well developed, thus helping facilitate deep understanding [44]. Thick description then allows us to evaluate to what extent our conclusions may be transferable [44] and to illustrate the theory and how we came to assess if and that UWTSD met the CFW spirit.

Table 4. Overview of data sources.

Document Type	Number
Academic papers (incl. conference presentations)	8
Reports (e.g., technical policy reports)	3
External UWTSD documentation	1
Internal UWTSD documentation	2
Other resources (e.g., project websites and blogs)	2

The authors started collaborating on the research project in early January 2021 by designing the research and dividing work. We held on-line hour-long meetings every 10–14 days where we discussed the process and reflected on and responded to what was emerging. The data collection and analysis was conducted iteratively as we first scanned major documents (such as the CFW and module handbooks) looking for signs of answers to our research questions. As we started writing up, we consulted the documents and added others that helped to achieve a clearer picture (e.g., blogs, reports, and academic publications). We reported regularly to each other what results we were producing and asked the other to reflect and comment on emerging findings. We kept notes of those meetings to go back to where necessary. Furthermore, we benefitted from feedback from UWTSD researchers and authors of some of the papers we analysed where we encountered issues.

Both authors have experience of teaching entrepreneurship education and have taken part in European projects in collaboration with teachers from UWTSD. The second author specialises in entrepreneurial education and has worked on research for the last decade in that area, mainly with qualitative methods. The first author has contributed to publications on entrepreneurial education for, among others, the European Union and published in the field. Our views are positive towards entrepreneurial education and to the quality of the work we got to know constructed by UWSTD in this area. We were aware of our attitudes and regularly reminded ourselves of the impact they might have on our results.

Both researchers involved in this research project have in the past collaborated with UWTSD entrepreneurial education researchers and educators on different projects. We were thus able to draw upon our experience and knowledge in terms of publications that outline the UWTSD approach. We were also able to identify where projects (those we participated in as well as others) were useful illustrations of different educational theories, e.g., "glorious failure" [45]. Additionally, we benefitted from support from personal contacts to UWTSD researchers to point out to us other publications and projects that could be of value. This support allowed us to achieve data saturation as we were able to find high-quality examples in the public domain that could be of interest to illustrate how UWTSD supports education for social change along the understanding of the CFW.

4. Results

The following sections will show how UWTSD already implements the dictates of the CFW in its teacher training and wider education approach. Examples are provided to highlight good practice(s) of an entrepreneurial education approach for social change and sustainability.

4.1. Delivery

Several core aspects of learning and teaching at UWTSD reflect various tenets of the new CFW delivery.

First, the CFW integral skill "creativity and innovation" is deeply rooted in UWTSD's entrepreneurial education. UWTSD has a long history of drawing on design education for entrepreneurial education [46–50]. Its learning and teaching approaches borrow methodologies from design to create value for others through seeing multiple perspectives. As early as 2008, researchers at the International Institute for Creative Entrepreneurial Development (IICED) at UWTSD proposed "curiosity-based learning" [46] as a strategy for learners to recognise problems and generate solutions. As the authors wrote:

> Reflecting the inquisitive entrepreneur, learners become aware of their shortfall in knowledge through their own experience, rather than simply being told it. They also learn to look around a problem and not just to see it at face value, or are encouraged to find problems within scenarios presented to them as a project assignment brief. Finding these problems is a necessity—failure to do so results in learners not being able to engage with the scenario [46] (p. 405).

UWTSD has taken this focus on creativity to an international forum, having contributed to United Nations education programmes, among others. Together with the United Nations Conference on Trade and Development, IICED developed a curriculum that aimed to enhance innovative capacity among learners in educational programmes [30]. For the OECD, their researchers authored a "Thematic Paper on Entrepreneurial Education in Practice" arguing that "retaining the creative thinking of the young mind is important and real world relevance and levels of connectivity will help to bring invaluable insights to our schools" [29] (p. 7).

Relatedly, real-world context plays an important role in UWTSD's teaching and learning, which is collaborative both within and beyond the classroom. UWTSD has a long history of engaging alumni as sources of information for evaluation and as external practitioners [28]. Within the classroom, learners engage in project work to develop their knowledge and ideas with both peers and teaching staff [47]. At the same time, engagement with these speakers ensures that learning and teaching take place in authentic contexts. As part of the Postgraduate Certificate in Education (PGCE), Professional Certificate in Education (PCE), and Professional Certificate in Education for Post Compulsory Education and Training (PCET) programmes, guest speakers (including students in the Education Doctorate programme, EdD) are invited to inform trainee teachers of recent developments within the field of post compulsory education and to enable learners to view education in relation to different contexts. Among others, Wales has an established method of providing Entrepreneurship Champions, who before they join any classes have to attend a short course as an introduction to learning and teaching.

Equally, learning and teaching are focussed on skills development for learners (including trainee teachers) to appreciate that value creation can go beyond economic value.

Table 5 outlines in detail the CFW requirements in relation to delivery, shows how UWTSD implements them, and provides examples.

Table 5. Delivery as outlined in the CFW guidance mapped against UWTSD implementation and visualised by examples.

CFW Guidance on Delivery	UWTSD Implementation	Example
Learning and teaching should be collaborative and cross-disciplinary	Trainee teachers at UWTSD are engaged in collaborative teaching and peer support to extend the range of strategies and methods they employ within their teaching and to seek to continually improve themselves to the benefit of their learners.	As part of the Professional Graduate Certificate in Education, trainee teachers are engaged in workshops to practice and develop their skills in teaching, research and critical analysis, resource development, experimentation with traditional and creative pedagogies, application of digital technology skills, self-reflection, and peer and self-evaluation; they also engage in project work to develop their knowledge and ideas with both peers and teaching staff [50].
Learning and teaching should make use of external practitioners	UWTSD has a long history of engaging alumni as sources of information for evaluation and as external practitioners.	UWTSD alumni have contributed to programme development. Especially, their "ideas, perceptions and experiences, networks and contacts, have provided particularly rich empirical evidence that enabled comprehensive and detailed consideration and evaluation" [37] (p. 234).
Learning and teaching should be delivered using a range of teaching and learning approaches	UWTSD designs "fit for purpose" learning and assessment, ensuring that is it constructively aligned. Cognition research into insightful as well as analytical thinking theoretically underpins this approach.	Penaluna et al. [46] explored practical measures of how student performance can be assessed and argued that inappropriate assessment strategies can significantly inhibit the creativity of students and teachers.
Learning and teaching should take place in a range of contexts and be cross-curricular	UWTSD's approaches borrow learning methodologies from design education, which seeks to create value for others through seeing multiple perspectives within wicked problem-solving scenarios.	For example, UWTSD makes use of pedagogies such as curiosity-based learning that distinguishes between, at first, focussing on divergent thinking (opening minds and synthesis), and then converging (analytical and solution-focused) to generate ideas and explore possible solutions [47–49].
Learning and teaching should allow for learners to develop their skills (e.g., critical thinking, problem solving, decision-making) and for them to generate different types of value (financial, cultural, social, and learning)	For over 10 years, UWTSD has been involved in various international innovation projects focussed on mainstreaming entrepreneurial education skills among teachers.	UWTSD was involved in the EU-level ADEPTT (http://adeptt.blogspot.com/, accessed on 4 June 2021), Eco System App (https://ecosystemapp.net/, accessed on 4 June 2021), EntreAssess (http://entreassess.com/, accessed on 4 June 2021), and EntreCompEdu (https://entrecompedu.eu/, accessed on 4 June 2021) projects. Dr. Jan Barnes, UWTSD Senior Lecturer in Cross-curriculum close to practice enquiry and research, described for the EntreAssess project in 2018 how she develops trainee teachers' entrepreneurial skills [42]. The EU-funded policy reform project EntreCompEdu, led by UWTSD, developed a professional skills framework of entrepreneurial education and ran a teacher training course with over 400 teachers globally [41].
Learning and teaching should ensure exposure to local, national, and international contexts at different stages of development	UWTSD has been involved in international teaching and research projects to develop teacher education for many years. Learners participate regularly in these opportunities.	Educators engaged in the development of the EntreCompEdu programme. As part of this training programme, they engage with educators from other countries, learn together, and exchange teaching ideas [41]. UWTSD also co-hosts international events, e.g., as part of Global Entrepreneurship Week, for their students [51]. For example, they hosted Fiorina Mugione, previously the Chief of Entrepreneurship at the United Nations.
Learning and teaching should take place in authentic context(s)	UWTSD makes extensive use of guest speakers (including alumni) in their teacher training.	For instance, guest speakers are invited to inform trainee teachers of recent developments within the field of post-compulsory education and to enable learners to view education in relation to different contexts. This also includes EdD students. See also this blog reflecting on an alumni-led event and [37,49] (https://www.lancaster.ac.uk/users/enterprisecentre/stepping-back-to-go-forward-alumni-voices/, accessed on 31 May 2021)

4.2. Assessment

In relation to assessment, UWTSD supports students extensively in reflecting on their learning progress over time. This is core to the arts and design education that much of UWTSD's work is built upon [29,46,50,51]. Students are evaluated through summative and formative assessment methods such as project work, presentations/pitches, self-reflection, as well as self-evaluation, peer evaluation, and external expert review/feedback [45]. They are encouraged to "fail fast" and to make mistakes and learn from them ("glorious failures") [29,45]. For instance, students take part in "The Crit", an arts-based discussion where students critique each other's work. Penaluna and Penaluna described this as follows:

> Students are expected to communicate and debate their thinking processes and enter into a discussion of their work with tutors and peers and in later studies when appropriate with external stakeholders such as industry practitioners, clients and community members [49] (p. 7).

The approach forces students to explain their decision-making in envisioning new futures as well as to build on feedback [49]. Across their learning, progress is mapped by charts and evidence that showcase which connections the student has made in their mind. The more complex and numerous these connections are, the better:

> The highest grades are given to those who can argue for a range of distinctly different yet justifiable solutions. The number of alternative solutions required will be determined by the educator, who will consider the developmental stage of the learners. New students may be asked to present only two alternatives, whereas more accomplished students will be more challenged, with 6 to 12 alternatives [49] (p. 7).

In addition, UWTSD has become adept at involving alumni in assessment approaches based on the long-standing "Continuous Conceptual Review" model [47]. Among other forms of participation, alumni join classes to reflect on their learning in their own contexts.

UWTSD is also involved in research and policy projects to foster entrepreneurial education assessment approaches. UWTSD educators have been part of the EU-funded "EntreAssess" project, which published assessment methods, tools, and examples to help educators assess entrepreneurial teaching and learning [52]. This project collated entrepreneurial education assessment methods, tools, and examples from across the world to develop a self-assessment tool and model to help educators understand their own assessment approaches and grow in their use of more creative and complex methods and tools. The different stages range all the way up to a whole-school approach. On the project blog, UWTSD educators provided insights from their own classrooms and how they encourage creative assessment approaches. For instance, Tom Cox, UWTSD Senior Lecturer in Creative and Innovative Teaching and Learning, outlined how he uses the EntreAssess tools in helping primary school teachers to find specific assessment methods for specific skills development [52]. Case studies of Welsh schools are also provided. Craigfelen Primary School in Swansea was among the first schools in Wales that participated in UWTSD's teacher training programmes. In a blog post on the EntreAssess website, Andy Penaluna narrated a visit he undertook to Craigfelen following an invitation to their year 1 and 2 learners, explaining how the learners (5–7 years old) ran a project to revitalise the local post office and summarising the learnings he drew from the visit. The learners opted to run a pop-up shop in the space and involved their parents as well as the local community and UWTSD in this project.

Table 6 outlines in detail the CFW requirements in relation to assessment, shows how UWTSD implements them, and provides examples.

Table 6. The CFW requirements on assessment mapped against UWTSD implementation and visualised by examples.

CFW Guidance on Assessment	UWTSD Implementation	Example
Assessment should enable reflection on learner progress over time (e.g., it should inform a learner on their strengths and achievements, as well as areas for improvement and, if relevant, barriers to learning)	UWTSD supports learners extensively via formative feedback.	An adaptation of the Art and Design "Crit" is used to enhance peer to peer learning—through the justification of multiple solutions, which are expected to be argued for their distinctiveness from each other (Forced Divergent Thinking) [37,49].
Assessment should enable reflection on group progress over time too (e.g., at school level)	Alumni join classes to reflect on their learning in their own teaching contexts.	"Glorious Failure" is a teaching/assessment approach in which students are allowed to "fail" if they reflect upon the why and articulate the reasoning [45].
A wide range of assessment approaches should be used to provide a holistic picture of learners' development	Students are subject to various summative and formative assessment methods such as project work, presentations/pitches, self-reflection, as well as self-evaluation, peer evaluation, and external expert review/feedback. In addition, UWTSD has been and is actively involved in research in entrepreneurial education assessment to help educators develop progress in their assessment challenge.	UWTSD educators have been part of the EU-funded EntreAssess project that published assessment methods, tools and examples to help educators with assessing entrepreneurial teaching and learning. The project focussed on practical and easy-to-use assessment methods and aimed to help enhance students' learning in entrepreneurial education and support the quality of education and outcomes in European contexts [52].
There should be engagement between the learner and the world outside of school, incl. parents or carers, and practitioners	UWTSD has a long history of engaging alumni as sources of information for evaluation and as external practitioners.	If an educator is innovative, they can "only be realistically evaluated and validated by their learners", in other words: alumni [45] (p. 29). Consequently, alumni engagement and the empowerment of alumni to return to university and provide advice as well as share their experience of their education, for instance, preparing them for their work, are crucial. See also [47].
Learners should participate in the assessment process (e.g., reflect on their learning journey)	Students are encouraged to "fail fast" and to be comfortable making mistakes and learning from them ("glorious failures").	"Glorious failures" denotes a concept that accepts that what is new will likely be a prototype that is improved with testing and feedback. In education, it means understanding and accepting that interventions will be prototypes, which means for both students and educators, the experience itself will be as valid as the immediate outcome [45].
As learners progress, they should become increasingly effective. This includes increasingly successful approaches to self-evaluation, identification of their next steps in learning and more effective means of self-regulation	UWTSD is continuously involved in developing training programmes focussed on teachers' continuous development.	UWTSD has been the project lead for the EU-funded policy reform project EntreCompEdu (2018–2020). It supports educators in developing their entrepreneurial education skills. The EntreCompEdu framework builds on good pedagogy in the field of entrepreneurial education. The framework rests on six pedagogical principles: (i) think creatively, (ii) look to the real world for inspiration, (iii), promote collaboration with a purpose, (iv) create something of value for others, (v) stimulate reflection, flexible thinking and learning from experience, and (vi) make entrepreneurial learning visible [41].

4.3. Social Change, Sustainability, and Entrepreneurial Education

The University's sustainability statement commits UWTSD to deliver "meaningful and relevant educational pathways". This includes promoting learning and social responsibility, which supports what the Brundtland Commission in 1987 described as "development that meets the needs of the present without compromising the ability of future generations to meet their own needs" [50]. UWTSD thus empowers learners to become active agents in building a socially just and sustainable society. In detail, the PGCE/PCE programmes aim to produce learners who "understand their professional responsibilities in relation to Education for Sustainable Development (ESD), with particular regard to the development of practice and engagement within the classroom and the ability to understand, critically evaluate, and adopt thoughtful sustainability values." Teacher trainees are encouraged to be experimental in their teaching and "experiment with pedagogies that embed ESD and consider sustainability through critical reflective practice and evaluation" [50] (p. 31).

Learning is delivered in authentic contexts. For UWTSD, collaboration with industry is a key focus. This facilitates "more value creation opportunities for students" while also augmenting learners' employability prospects [51]. At the same time, industry collaboration allows UWTSD learners to explore value creation opportunities for the development of new sustainable businesses, products, and services. In 2018–2019, UWTSD was ranked first in Wales and second in the UK by the Higher Education Statistics Agency (HESA) for the number of graduate businesses running for three years or more [51]. Over 550 alumni are enhancing and supporting UWTSD's entrepreneurial education ambitions. This demonstrates how UWTSD, in teaching about (and through) entrepreneurial education, focuses on value creation that is ecological, humane, and social, in addition to creating economic value. In 2020, UWTSD teaching staff launched the "Harmonious Entrepreneurship Society" to "set up and advance harmonious approaches to entrepreneurship to address the sustainability challenge facing our planet" [51].

Table 7 outlines in detail the CFW expectations in relation to social change and sustainability, shows how UWTSD implements them through entrepreneurial education, and provides examples.

Table 7. The CFW expectations on social change mapped against UWTSD implementation and visualised by examples.

CFW Guidance on Social Change	UWTSD Implementation	Example
Learners should be empowered to become active agents of building a socially just and sustainable society	The University's sustainability statement commits UWTSD to deliver "meaningful and relevant educational pathways." This includes promoting learning and social responsibility, which supports what the Brundtland Commission in 1987 has described as "development that meets the needs of the present without compromising the ability of future generations to meet their own needs" [53].	The PGCE/PCE programmes aim to produce learners who "understand their professional responsibilities in relation to ESD, with particular regard to the development of practice and engagement within the classroom, and the ability to understand, critically evaluate and adopt thoughtful sustainability values." Teacher trainees are encouraged to "experiment with pedagogies that embed ESD and consider sustainability through critical reflective practice and evaluation" [50] (p. 31).
Learner engagement is emphasised. Learning should take place in authentic contexts across curricular areas	For UWTSD, collaboration with industry is a key focus. It facilitates "more value creation opportunities for students" while also augmenting learners' employability [53]. At the same time, industry collaboration allows UWTSD learners to explore value creation opportunities for the development of new sustainable businesses, products, and services.	UWTSD has been ranked 1st in Wales and 2nd in the UK in 2018/19 by the HESA for the number of graduate businesses running for three years or more [51]. Over 550 alumni are enhancing and supporting UWTSD's entrepreneurial education ambitions.
Enterprising spirit and action competence. Being able to create value of different kinds—financial, cultural and social	UWTSD in teaching about (and through) entrepreneurial education focuses on value creation that is ecological, humane and social, in addition to the traditional economic value creation.	In 2020, UWTSD teaching staff launched the "Harmonious Entrepreneurship Society" to "set up and advance harmonious approaches to entrepreneurship to address the sustainability challenge facing our planet" [51]. All units in the PGCE/PCET/PCE programmes are also mapped against the university's "Education for Sustainable Development" plan, which outlines skills developed in the teacher trainees [50].

Table 7. *Cont.*

CFW Guidance on Social Change	UWTSD Implementation	Example
Learners should become enterprising in managing their own and others' resources, valuing failure as a part of the creative process, and relatedly strengthening their employability skills	Creativity and innovation are at the heart of UWTSD's mission to enhance graduate employability and the number of graduate start-ups.	UWTSD has been ranked 1st in Wales and 2nd in the UK in 2018/19 by the HESA for the number of graduate businesses running for three years or more [51]. Over 550 alumni are enhancing and supporting UWTSD's entrepreneurial education ambitions.
Learners should become sustainable citizens through a sustainable education, and should be able to respond to challenges of a social, economic and environmental nature	UWTSD's Sustainability Statement focuses on providing meaningful education that considers social responsibility and the needs of future generations.	The university aims to "utilise our collective skills, knowledge and technology to enable the University and its graduates to offer solutions to the most urgent societal challenges—in Wales and further afield. We are also committed to building a sustainable society driven through enterprising innovation and entrepreneurship" [51] (p. 3).
Learners should be able to make responsible decisions, to act as caring, participative citizens of their local, national, and global communities, committed to justice, diversity and the protection of the environment	Entrepreneurial value creation—where value may be cultural, social, or environmental, in addition to economic—is well understood in UWTSD teaching.	UWTSD's "Harmonious Entrepreneurship Society" was set up to advance entrepreneurial approaches that have sustainability at their heart [51].

5. Discussion

This study sought to present how a university works towards embedding sustainability thinking and actions for social change through an entrepreneurial education approach. It analysed the very forward-looking curriculum demands that Wales makes towards its educators in the "Curriculum for Wales", and mapped against these demands how a specific university with a track record in delivering entrepreneurial education enables education for sustainability and social change. The results highlight some noteworthy practices. Through some of its core entrepreneurial education approaches, UWTSD manages to seamlessly embed sustainable thinking and education for social action in their learning and teaching.

First, through their long history of making use of external practitioners, UWTSD is able to foster an enterprising spirit and action competence in learners. Learners are then able to create value of different kinds. UWTSD, in teaching about—and importantly, through—entrepreneurial education allows broad thinking about value creation that goes beyond the creation of economic value and is ecological, humane, and social. For instance, for UWTSD, collaboration with industry is a key focus. It facilitates "more value creation opportunities for students" while also augmenting students' employability prospects [53]. At the same time, industry collaboration allows UWTSD students to explore value creation opportunities for the development of new sustainable businesses, products, and services. Equally, trainee teachers at UWTSD engage with guest speakers, including alumni, who become participants in UWTSD's teaching approach. Specifically in teacher education, UWTSD includes in-house (budding) experts such as EdD students to inform trainee teachers of recent developments in the field of education. Such approaches aim for learners to be able to view education in relation to different contexts. This kind of contextualised approach to teaching and learning is important for sustainability education, connecting the local and global, and acquiring a sense of *cynefin*.

In practice, UWTSD presents impressive numbers in learner engagement through entrepreneurship, as demonstrated by its HESA ranking and extensive alumni participation. Within the teacher training programmes delivered at UWTSD, all are mapped against the "Education for Sustainable Development" plan. The aim is for trainee teachers to "understand their professional responsibilities in relation to ESD, with particular regard to the development of practice and engagement within the classroom, and the ability to understand, critically evaluate and adopt thoughtful sustainability values" [51].

This is supported in UWTSD's approach to assessment, which encourages educators to become increasingly effective in their assessment approaches and methods in relation to measuring—and fostering—entrepreneurial skills. This includes for learners to be increasingly successful to self-evaluate and identify their next steps in learning and more effective means of self-regulation. For example, UWTSD has been involved in a European project to gather, and make available in an accessible way, assessment methods, tools, and examples of increasingly differentiated entrepreneurial education assessment approaches for educators. Relatedly, there is a strong focus in UWTSD's own work on assessment for learning. The elective course "Enterprise Educators" is built around different rounds of formative feedback. Similarly, UWTSD learners participate in the assessment process, for example, through reflection. UWTSD has made the EU's EntreComp framework work for itself in this regard. Such assessment methods can enhance learner engagement and agency and support the cultivation of action competence in the spirit of sustainability education including the potentials for social change.

Furthermore, UWTSD is focussed on fostering entrepreneurial skills in their learners. Learning and teaching thus foster skills such as critical thinking, problem solving, and decision-making, as well as the ability to see how value of different kinds (financial, cultural, social, and learning) may be generated. Teacher trainees in particular are encouraged to put these skills into action and be experimental in their teaching. For over 10 years, UWTSD has been involved in different international (education) innovation projects focussed on mainstreaming entrepreneurial education among teachers. In a similar but wider manner, creativity and innovation are at the heart of UWTSD's mission to graduate employment. This allows all learners to become enterprising and be able to manage their and others' resources, giving them agency and skills to influence their social, economic, and environmental conditions, both locally and globally. UWTSD's Sustainability Statement focuses on providing meaningful education that considers social responsibility and the needs of future generations. The university aims to "utilise our collective skills, knowledge and technology to enable the University and its graduates to offer solutions to the most urgent societal challenges—in Wales and further afield. We are also committed to building a sustainable society driven through enterprising innovation and entrepreneurship" [54] (p. 3).

6. Conclusions

The findings are not generalisable and were not meant to be. They are limited to a case of how the work of UWTSD in the area of entrepreneurial education emerges as practice in the spirit of sustainable education and has the potential to support social change. It is a descriptive and analytical study where core elements in the progressive CFW are illustrated against the work of UWTSD to exemplify how entrepreneurial education can provide affordances to support education for social change.

A novelty of this research is that the aspects of entrepreneurial education are highlighted as a curriculum ideology (CFW) and as examples in practice at the university level in teacher education. Tables 4–6 illustrate the essence of the research and can be used to develop similar frameworks to scrutinise other curricula focusing on social change and entrepreneurial education. The presented examples are related to the professional context and practical implementation is illustrated plastically, and they show how entrepreneurial education approaches can embed sustainable thinking and education for social change. The findings can be informative and explanatory for teachers and teacher educators looking for ways to enhance sustainability education and social change.

Further research could deepen our findings and contextualise how a progressive curriculum can pan out in practice; for example, research on how entrepreneurial education emerges in school practice and other educational settings in Wales, looking for signs of the elements presented in the framework (Tables 4–6). Similar research in other countries focusing on how entrepreneurial education can support sustainability education and social change could also benefit from applying the framework or adjusting it accordingly.

Author Contributions: R.W. had a leading role in gathering data and designing the approach and S.R.J. handled the overall supervision. Both authors worked in tight collaboration on all other parts of the research and writing up the article. Both authors have read and agreed to the published version of the manuscript.

Funding: This research was funded by the University of Iceland Research Fund.

Institutional Review Board Statement: Not applicable.

Informed Consent Statement: Not applicable.

Data Availability Statement: Most data is available online and cited in the reference list, except course handbooks from the UWTSD that we have not permission to share publicly.

Acknowledgments: We want to thank the University of Iceland Research Fund for the grant allocated to this research and the Educational Research Institute at the University of Iceland for advice and administrative support. The authors are grateful to Andy Penaluna, the editors of this issue, as well as anonymous referees for feedback on earlier versions of this paper. We also want to thank our language editor Jakob Maas for his excellent and swift services.

Conflicts of Interest: The authors declare no conflict of interest. The funders had no role in the design of the study; in the collection, analyses, or interpretation of data; in the writing of the manuscript, or in the decision to publish the results.

References

1. McCallum, E.; Weicht, R.; McMullan, L.; Price, A. *EntreComp into Action: Get Inspired, Make It Happen*; Publications Office of the European Union: Luxembourg, 2018.
2. European Commission. Effects and Impact of Entrepreneurship Programmes in Higher Education. 2012. Available online: https://ec.europa.eu/growth/content/effects-and-impact-entrepreneurship-programmes-higher-education-0_en (accessed on 4 June 2021).
3. European Commission. *European Skills Agenda. Sustainable Competitiveness, Social Fairness and Resilience*; Publications Office of the European Union: Luxembourg, 2020.
4. Gibb, A. In pursuit of a new "enterprise" and "entrepreneurship" paradigm for learning: Creative destruction, new values, new ways of doing things and new combinations of knowledge. *Int. J. Manag. Rev.* **2002**, *4*, 233–269. [CrossRef]
5. Seikkula-Leino, J.; Salomaa, M. Entrepreneurial Competencies and Organisational Change—Assessing Entrepreneurial Staff Compete ncies within Higher Education Institutions. *Sustainability* **2020**, *12*, 7323. [CrossRef]
6. Weicht, R. World Economic Forum. 2018. Available online: https://www.weforum.org/agenda/2018/04/education-systems-can-stifle-creative-thought-here-s-how-to-do-things-differently/ (accessed on 1 June 2021).
7. David, K.; Penaluna, K.; McCallum, E.; Usei, C. Embedding entrepreneurial skills development in teacher education. In *Entrepreneurial Learning City Regions: Delivering on the UNESCO 2013, Beijing Declaration on Building Learning Cities*; Springer: Cham, Switzerland, 2018; pp. 319–340.
8. Jónsdóttir, S.R.; Þórólfsson, M.; Karlsdóttir, J. Að Uppfæra Ísland: Sýn Stjórnenda Íslenskra Framhaldsskóla á Nýsköpunar- og Frumkvöðlamennt og Framkvæmd Námssviðsins í Námskrárfræðilegu Ljósi. [Upgrading Iceland: Icelandic Secondary School Administrators' Views about Innovation and Entrepreneurial Education (IEE) within the Context of Curriculum Studies. In Icelandic, Long Abstract in English]. Netla—Veftímarit um Uppeldi og Menntun 2014. Available online: https://netla.hi.is/greinar/2014/ryn/011.pdf (accessed on 29 July 2021).
9. Jónsdóttir, S.R. *The Location of Innovation Education in Icelandic Compulsory Schools*; University of Iceland: Reykjavík, Iceland, 2011.
10. Jónsdóttir, S.R.; Gunnarsdóttir, R. *The Road to Independence: Emancipatory Pedagogy*; Sense: Rotterdam, The Netherlands, 2017.
11. Huckle, J. Education for Sustainable Development: A Briefing Paper for the Teacher Training Resource Bank. 2006. Available online: https://huckleorguk.files.wordpress.com/2016/10/huckle2006.pdf (accessed on 4 June 2021).
12. Pálsdóttir, A. *Sustainability as an Emerging Curriculum Area in Iceland. The development, Validation and Application of a Sustainability Education Implementation Questionnaire*; University of Iceland: Reykjavík, Iceland, 2014.
13. Hannay, L.; Earl, L. School districts triggers for reconstructing professional knowledge. *J. Educ. Chang.* **2012**, *13*, 311–326. [CrossRef]
14. Weicht, R.; Ivanova, I.; Gikopoulou, O. The CRADLE teaching methodology: Developing foreign language and entrepreneurial skills in primary school pupils. *Entrep. Educ.* **2020**, *3*, 265–285. [CrossRef]
15. Curran, M.A. Wrapping Our Brains around Sustainability. *Sustainability* **2009**, *1*, 5–13. [CrossRef]
16. Rosen, M.A. Sustainability: A Crucial Quest for Humanity. *Sustainability* **2009**, *1*, 1–4. [CrossRef]
17. Šlaus, I.; Jacobs, G. Human Capital and Sustainability. *Sustainability* **2011**, *3*, 97–154. [CrossRef]
18. Hameed, I.; Khan, M.B.; Shahab, A.; Hameed, I.; Qadeer, F. Science, Technology and Innovation through Entrepreneurship Education in the United Arab Emirates (UAE). *Sustainability* **2016**, *8*, 1280. [CrossRef]
19. European Commission. A New Skills Agenda for Europe. 2016. Available online: https://ec.europa.eu/transparency/regdoc/rep/1/2016/EN/1-2016-381-EN-F1-1.PDF (accessed on 13 May 2021).

20. European Commission. Council Recommendation of 22 May 2018 on Key Competences for Lifelong Learning. 2018. Available online: https://eur-lex.europa.eu/legal-content/EN/TXT/?uri=uriserv%3AOJ.C_.2018.189.01.0001.01.ENG&toc=OJ%3AC%3A2018%3A189%3ATOC%20.%20Accessed%2013%20August%202020%20https://eur-lex.europa.eu/legal-content/EN/TXT/%3furi%3duriserv%253AOJ.C_.2018.189.01.0001.01.E (accessed on 13 May 2021).
21. Eurydice. Entrepreneurship Education at School in Europe. 2016. Available online: https://op.europa.eu/en/publication-detail/-/publication/74a7d356-dc53-11e5-8fea-01aa75ed71a1/language-en (accessed on 13 May 2021).
22. Penaluna, A.; Penaluna, K.; Polenakovikj, R. Developing entrepreneurial education in national school curricula: Lessons from North Macedonia and Wales. *Entrep. Educ.* **2020**, *3*, 245–263. [CrossRef]
23. Margherita, B.; Pan, K.; Yves, P.; Lieve, v.d.B. *EntreComp: The Entrepreneurship Competence Framework*; Publications Office of the European Union: Luxembourg, 2016.
24. European Commission. Final Report of the Thematic Working Group on Entrepreneurship Education. 2014. Available online: https://ec.europa.eu/assets/eac/education/experts-groups/2011-2013/key/entrepreneurship-report-2014_en.pdf (accessed on 13 May 2021).
25. Power, S.; Newton, N.; Taylor, C. 'Successful futures' for all in Wales? The challenges of curriculum reform for addressing educational inequalities. *Curric. J.* **2020**, *31*, 317–333. [CrossRef]
26. OECD. *Improving Schools in Wales: An OECD Perspective*; Organisation for Economic Co-Operation and Development: Paris, France, 2014.
27. Taylor, C.; Rees, G.; Davies, R. Devolution and geographies of education: The use of the Millennium Cohort Study for 'home international' comparisons across the UK. *Comp. Educ.* **2013**, *49*, 290–316. [CrossRef]
28. Welsh Government. *Curriculum for Wales Guidance*; Welsh Government: Cardiff, UK, 2020.
29. Penaluna, K.; Penaluna, A.; Usei, C.; Griffiths, D. Enterprise education needs enterprising educators: A case study on teacher training provision. *Educ. Train.* **2015**, *57*, 948–963. [CrossRef]
30. Penaluna, A.; Penaluna, K. *Entrepreneurial Education in Practice; Part. 2-Building Motivations and Competencies*; OECD: Paris, France, 2015.
31. Mugione, F.; Penaluna, A. Developing and evaluating enhanced innovative thinking skills in learners. In *Entrepreneurial Learning City Regions*; Springer: Berlin/Heidelberg, Germany, 2018; pp. 101–120.
32. Penaluna, A. Quality Assurance Agency. 2018. Available online: https://www.qaa.ac.uk/docs/qaa/about-us/enterprise-and-entrpreneurship-education-2018.pdf?sfvrsn=20e2f581_10 (accessed on 13 May 2021).
33. Donaldson, G. *Successful Futures; Independent Review of Curriculum and Assessment Arrangements in Wales*; Welsh Government: Cardiff, UK, 2015.
34. Furlong, J.; Hagger, H.; Butcher, C.; Howson, J. *Review of Initial Teacher Training Provision in Wales; A Report to the Welsh Assembly Government*; University of Oxford: Oxford, UK, 2006.
35. Tabberer, R. *A Review of Initial Teacher Training in Wales*; Welsh Government: Cardiff, UK, 2013.
36. Yin, R.K. *Case Study Research; Design and Methods*, 5th ed.; Sage: London, UK, 2014.
37. Sinnema, C.; Nieveen, N.; Priestley, M. Successful futures, successful curriculum: What can Wales learn from international curriculum reforms? *Curric. J.* **2020**, *31*, 181–201. [CrossRef]
38. Newton, N.; Power, S.; Taylor, C. *Successful Futures for All: Explorations of Curriculum Reform*; Wales Institute of Social & Economic Research, Data & Methods: Cardiff, UK, 2019.
39. Hughes, S.; Lewis, H. Tensions in current curriculum reform and the development of teachers' professional autonomy. *Curric. J.* **2020**, *31*, 290–302. [CrossRef]
40. Penaluna, A.; Penaluna, K. Stepping Back to Go Forward: Alumni in Reflexive Design-Curriculum Development. In Proceedings of the Internationalising Entrepreneurship Education and Training Conference, Sao Paulo, Brazil, 9–12 July 2006.
41. Grigg, R. EntreCompEdu, a professional development framework for entrepreneurial education. *Educ. Train.* **2020**, *1*. [CrossRef]
42. Barnes, J. EntreAssess Blog. 2018. Available online: http://entreassess.com/2018/05/30/entrepreneurial-or-innovation-not-just-another-buzz-word-thoughts-from-a-teacher-educator/ (accessed on 4 June 2021).
43. EntreCompEdu. EntreCompEdu Blog. Available online: https://entrecompedu.eu/phase-1-educators-in-wales-report-on-help-with-covid-19-challenges/ (accessed on 4 June 2021).
44. Lincoln, Y.S.; Guba, E.G. *Naturalistic Inquiry*; Sage Publications: Newbury Park, CA, USA, 1985.
45. Komarkova, I.; Gagliardi, D.; Conrads, J.; Collado, A. *Entrepreneurship Competence: An Overview of Existing Concepts, Policies and Initiatives-Final Report*; Publications Office of the European Union: Luxembourg, 2015.
46. Penaluna, A.; Penaluna, K.; Jones, C.; Matlay, H. When Did You Last Predict a Good Idea?: Exploring the Case of Assessing Creativity through Learning Outcomes. *Ind. High. Educ.* **2014**, *28*, 399–410. [CrossRef]
47. Penaluna, A.P.K. Business Paradigms in Einstellung: Harnessing Creative Mindsets, A Creative Industries. *J. Small Bus. Entrep.* **2008**, *21*, 231–250. [CrossRef]
48. Penaluna, A.; Coates, J.; Penaluna, K. Creativity-based assessment and neural understandings: A discussion and case study analysis. *Educ. Train.* **2010**, *52*, 660–678. [CrossRef]
49. Penaluna, A.; Penaluna, K. In search of entrepreneurial competencies: Peripheral vision and multidisciplinary inspiration. *Ind. High. Educ.* **2020**, 0950422220963796. [CrossRef]
50. UWTSD. *Programme Titles PGCE, PCET, PCE*; Centre for Continuing Professional Learning and Development: Swansea, UK, 2016.

51. UWTSD. UWTSD Annual Review 2020. 2020. Available online: https://www.uwtsd.ac.uk/media/uwtsd-website/content-assets/documents/ries/WGAnnualReview2020.pdf (accessed on 4 June 2021).
52. Bowe, G.A. Document Analysis as a Qualitative Research Method. *Qual. Res. J.* **2009**, *9*, 27–40.
53. Jónsdóttir, S.R. Developing practical assessment methods for entrepreneurship education. In Proceedings of the 46th Nordic Educational Research Association (NERA) Congress, NERA 2018-Boundaries, Breaches and Bridges, Oslo, Norway, 8–10 March 2018.
54. UWTSD. *Strategic Planning and Engagement Document, W14/36HE Annex A*; UWTSD: Swansea, UK, 2014.

Case Report

Exploring the Connections of Education for Sustainable Development and Entrepreneurial Education—A Case Study of Vocational Teacher Education in Finland

Eveliina Asikainen * and Annukka Tapani

Professional Teacher Education, Tampere University of Applied Sciences, Kuntokatu 3, 33520 Tampere, Finland; annukka.tapani@tuni.fi
* Correspondence: eveliina.asikainen@tuni.fi

Abstract: Education for Sustainable Development (ESD) and Entrepreneurial Education (EE) are quite abstract and demanding concepts for teacher students. Yet, Key Sustainability Competences and Entrepreneurial Competences entail important qualities of future citizens and workers, and teacher students should become prepared to accommodate education for these competencies in their teaching practice. This paper explores teacher students' process of sense-making of sustainable development and how becoming a teacher who practices ESD connects with entrepreneurship. EE serves as a good mirroring surface to ESD as they both have their roots in Transformative Learning (TL) but pursue transformation towards different goals. The case study follows the vocational teacher education (VTE) students' sensemaking of Sustainable Development as a part of teacher's work during one semester which included integrated Thematic Studies of Sustainable Development. The qualitative content analysis of students' texts focused on signs of transformative learning and was guided by the dimensions of sustainable development and learning goals set for teacher's sustainability competences in the VTE curriculum. The results indicate that transformative learning is possible. Furthermore, they address the importance of certain entrepreneurial capabilities in the actualization of change agency.

Keywords: education for sustainable development; entrepreneurial education; key sustainability competencies; transformative learning

Citation: Asikainen, E.; Tapani, A. Exploring the Connections of Education for Sustainable Development and Entrepreneurial Education—A Case Study of Vocational Teacher Education in Finland. *Sustainability* **2021**, *13*, 11887. https://doi.org/10.3390/su132111887

Academic Editors: Jaana Seikkula-Leino, Mats Westerberg, Priti Verma, Maria Salomaa and Adela García-Aracil

Received: 31 August 2021
Accepted: 21 October 2021
Published: 27 October 2021

1. Introduction

Education is considered one of the strongest promoters of Sustainable Development (SD), and the need for teacher education to promote teachers' change agency has been identified as one of the critical prerequisites for ecological and societal transition [1,2]. The demand for teachers' capacities to facilitate the development of change making capacities in their students is also clearly stated in the Goal 4.7 of Agenda 2030, saying, "all learners should acquire the knowledge and skills needed to promote sustainable development by 2030." The achievement of this goal should manifest as actions of both teachers and students [3]. Developing educators' capacities and competencies in Education for Sustainable Development (ESD) has been recognized as one of the priority action areas required to address climate change and sustainability challenges in the UNESCO publication *Education for Sustainable Development: A Roadmap* [4].

Competencies consist of cognitive, affective and motivational elements and a voluntary will and intention of doing things. Each competency is an interplay of knowledge, capacities and skills, motives, and affective dispositions, and their interaction is what matters in the actualization of the competency [5]. Competencies are also situational and context-specific, but transversality can be developed. Key competencies can be understood as multifunctional and context-independent competencies, which are considered to be particularly crucial for implementing societal goals important in a normative framework (e.g., sustainability) and which are important for all individuals [5]. In the education profession, competencies of teaching are about promoting learning by combining a teacher's substantial knowledge about an issue, concept, or phenomena with pedagogical knowledge [5,6].

Cebrían and Junyent state ESD competencies focus on the competencies that teachers and educators need to put in place in educational settings to promote sustainability competencies amongst their students [7]. Rieckmann [8] defines teachers' competencies for Education for Sustainable Development as "teacher's capacity to help people develop sustainability competencies through a range of innovative teaching and learning practices" [8] p. 38. The definitions draw attention to the pedagogical competencies and emphasize the competence to transform, which lies in the heart of many other definitions as well (for example, [9] p. 204 and [10]). The transformative dimension of ESD can be stated as follows: "ESD draws on the experience of learners and creates opportunities for participation and for the development of creativity, innovation and the capacity to imagine alternative ways of living. It encourages learners to reflect on the impact of their everyday choices in terms of sustainable development", e.g., [11] p. 17. Thus, the ultimate aim of ESD is to facilitate transformative learning so that people are empowered to take sustainable actions in complex situations [3,12]. To achieve competency to encourage learners to reflect and change their behaviors, teachers should learn about the contents of SD and the concept of ESD, develop in Key Sustainability Competences, and learn how to facilitate all this as educators [13]. Still, a recent review on initial teacher education and ESD discovered a deficit in the development of the professional skills needed to implement ESD, even the students showed positive attitudes towards sustainable development [2]. Often, students do not feel ready or prepared to teach SD topics [14] or have only a partial understanding of SD, often comprising mostly of environmental sustainability [7] or having strong emphasis on social sustainability [15]. Moreover, students' understanding of their possibilities to influence outside school can be weak [15], and they do not consider the competence to teach for a sustainable society as a part of teacher professionalism [14]. One part of the phenomenon seems to be that ESD is often a separate course, and the goal for a teacher to be a promoter of change for a more sustainable world is not made visible in the professional competences or goals of the curriculum [2,14,16,17].

Transformative Learning (TL) theory refers to a qualitative shift in perception and meaning making on the part of the learner, especially when the learner reframes or questions his/her assumptions or habits of thought and becomes critically reflective of those beliefs that become problematic [18]. The broad understanding and general aims of transformative learning are to contribute to a more significant social change (or transformation) through education, which has attracted many ESD scholars to use TL as a theoretical background [19]. For example, Blake et al. [20] p. 5348 describe the need for "that type of learning that is consistent with and helps manifest individual, organizational and social change towards more sustainable practices".

Sterling [21] connects TL with Bateson's [22] models of three orders of learning and change: first-order learning is learning more of the same; second-order learning refers to a significant change in thinking or actions as a result of examining assumptions and values, and it demands the negotiation of meaning; and, finally, third-order learning involves a shift in the operative way of knowing and thinking, in the world view. The levels of learning are connected: second-order or third-order learning also affects our interpretations of our first-order learning, which can show as transformative actions [21].

Even though TL has been used widely as a background theory in ESD [19], the relationship of ESD and TL is not unproblematic [23]. First, the scale of transformation is often blurred in ESD: Any change can be called transformation. Secondly, it is always not at all clear what needs to be sustained and what transformed in transformative ESD [23] p. 306. We apply the strict interpretation of transformative learning, according to which only learning that involves a profound shift in the operative way of knowing and thinking that frames people's perception of, and interaction with, the world—an epistemic change—is transformative [21]. It is about being challenged and about becoming conscious of one's worldview. It is not easy or comfortable and demands trusted peers and capable mentors or facilitators [20].

Entrepreneurial education (EE) may seem at first hand a counterpart of ESD, as it is often connected to capitalism and economic growth. On the other hand, there are many ways to define entrepreneurial education [24], and some definitions contain notions of new entrepreneurship that are more connected to ethics, sustainability, sensitivity, and intuitiveness [25]. The connections of EE and Science education are a quite popular research topic [24], but the connections between EE and ESD have not been researched much. For example, the extensive review of Deveci and Seikkula-Leino [24] does not mention any such studies. However, many similarities can be observed easily: making change, the re-examination of assumptions and values, critical thinking, and new creativity can all be found in the learning objectives or competences of ESD and EE [11,26].

Allan Gibb [26] used the term enterprise to mean "a set of behaviors, skills and attributes which may be exhibited by a person". Enterprising behavior has largely been regarded as a vehicle for a change in society with a growing emphasis of managing risks and uncertainty [27]. Gibb's underlying proposition is that there is some enterprise inside every student, and that the inner enterprise can be developed by means of education and training—through EE—which gives freedom, provides ownership, allows control, gives responsibility, allows mistake taking, keeps informal, and provides flexibility in learning situations [26]. Still, clear learning goals are needed, and the teacher's role is essentially to guide, facilitate, and to be a partner in the learning process by focusing on the different ways people learn [19] following the principles of TL. EE also enables students to forecast the future and the changing needs of society and to understand the substance deeply and figure out how it is related to other issues in the society—see "all-overness" [28]. This is clearly key to sustainability competences of systems thinking and anticipatory competency [13] p. 10.

For us, the transformative ambition of ESD and EE provides a lens to examine teachers' competences for ESD. Sustainability and entrepreneurship are both quite abstract ideas as part of vocational teacher's work, even though entrepreneurship has been included in the VTE curricula since the first decade of the 21st century [28] and the ways of proceeding SD in educational institutions has been nationally assessed for 20 years [29,30]. For example, in a recent study on the skills and competencies of Finnish vocational teachers, neither of them was among the 53 identified skills, but the skill of transforming society was a part of Innovator Competency [31].

Vocational teachers are an interesting group of educators for SD. Firstly, they have a professional identity when they start teacher education, and secondly, they will be educating future professionals who should acquire competences mentioned in SDG 4.7, also in relation to their work and professional identity. We find that the vocational teacher students' perceptions of ESD and SD and their perceptions of their professional competencies are particularly important as they affect their future teaching practice and the way they prepare the future professionals who should be able to build a professional identity of being part of the solution in ecological transition.

The concept of sensemaking is useful to describe the process in which the teacher students are trying to grasp concepts which are hard to make concrete: As sensemaking deals, for example, with identity, it is an ongoing and social activity. In addition to that, it offers people extracted cues that provide points of reference for linking ideas for broader networks of meaning [32]. In this paper, we search for answers to the following questions: (1) how students in vocational teacher education make sense of sustainable development as a part of vocational teachers' work and (2) what kinds of connections can be found between education for sustainable development and entrepreneurial education as teachers' competencies.

We explore these connections via following the development of vocational teacher students' thinking as Thematic Studies in Sustainable Development (TSSD) were integrated to the Vocational Teacher Edcuaction (VTE) programme in Tampere University of Applied Sciences (TampereUAS). The idea of TSSD was not to organize a separate course but to produce materials with pedagogical insights and assessment frames, which the teacher educators could use according to their choice as part of the VTE programme. We take a detailed look at one teacher student groups' learning to understand the students' sense-making of ESD and competencies related to it and make abstract concepts more concrete [32]. To capture the change in the students' perceptions, to identify signs of transformative learning, and to establish how the learning goals connected with transformative learning were achieved, we present the students' thoughts and opinions of teacher's role as a promoter of sustainable development at two points of time. In discussion, we relate the findings to entrepreneurship on conceptual level.

2. The Case: Introducing ESD Competencies in Vocational Teacher Education Curriculum

The Finnish VTE programme (60 credits) consists of vocational pedagogical studies, teaching practice, basic studies in education, and elective pedagogical studies. The education is research-based and aims at educating vocational teachers who are able to justify their decisions and actions based on both experience and theory [28]. Most students in vocational teacher education hold a master's degree and at least three years' work experience in the field in which they aim to teach [31]. VTE curriculum in TampereUAS is competency-based [33]. The main pedagogical approach is participatory pedagogy, meaning that the teacher students and their communities are the main actors, and attention is paid to individual learning possibilities and possibilities to demonstrate the skills and competencies the teacher student already has.

The sustainability commitments of TampereUAS have served as drivers for the introduction of ESD competencies in the curriculum. Since 2016, TampereUAS has committed to educating professionals able to further SD in working life and thus stated the will to act out the SDG Goal 4.7. More specifically, the School of Vocational Teacher Education has signed the Climate Challenge of the Teacher Student Union of Finland in 2019 and committed to include education on the competences of promoting climate actions and sustainable practices in vocational colleges in the VTE curriculum [34].

The work for TSSD started by creating a common understanding of sustainable education by a team, which was established in 2020 to promote the integration of sustainability themes in the VTE curriculum and to facilitate their implementation. The common understanding is based on critical thinking and the capability to challenge values and norms as the foundations of sustainable education [7,35]. Competences were defined broadly, not only as performance and professional growth but also as human growth, large knowledge foundations, and theoretical thinking: as understandings of principles, moralities, and responsibilities. Acting as a change agent was set in the core of sustainable education and the role of sustainable community culture of an educational institution is addressed; the importance to connect sustainability with the growth of teacher identity and the need to collaborate with working life were emphasised. This served as a basis for describing the core sustainability competences of vocational teachers in relation to the study modules of the present curriculum (Table 1).

Table 1. Vocational Teachers Sustainability Competencies as defined in Tampere University of Applied Sciences Vocational Teacher Education in 2020 (originally in Finnish, translated by EA).

Vocational Teacher's Sustainability Competences Core competence and conceptual base: history and philosophy of sustainable development (SD), ecological, social, cultural, and economic sustainability, Agenda 2030			
Level of competence / Competence areas	Teachers in learning environments and networks	Teachers' professional identity and pedagogical skills	Teachers in society
Descriptive	Student can describe features of a sustainable educational institution. Knows certification systems of educational institutions.	Recognizes how the teaching profession connects with SD. Recognizes the relations of SD and pedagogical choices.	Knows the agreements and goals related to SD
Applying	Furthers active sustainability in an educational institution. Implements the principles of sustainability at work	Connects professional knowledge, practices, and professional identity to form a sustainable professional identity. Uses pedagogical solutions which further SD.	Acts in order to promote SD. Uses different methods of influencing and communication.
Integrated	Develops sustainable practices of an educational institution. Creates networks.	Acts as a role model for a sustainability promoting professional. Develops pedagogics and contents.	Uses theoretical knowledge and practical experience to create a more sustainable world.

The ESD competences were described through three competence areas, which matched the study modules of the VTE curriculum: Teachers in learning environments and networks, Teachers' professional identity and pedagogical skills, and Teachers in society. This structure responds to the CSTS model's structuring of competences for ESD [10]. For each competence, the aspects of knowing, doing, and being are described. Mastering the competences was described in three different levels—Descriptive, Applying, and Integrated. The descriptive level goal in each of the three competence areas is based on Knowing about norms and systems. Quite systematically, Doing is emphasized in the descriptions of applied-level mastering and Being in the integrated mastering. Only in the professional development and pedagogical skills is doing already part of the descriptive and being part of the descriptive. This is due to the central role of identity work in this competence area.

The three levels in our learning outcomes resemble the idea behind Stephen Sterling's interpretation of transformative learning [21] based on Bateson's three levels of transformative learning [22]. Students should first learn (more) about SD and its applications in education, then they can change their practices and attitudes. Mastering in the integrated level has elements of epistemic change: "develops sustainable practices", "acts as a role model of a sustainability promoting professional", and "uses theoretical knowledge and practical experience to create a more sustainable world".

We want to emphasise that TSSD was not a separate course. It was designed to be an integrative theme and viewpoint to be discussed in different phases of teacher education and to be independently carried out by the teacher educators of TampereUAS. To ensure the even quality of implementations, learning materials with pedagogical suggestions were produced comprising of an introductory video lecture with slides and linked further readings and suggestions on how to relate the themes into the courses. Materials were made available for all teacher educators through a shared Moodle platform.

3. Analysis

We followed the learning process of one VTE student group (n = 28) during their first semester, autumn 2020. Most of the students (n = 25) had no prior experience as a teacher in vocational education. Their backgrounds were diverse: from social and health care, culture, engineering, agriculture, and business management to communication. We collected students' thoughts and opinions of teacher's role as a promoter of sustainable development in two occasions.

The first collection of students' ideas of sustainable development and teachers' work took place in September 2020 as the students were introduced to the Thematic Studies in Sustainability. The students had studied the introductory material beforehand and discussed their insights of the phenomenon during the contact day. At the end of the contact day, the students were asked to write a short text (max. one page) with the following instruction: "What does sustainability mean in vocational teacher´s work? In what ways it is/it should be visible in vocational teacher´s work and in the practices of teaching communities?"

During the semester, the students participated in teaching practice and vocational pedagogical studies. The second collection of students' thoughts on sustainable development and teachers' work took place in November 2020. The students were asked to reflect on their first stories and write shortly about the change. It was expected that seeing the reality of vocational institutes would create a reflective surface for the initial ideas of SD. Ten stories were received (four pages in total). The texts were analysed modifying the grounded theory processes [11,36]: We recognized the role of prior knowledge as mini-frameworks, but analysis involved both theoretical and empirical sensitivity. We attempted to identify specific features of the texts: the context in which the phenomenon was embedded and the conditions that gave rise to it. Although we had prior knowledge of the phenomenon, we tried to be as open as possible to the texts and their voice and reviewed the texts multiple times. The ideas and thoughts presented in the texts were identified to understand the differences and connections between them [12,37].

In our analysis, we combined concept-driven and data-driven qualitative content analysis [38]. Following concept-driven qualitative content analysis [13,38], the expressions were grouped in categories drawn from the theoretical background: the pillars of sustainable development. Table 2 presents the data according to the themes and occasions of data collection. All the teacher students' notions did not, however, fit clearly to the pillars of sustainable development. Some were quite general and focused on the role of education in sustainable development. Here, we applied data-driven analysis [38] and created a category "education for sustainable development", presented in Table 3.

Table 2. The notions of the pillars of sustainable development in students' texts grouped according to the themes and occasions of data collection (1 and 2). Translation by AT.

Ecological 1	Ecological 2
• paying attention to materials used, how to recycle them and save energy • using more electric tools in teaching, trying to reduce the waste of paper, reusing books • electric tools are good in teaching but there is still a need for encounterings: students need to be seen and heard, and net-based teaching in not the only good way • do we need such large spaces in educational institutes	• transport: using public transport or bicycling is supported in the institutes • respecting nature, using nature as a learning possibility; more attention should be paid to aesthetic views of environments • paying attention to habits in the educational institutes: for example, how the waste food is reduced • I pay a lot of attention to discussions on the theme

Table 2. *Cont.*

Social 1	Social 2
• comfortable space is important in feeling welcomed • supportive working community • paying attention to special needs students and how to guide them • supportive culture for learning for all kinds of students • no bullying • interaction is important (between teacher and student but also in teacher community) • justice and equality: for example, trying to help the immigrant students to overcome the language barriers in employment • open-minded attitudes to different cultures: what we have in common is a good starting point • there is a need to be unselfish and pay attention to future generations	• I should concentrate on my doings, doing one thing at a time, that would be sustainable • for me, social and cultural aspects are easier to think about
Cultural 1	**Cultural 2**
• knowing one´s own cultural habits, knowing the cultures nearby and also other cultures • there is a lot of good things also in the past • the base elements in life should be arranged before there is a possibility to promote cultural sustainability	• for me, social and cultural aspects are easier to think about
Economic 1	**Economic 2**
• teachers´ work is promoting the economical sustainability while they educate future employees who commit to society and give their work efforts and pay taxes, and so these employees will keep the society going • all students are not in the same position as far as learning possibilities are concerned—students have different status as financial backgrounds • durable teaching materials (not only materials that are used once), possibly recycled (has to do with the ecological sustainability, too) • teacher should stress the proactive way of teaching the skills needed in the future	• while working in the company, sustainability is one of our business strategies • some themes that we do in the companies have not landed to educational institutes in a bigger scale yet (such as recycling materials)

Table 3. Notions of education for sustainable development students' texts grouped according to the occasions of data collection (1 and 2).

Education for sustainable development 1	Education for sustainable development 2
• education plays a big role in supporting sustainability • sustainability should be a subject in every curriculum • mutli-professional cooperation, sharing one´s expertise • grouping students and making the environments as collaborative as possible	• I have noticed that the phenomena is "all-over" • should be taken into account everywhere hard to grasp • what are my real possibilities to promote sustainability as a teacher? • I would love to work as a teacher and discuss the sustainability questions with my students. Teachers can be a huge value influencer and I would like to be one.

3.1. Sustainable Development as a Part of Vocational Teachers' Work

All the pillars of sustainable development were identified in the first texts. More than half of the notions in the first texts were on socially sustainable development: ideas on safety in learning environments, supporting the special needs students, supportive teacher communities, sharing the expertise with colleagues, zero tolerance against bullying, notifying meaningful encounterings, and having the future generations in mind. Ecological views concerned the material and recycling issues but also alternative transportation and thoughts on the needs for huge buildings and spaces of educational institutions. The educational profession was seen to be economical when they have a possibility to train future employees and while using sustainable products as teaching materials. Cultural views concerned knowing one's personal life, history, and surroundings. Knowing one's own and local history was regarded as equally important in relation to understanding other cultures. This category was present in the first texts only. In the second texts, the notions were divided more equally between the different pillars of sustainable development and contained thoughts of changing one's own behavior.

The proportion of notions concerning teachers' work was larger in the second collection of data.

In the first texts, students presented the need to include sustainability perspectives in all curricula and identified the importance of education in supporting sustainable development. They also described some pedagogical arrangements which could help in furthering sustainable development such as paying attention to collaboration and multi-professional cooperation in teaching. The second texts tell about beginning to understand the "overallness" of the phenomenon and, at the same time becoming uncertain of teachers' possibilities for promote sustainability. On the other hand, some students express enthusiasm and write about realizing the possibilities a teacher has in influencing future students.

The results of Tables 2 and 3 can be summarized into three main perspectives to make a holistic picture of how the students' understanding of sustainable development as a part of teachers' work develops: the meaningfulness of education, teachers as an example, and understanding the connections.

Concerning the *meaningfulness of education,* a change in tone can be noticed. The first texts were written more in the style of what *ought* to be done: There should be sustainability as a core competence for all vocational students and that should be obligatory for all to master in addition to the vocational skills. There was also an understanding that education brings along health and well-being. In the second texts, the students wrote more about the concrete ideas to carry out sustainable things, such as how educational institutions try to support their employees in using public transport or how the waste of food is trying to be avoided. There was also an example of how to bring some nature elements into the classrooms and referring to that it is said to increase the ecological thinking in addition to that it could be used as a teaching tool. At the same time, students presented their individual feelings towards teachers' possibilities: both hesitation and enthusiasm.

In the first texts, *teachers as an example* was more about knowing how, organizing the learning environments to act, and knowing about the facts, for example, sharing expertise and making environments as collaborative as possible. In the second texts, understanding was more on affecting the students: willingness to discuss the themes with students and to be more coherent of one's own doings, such as trying not to concentrate on multiple things at the same time.

Understanding the connections was described in the first texts more in terms of "hope to do these things". The students described understanding of different cultures and beautiful learning environments. They wanted to encourage students to think on their own behavior and its effects. In the second stories, there was more understanding of sustainability as a phenomenon everywhere. Their eyes were opened to it as being "all over the world". They were a little concerned about how it is everywhere, but it is difficult to grasp and make concrete.

The results indicate that students' first texts were written usually from a viewpoint of an *observer* and the second texts more from a viewpoint of a *member of the teacher community.* This we consider as the beginning of sensemaking, remarkable transformative learning in three months, and as an initiation of the development of teacher identity.

3.2. Meeting the Learning Objectives of the Course

To reflect on how the goals set when planning the TSSD were achieved, we connected the teacher students' thoughts and observations with the expected learning outcomes (Table 4). This analysis was carried out as a concept-driven analysis using the descriptions of learning outcomes as the basis of the categories. This analysis shows that teacher students' thoughts and reflections meet the applying and even the integrated level of mastering in the latter texts. The students seemed to have encountered new thoughts which have changed their way of looking at teachers' work and their understanding of the power and potential to make change through teachers' work. Even though the students were not working as teachers, some of them have reached the integrated level in their writings on how teachers can act as examples of sustainably acting and responsible professionals. Altogether, students were able to recognize practical examples of actions for sustainable development in educational institutions. However, they did not produce notions of the certification systems or descriptions of the agreements in the background of sustainable development.

Table 4. Learning on sustainability competences shown in the students' texts (*in italics*) and Vocational Teacher's Sustainability Competences as defined in Tampere University of Applied Sciences Vocational Teacher Education in 2020.

Vocational Teacher's Sustainability Competences Core competence and conceptual base: history and philosophy of sustainable development (SD), ecological, social, cultural, and economical sustainability, Agenda 2030				
Level of compe-tence	Frame of Competence according to the study modules	Teachers in learning environments and networks	Teachers' professional identity and pedagogical skills	Teachers in society
Descriptive		Students can describe features of a sustainable educational institution. *Students could recognize practical examples in educational institutions.* Knows certification systems of educational institutions. *(not visible in the data)*	Recognizes how the teaching profession connects with SD. Recognizes the relations of SD and pedagogical choices. *Reflections on the impact of education.*	Knows the agreements and goals related to SD *(not visible in the data)*
Applying		Furthers active sustainability in an educational institution.Implements the principles of sustainability at work *Students seem to pay more attention to sustainability in vocational teacher's work.*	Connects professional knowledge, practices, and professional identity to form a sustainable professional identity. Uses pedagogical solutions which further SD. *Especially the social points are recognized—the core of being a vocational teacher.*	Acts in order to promote SD. Uses different methods of influencing and communication. *The power of teacher being an example is recognized.*
Integrated		Develops sustainable practices of an educational institution. Creates networks. *Most of the teacher students are not working as teachers yet. Clear reflections outside the teacher communities.*	Acts as a role model for a sustainability promoting professional. Develops pedagogies and contents. *Well-defined the idea of the teacher being an example. During studies, broadened views to the phenomenon.*	Uses theoretical knowledge and practical experience to create a more sustainable world. *The journey has started by thematic studies. It is the responsibility of teacher educator to keep discussions alive during the study process.*

As we took a critical view on the course implementation, we realized that it was no surprise that students did not take the documents of sustainability into account: There were no lectures or documents available for that. Some teacher students were aware of the documents in their previous working places but have not seen any documents when visiting the educational institutes. Furthermore, the students who had no prior experience in teaching were quite keen on becoming acquainted with the teaching and learning environments. Encounterings—that is, many kinds of situations of close and authentic interaction—have played a big role during the studies and discussions on how to keep every student along, so it was no wonder that they commented most on the social dimensions of sustainability.

4. Discussion

Separate courses on ESD have been reported to be somewhat inefficient in creating change agency in teachers [2,7,14,15]. For example, [14–17] call for a more holistic inclusion of sustainability in strategies and curricula of teacher education. Especially [14] addresses the power of making sustainability competency a visible professional goal in the curriculum. Thematic studies were an attempt to answer these calls by providing a sustainability lens to the teaching profession instead of a separate course. The reflections of the students point to the direction that strong integration with the identity process provides an important ingredient for enabling transformative learning towards ESD.

As promotion of SD is not commonly identified as a part of vocational teachers' work [31], it is not surprising that the students' first texts were generally quite descriptive and resembled answers to a test or described their expectations. The second ones were written with more subjective tones, presenting observations of how sustainability relates with education and describing teachers' roles in promoting sustainability. All this speaks about ongoing sensemaking [32].

The sensemaking process started with grasping what SD is and, at the same time, trying to understand what is a teacher's role in this. The latter texts contain thoughts, both inspired and hesitant, about teachers' possibilities. We interpret the inspired thoughts as becoming aware of the transformative power of a teacher in the context of SD, which is a step towards acting for such transformation as a teacher. Beyond the possibilities of an individual teacher, students also described the practicalities of educational institutions and possibilities to make change in them. Students have become aware of sustainability and its "all-overness" and can identify teachers' transformative role in the educational institution [39].

The hesitant notions are reflections of teacher students becoming more aware of the "doing and being" potentials of teachers. This is related to becoming aware of how huge and difficult a phenomenon sustainability is [15]. The next step would be to start thinking about how to realize sustainability in their teaching practice. Whether subjective observations on practices, teachers' potentials and possibilities to promote SD on the level of educational institutions are actualized as change agency and depend greatly on the interplay of all aspects of the ESD competency but especially of motives and affective dispositions [5] during teaching practice or in the school community. This is the phase where students' courage for transformative learning can fail if they are not around trusted peers and a teacher educator who is a capable mentor or facilitator [20] who guides, facilitates, and partners [2,26] with them in the learning process. Starting to practice ESD demands competencies of risk taking and withstanding uncertainty and ambiguity, which can be described as entrepreneurial [40]. Here, entrepreneurial education seems to have something to offer for ESD as the pedagogies for ESD address poor tolerances for ambiguity and uncertainty [41], and do not even recognize risk taking as it is not considered a Key Sustainability Competence, while these competences are in the heart of entrepreneurial education [27].

Finally, some students wrote about changes in their world view. The introductory materials on sustainability, instructed ponderations, and other studies steered the students to observe ecological and social sustainability in educational institutions and curricula. This directed some students to choose sustainability-inclined topics for Development Works (5 ect), such as "how to teach regenerative farming" or "trying to figure out eco-sociological social work or studying how garden work could be implemented as many possible studies". In these students, the studies have ignited the will to act as change agents in educational institutions, and they show the actualization of change, which is coming close to third-order learning [22].

As some of the students have actually made choices which alter their professional development towards a change agent, we find that we have been able to create studies which can open students' eyes to the meaning of their choices concerning the sustainable future and sustainable life, thus providing some evidence of third-order transformative learning [21]. Some of this is undoubtedly due to the teacher education process being an identity-building process with strong group dynamics. It seems that introducing the theme of ESD in the beginning of studies directed the students' attention to the sustainability themes in an early phase of making sense of the whole profession and facilitated integrating the mission of sustainability as part of the professional growth [17]. In our opinion we have developed one way to incorporate ESD in teacher education in a way that meets the challenge of establishing "professional development approaches and opportunities that would enable teaching colleagues to prepare students … to understand and apply their professional and global responsibilities in sustainability [2]."

We have followed only one student groups' learning and only for a short time. Thus, we cannot generalize our results. Still, we find this case and its results quite promising. Transformative learning processes can take a long time, and the seeds sown in teacher education may manifest later in during the teaching career. More initiatives and more research on the connections of ESD and EE would be needed to better understand how their transformative capacities can best be combined for a more sustainable future through teacher education.

5. Conclusions

In the beginning of the paper, we stated that transformative learning connects ESD and EE and noted that many of the competences are related. On the other hand, some contents and commitments are contradictory, most visibly the commitment to economic growth in mainstream definitions of EE [24]. Through our case study, and especially through students' learning outcomes, we propose that it would be important to the recognize Gibb's [26] inner enterprise as a crucial component of an active educator striving to transform educational practices and institutions. Especially the entrepreneurial competencies of risk-taking and standing uncertainty and ambiguity could add to the agency of student when connected with the often-emphasized intra-personal competencies such as collaboration and empathy [3]. Addressing these competencies demands the same capacities also from the teacher educators, which addresses how demanding it is to design transformative learning processes.

We find that the developed model for integrating the sustainability theme into VTE offers teacher students a trigger to start making sense of the teacher profession and educational institutions through the sustainability lenses, to develop their sustainability competences. It is important to note that the students' transformative learning process does induce ambiguity and doubts. Here, some perspectives of entrepreneurial education can be useful: especially paying attention to the encouragement, risk-taking, and withstanding the uncertainty of change making.

It is also important to understand that the integrated mastering of transformative sustainability competences can develop only in encounters with authentic teaching and learning situations and with working life. For this to take place, the sustainability theme has to live all through the curriculum, and the inner enterprise of students has to be strengthened through the pedagogical choices of guiding, facilitation, encouraging to take risks and to try new teaching methods, and by partnering in the learning process by focusing on the different ways people learn. These solutions would allow for the transformative learning and growth of vocational teachers who strive for educating professionals who are willing and capable to take action to carry out sustainable transitions.

As a more general conclusion, we highlight the importance of making the sustainability competencies visible as part of the education profession in teacher education curricula and providing teacher educators with the capacities which facilitate transformative learning.

Author Contributions: Concepts and background, E.A. and A.T.; curriculum development, E.A.; case study with students, A.T.; writing, E.A. and A.T. All authors have read and agreed to the published version of the manuscript.

Funding: This research received no external funding.

Institutional Review Board Statement: Ethical review and approval were waived for this study, due to that the research does not deal with vulnerable groups or sensitive issues.

Informed Consent Statement: Informed consent was obtained from all subjects involved in the study. Preliminary results of the study were presented to the students. Individual students cannot be identified from the study.

Data Availability Statement: Data available on request due to restrictions of privacy.

Acknowledgments: We want to thank the teacher students for versatile, honest and inspiring answers.

Conflicts of Interest: The authors declare no conflict of interest.

References

1. UNESCO. *Progress on Education for Sustainable Development and Global*; UNESCO: Paris, France, 2018.
2. Pegalajar-Palomino, M.D.C.; Burgos-García, A.; Martinez-Valdivia, E. What does education for sustainable development offer in initial teacher training? A systematic review. *J. Teach. Educ. Sustain.* **2021**, *23*, 99–114. [CrossRef]
3. Giangrande, N.; White, R.M.; East, M.; Jackson, R.; Clarke, T.; Coste, M.S.; Penha-Lopes, G. A competency framework to assess and activate education for sustainable development: Addressing the UN sustainable development goals 4.7 challenge. *Sustainability* **2019**, *11*, 2832. [CrossRef]
4. UNESCO. *Education for Sustainable Development: A Roadmap*; UNESCO: Paris, France, 2021.
5. Rieckmann, M. Future-oriented higher education: Which key competencies should be fostered through university teaching and learning? *Futures* **2012**, *44*, 127–135. [CrossRef]
6. Shulman, L.S. Those who understand: Knowledge growth in teaching. *Educ. Res.* **1986**, *15*, 4. [CrossRef]
7. Cebrián, G.; Junyent, M. Competencies in education for sustainable development: Exploring the student teachers' views. *Sustainability* **2015**, *7*, 2768–2786. [CrossRef]
8. Rieckmann, M. Education for sustainable development in teacher education. An international perspective. In *Environmental Education*; Lahiri, S., Ed.; Studera Press: Delhi, India, 2019; pp. 33–48.
9. Wiek, A.; Withycombe, L.; Redman, C.L. Key competencies in sustainability: A reference framework for academic program development. *Sustain. Sci.* **2011**, *6*, 203–218. [CrossRef]
10. Sleurs, W. Competencies for ESD (Education for Sustainable Development) Teachers: A Framework to Inte-Grate ESD in the Curriculum of Teacher Training Institutes. Available online: https://unece.org/fileadmin/DAM/env/esd/inf.meeting.docs/EGonInd/8mtg/CSCT%20Handbook_Extract.pdf (accessed on 31 January 2008).
11. UNECE. *Learning for the Future: Competences in Education for Sustainable Development*; United Nations Economic Commission for Europe: Geneva, Switzerland, 2012.
12. Sterling, S.; Glasser, H.; Rieckmann, M.; Warwick, P. competencies. In *Envisioning Futures for Environmental and Sustainability Education*; Wageningen Academic Publishers: Wageningen, The Netherlands, 2017; pp. 153–168.
13. UNESCO. *Education for Sustainable Development Goals: Learning Objectives*; UNESCO: Paris, France, 2017.
14. Dahl, T. Prepared to teach for sustainable development? Student teachers' beliefs in their ability to teach for sustainable development. *Sustainability* **2019**, *11*, 1993. [CrossRef]
15. Koskela, T.; Kärkkäinen, S. Student teachers' change agency in education for sustainable development. *J. Teach. Educ. Sustain.* **2021**, *23*, 84–98. [CrossRef]
16. Wolff, L.-A.; Sjöblom, P.; Hofman-Bergholm, M.; Palmberg, I. High performance education fails in sustainability?—A reflection on Finnish primary teacher education. *Educ. Sci.* **2017**, *7*, 32. [CrossRef]
17. Raus, R. Student teacher ecological self in the context of education for sustainable development: A longitudinal case study. *J. Educ. Sustain. Dev.* **2017**, *11*, 123–140. [CrossRef]

18. Mezirow, J. *Learning as Transformation: Critical Perspectives on a Theory in Progress*; Jossey-Bass Publishers: San Fransisco, CA, USA, 2000.
19. Aboytes, J.G.R.; Barth, M. Transformative learning in the field of sustainability: A systematic literature review (1999-2019). *Int. J. Sustain. High. Educ.* **2020**, *21*, 993–1013. [CrossRef]
20. Blake, J.; Sterling, S.; Goodson, I. Transformative learning for a sustainable future: An exploration of pedagogies for change at an alternative college. *Sustainability* **2013**, *5*, 5347–5372. [CrossRef]
21. Sterling, S. Transformative learning and sustainabiity: Sketching the concpetual ground. *Learn. Teach. High. Educ.* **2011**, *5*, 17–33.
22. Bateson, G. *Steps to an Ecology of Mind*; Chandler: San Fransisco, CA, USA, 1972.
23. Selby, D.; Kagawa, F. Teetering on the brink: Subversive and restorative learning in the times of climate turmoil and disaster. *J. Transform. Educ.* **2018**, *16*, 302–322. [CrossRef]
24. Deveci, I.; Seikkula-Leino, J. A review of entrepreneurship education in teacher education. *Malays. J. Learn. Instr.* **2018**, *15*, 105–148. [CrossRef]
25. Rae, D. Universites and enterprise-education: Responding to the challenges of the new era. *J. Small Bus. Enterp. Dev.* **2010**, *17*, 591–606. [CrossRef]
26. Gibb, A.A. Enterprise culture and education: Understanding enterprise education and its links with small business, entrepreneurship and wider educational goals. *Int. Small Bus. J.* **1993**, *11*, 11–34. [CrossRef]
27. Gabrielsson, J.; Hägg, G.; Landström, H.; Politis, D. Connecting the past with the present: The development of research on pedagogy in entrepreneurial education. *Educ. Train.* **2020**, *62*, 1061–1086. [CrossRef]
28. Seikkula-Leino, J.; Ruskovaara, E.; Hannula, H.; Saarivirta, T. Facing the changing demands of Europe: Integrating entrepreneurship education in Finnish teacher training curricula. *Eur. Educ. Res. J.* **2012**, *11*, 382–399. [CrossRef]
29. Rajakorpi, A.; Salmio, K. *Toteutuuko Kestävä Kehitys Kouluissa Ja Oppilaitoksissa*; National Board for Education: Helsinki, Finland, 2001.
30. Räkköläinen, M. *Kestävän Kehityksen Osaaminen, Opetus Ja Koulutuksen Järjestäminen Ammatillisissa Perustutkinnoissa*; Kansallinen koulutuksen arviointikeskus: Helsinki, Finland, 2017.
31. Tapani, A.; Salonen, O.A. Identifying teachers' competencies in Finnish vocational education. *Int. J. Res. Vocat. Educ. Train.* **2019**, *6*, 243–260. [CrossRef]
32. Weick, K.E. *Sensemaking in Organisations*; Sage: London, UK, 1995.
33. TAMK. *Vocational Teacher Education-Curriculum*; TAMK: Tampere, Finland, 2020.
34. SOOL-Teacher Students Union in Finland. Nämä Opettajankoulutusyksiköt Osallistuivat Soolin Ilmastohaasteeseen. Available online: https://www.sool.fi/uutiset/nama-opettajankoulutusyksikot-osallistuivat-soolin-ilmastohaasteeseen/ (accessed on 14 May 2021).
35. Asikainen, E.; Kukkonen, H.; Hakala, K.; Harju, E.; Kyhä, H.; Lahtinen, J.; Nevalainen, V.; Ranne, P.; Tapani, A. Kohti Kestävää Koulutusta ja Opettajuutta. TAMKJournal 2020. Available online: https://sites.tuni.fi/tamk-julkaisut/pedagogiset-ratkaisut/kohti-kestavaa-koulutusta-ja-opettajuutta/ (accessed on 3 September 2020).
36. Corbin, J.; Strauss, A. *Basics of Qualitative Research: Grounded Theory Procedures and Techniques*; Sage Publications: London, UK, 1990.
37. Kleinsasser, R.C.; Silverman, D. Interpreting qualitative data methods for analysing talk, text and interaction. *Mod. Lang. J.* **1997**, *81*, 136. [CrossRef]
38. Schreirer, M. *Qualitative Content Analysis in Practice*; Sage: London, UK, 2012.
39. Rauch, F.; Steiner, R. Competences for education for sustainable development in teacher education. *CEPS J.* **2013**, 3, 9–24.
40. EntreComp. The European Entrepreneurship Competence Framework. Available online: https://ec.europa.eu/social/main.jsp?catId=1317&langId=en (accessed on 18 August 2021).
41. Lozano, R.; Merrill, M.Y.; Sammalisto, K.; Ceulemans, K.; Lozano, F.J. Connecting Competences and Pedagogical Approaches for Sustainable Development in Higher Education: A Literature Review and Framework Proposal. *Sustainability* **2017**, *9*, 1889. [CrossRef]

MDPI

Article

Impact of Universities' Partnerships on Students' Sustainable Entrepreneurship Intentions: A Comparative Study

Shehnaz Tehseen * and Syed Arslan Haider

Department of Management, Sunway University Business School (SUBS), Sunway University, No 5, Jalan Universiti, Bandar Sunway, Selangor Darul Ehsan 47500, Malaysia; syed.h11@imail.sunway.edu.my
* Correspondence: shehnazt@sunway.edu.my

Abstract: This study investigated the impact of entrepreneurial attitude, perceived desirability, and perceived feasibility on sustainable entrepreneurship intentions under the moderating impact of entrepreneurial passion among undergraduate students of Malaysia. It was a quantitative study that compared two groups of students, i.e., Group A, comprised of students who have studied entrepreneurship modules and whose programmes did not offer any dual/triple award degrees and Group B, made up of students who have studied entrepreneurship modules and whose programmes offered dual/triple award degrees. Data were collected from 542 undergraduate students of universities located in Kuala Lumpur and Selangor through survey questionnaire. WarpPLS Software version 7.0 was used to analyse the data. The findings of this study revealed that Group B students' entrepreneurial attitude, perceived desirability, and perceived feasibility positively and significantly impacted the sustainable entrepreneurship intentions under the moderating impact of entrepreneurial passion. However, the impact of entrepreneurial attitude was found positive and significant on sustainable entrepreneurship intentions among students of Group A and entrepreneurial passion was found to be significant moderator to improve the impact of only entrepreneurial attitude and perceived desirability on sustainable entrepreneurship intentions but not for the impact of perceived feasibility on sustainable entrepreneurship intentions among these students. Moreover, the direct impacts of perceived desirability and perceived feasibility were also found non-significant on sustainable entrepreneurship intentions among Group A students. The findings reveal that universities having partnership with other overseas' universities may offer high quality entrepreneurship modules due to which their students have high entrepreneurial passion and develop more entrepreneurial attitudes, and are more willing and capable to start their own businesses as compared to students of other local universities who have no partnership with overseas' universities.

Keywords: universities' partnerships; entrepreneurial attitude; perceived desirability; perceived feasibility; entrepreneurial passion; sustainable entrepreneurship intentions

Citation: Tehseen, S.; Haider, S.A. Impact of Universities' Partnerships on Students' Sustainable Entrepreneurship Intentions: A Comparative Study. *Sustainability* **2021**, *13*, 5025. https://doi.org/10.3390/su13095025

Academic Editors: Jaana Seikkula-Leino, Mats Westerberg, Priti Verma and Maria Salomaa

Received: 25 February 2021
Accepted: 22 March 2021
Published: 30 April 2021

1. Introduction

Universities are using several strategies to promote the sustainable entrepreneurship intentions among undergraduate students as well as to provide quality entrepreneurship education to equip the students with the essential skills needed to run sustainable entrepreneurial businesses [1]. Academic entrepreneurship is becoming increasingly popular among scientific institutions, businesses, and local governments [2]. There are several reasons for the interest of academic entrepreneurship. For instance, increasing importance of knowledge for the economic development based on human capital entrepreneurship theory [3], research funding and prestige universities, as well as positive influence of entrepreneurship academic research to build competitiveness and international innovation of the economy [4]. The integration of scientific techniques within commerce leads towards the sustainable businesses. These factors have changed the ways of doing businesses. Therefore, universities are contributing towards the practical implications of their research findings due to which the entrepreneurial firms are improving their performances and are becoming more sustainable and innovative businesses [4]. Schumpeter has considered the entrepreneurship as an economic resource that also determines the effective use of material resources [5]. The existing studies have mentioned the importance of universities in promoting entrepreneurship education in every field of study [2–4]. This is because the universities' entrepreneurial cultures foster the entrepreneurial skills of students through their participation in entrepreneurship related research projects, new venture or start-ups projects, as well as through their participation in entrepreneurship classes conducted by business incubators [2]. An academic entrepreneur is an animator of scientific research, an organizer of the transfer of scientific research to the economy, an inspirer of the creation of innovative firms. Therefore, the entrepreneurship education can be enhanced due to the active cooperation between students and academic entrepreneurs.

Entrepreneurship has been considered as an attractive career option for students. Therefore, there is a real need to focus on developing an entrepreneurial mind-set among undergraduate students of universities [6]. Universities should focus more on producing graduates with entrepreneurial attitudes to create more jobs. Thus, universities play a key role in providing the ecosystem to foster entrepreneurship and encourage students to become entrepreneurs. Although universities are striving their best to implement diverse approaches to promote the entrepreneurship, however, not all entrepreneurship related programmes facilitate entrepreneurship as a potential career option for students [6,7]. Knowledge regarding the students' entrepreneurial intentions assists in understanding the factors that could be considered to develop entrepreneurship intention among them.

The entrepreneurship intention among university students is evident regarding the career alternative. The universities play an important role in developing the entrepreneurial skills among students and the focus of the existing entrepreneurship intention of the students has been on education of entrepreneurship [8]. The empirical studies on students' entrepreneurship have provided evidence regarding the positive influence of entrepreneurship courses on their decision to become entrepreneur but with a few studies of contrasting results [4,9,10]. Although the existing literature focused more on the importance of entrepreneurship courses in developing the entrepreneurship intention among students, however, very less or no attention has been paid on how the educational resources and outcomes provided in a university with its partnership university/ies at the overseas, can assist the students to choose entrepreneurship as a career choice. Even though some universities have partnership with other universities in other developed countries and are providing unique resources to support graduate entrepreneurship, however, the influence of these partnerships in promoting the entrepreneurial intention among the undergraduate students is not evident in the existing entrepreneurship literature.

Furthermore, the research on entrepreneurial intentions has tended to focus on one or two aspects of value creation in the context of sustainable entrepreneurship [11–14]. The term value creation differs the conventional entrepreneurship from sustainable entrepreneurship. Although entrepreneurs have earlier believed to pay attention on economic value creation. In these novel entrepreneurship schemes, economic value creation is perceived to an end or to blend different values [14–16]. Environmental entrepreneurship emphasises environmental value creation, while social entrepreneurship is about social value creation [14]. Sustainable entrepreneurship has been acknowledged to blend social, economic, and environmental value creation [12,17]. The sustainable entrepreneurship comprises both environmental and social entrepreneurship [12]. Thus, sustainable entrepreneurship intentions refer to the intentions of the individuals to create businesses by incorporating the elements of social, economic, and environmental factors. In other words, sustainable entrepreneurship intentions refer towards the intentions of individuals to engage themselves in the process of recognizing, assessing, and availing the entrepreneurial opportunities that could minimize a firm's influence on the natural environment and create the benefits for the whole society as well as for local communities by improving their living standards.

Entrepreneurial attitude, perceived desirability, and perceived feasibility have been considered the critical factors that have been studied widely to explore the entrepreneurship intention among students under various contexts [1,18,19]. The attitude towards a certain behaviour indicates the favourable or unfavourable assessments of the individual regarding that behaviour [18]. The perceived desirability is the personal attractiveness of initiating a business with respect to both extra personal and intrapersonal impacts [18]. Perceived desirability indicates the thoughts, enthusiasm as well as attractiveness towards starting the new venture. It refers to the degree of intensity to which a person is attracted to become an entrepreneur for behavioral success. Moreover, previous research observed that desirability is influenced by cultural influences and social norms. If one believes that individuals from their surrounding society accept the activity, this will improve one's attitude towards the behavior. Such social burdens are an obstacle to embarking on any business venture [19]. On the other hand, the perceived feasibility is the extent to which the individual feels capability to start business [1,18]. Perceived feasibility indicates the extent to which one feels personally more competent to start the new venture and refers to the extent to which one believes himself to be capable of carrying out a behaviour. Thus, the presence of mentors, guidance, and role models assist in developing one's perception towards feasibility and gives more confidence to believe that there is some potential and implementation of business idea is very much possible [19].

However, the determinants of sustainable entrepreneurship intentions have not yet been explored among those universities' students whose universities do offer dual or triple award degree programs e.g., [1,20–23]. Many internally-developed degrees level as well as diploma level programmes offered by Malaysian private HEIs are recognised and validated by various top foreign universities in the UK, USA, Canada and other developed countries. These partnerships bring a lot of opportunities for the students of Malaysian universities. For instance, superior quality of education is ensured through the process of external moderation of subject modules of local private universities by the professors of foreign universities and employers are also ensured about quality of degree due to dual/triple award degrees programmes. The students could perceive more value of their degrees due to validation of their local degree from prestige university/ies of abroad and thus strive hard to get success in their modules. The main purpose of any entrepreneurship module is to develop the entrepreneurial intentions among the students towards sustainable entrepreneurial businesses. This study argues that the students could feel more motivation towards sustainable entrepreneurship businesses in universities with dual/triple award degree programmes and are more passionate to start their own businesses due to the unique resources of entrepreneurship knowledge which the partner universities provide to local universities to teach the entrepreneurship modules.

Although the government of Malaysia has taken various incentive measures to attract entrepreneurial activities, however, it has not reached at desired level. The level of Malaysian's entrepreneurial activity is still at a low level compared to several other developed nations. For instance, Global Entrepreneurship Monitor (GEM) stated that only 4.9% Malaysians have entrepreneurial intentions, which ranked Malaysia 64th out of 65 countries. One of the best ways to increase the future entrepreneurial activities is to create entrepreneurial intentions among universities' students. Undergraduate students should develop sustainable entrepreneurship intentions to create social, economic, and environmental values. Unfortunately, not many studies have examined the sustainable entrepreneurship intentions among undergraduate students, e.g., [1,15,20,21,24,25]. Although many studies have explored the entrepreneurial intentions among universities' students, however, these studies have not investigated the influence of partnerships of local universities with overseas' universities on the sustainable entrepreneurship intentions among undergraduate students. Therefore, this study's focus is to investigate the effect of local universities' partnerships on sustainable entrepreneurship intentions among their students in the settings of an emerging country. A comparative study is planned to be conducted among Malaysian universities' students whose programs are affiliated with any other overseas' university/ies (with dual/triple award degree programs) and other students of Malaysian universities whose programs are not affiliated with any of other overseas' university/ies (without dual/triple award degree programs).

While the concept of sustainable entrepreneurship intention is achieving a significant attention in the field of entrepreneurship, prior research focuses mainly on different determinants on entrepreneurial intentions among universities' students in general in different countries [2,20,22,25]. However, the impact of different factors on sustainable entrepreneurial intentions of universities' students is poorly known. More importantly, although a few studies have investigated the impact of entrepreneurial attitudes, perceived desirability and perceived feasibility on entrepreneurial intentions of universities' students under different contexts [20,26]. These studies have found contradictory results regarding the impact of these predictors on developing the sustainable entrepreneurship intentions. For instance, some studies have found very strong influence of entrepreneurial attitude, perceived desirability, and perceived feasibility on entrepreneurial intentions among universities' students [1,27], while other researchers have found weak impacts of these variables on entrepreneurial intentions among universities' students [21]. Due to inconsistent findings regarding the impact of entrepreneurial attitude, perceived desirability, and perceived feasibility on entrepreneurship intentions among students in the existing studies under various contexts, this study argues that entrepreneurial passion could be the potential moderator that could improve the impact of entrepreneurial attitude, perceived desirability, and perceived feasibility on students' intentions towards the sustainable entrepreneurship. This is because the entrepreneurial passion is regarded to be the most observed factor in the entrepreneurial process and has been considered as the number one characteristic for any successful entrepreneur [5]. As mentioned earlier, since the existing studies did not examine the impact of universities' partnerships with overseas' universities on the sustainable entrepreneurship intentions among undergraduate students, which is one of the strategies to enhance the academic entrepreneurship to promote sustainable entrepreneurship intentions among students. Thus, it would be interesting to examine the impact of understudy variables on sustainable entrepreneurship intentions among universities' students with and without dual/triple award degree programs. Therefore, this study seeks to answer the following two questions:

(1) How is the impact of entrepreneurial attitudes, perceived desirability, and perceived feasibility on sustainable entrepreneurship intentions among universities' students with and without dual/triple award degree programs?
(2) How does the entrepreneurial passion moderate the positive impact of entrepreneurial attitude, perceived desirability, and perceived feasibility on sustainable entrepreneurship intentions among universities' students with and without dual/triple award degree programs?

This study provides useful insights for future development of sustainable entrepreneurship intentions among Malaysian universities' students and will reveal the impact of universities' brand image through their partnerships on students' sustainable entrepreneurship intentions.

This paper has been divided into several sections. For instance, after the Introduction section, the proposed model is presented which is followed by development of hypotheses based on existing literature. Methodology, data analysis and results are then presented in the next sections followed by discussion, study limitations and future recommendations, and finally conclusions.

2. Proposed Model

The entrepreneurial intention literature has emphasised five main themes; entrepreneurship education, the core entrepreneurial intention models, social and sustainable entrepreneurship, the entrepreneurial intention-behavior link, and the factors influencing entrepreneurial intentions including regional, cultural as well as institutional and individual-level variables [28]. The social and sustainable entrepreneurship themes of entrepreneurial intention have emerged more lately [28]. Only one or two features of value creation were emphasized in the research on entrepreneurial intentions under sustainable entrepreneurship's context [14,29]. This study's model consists of sustainable entrepreneurship intentions as the dependent variable. Moreover, entrepreneurial attitude, perceived entrepreneurial feasibility and desirability are proposed as drivers of sustainable entrepreneurship intention. These drivers are the constructs that describe the perception of individuals regarding their abilities to perform the given tasks [29]. This study has also used entrepreneurial passion as a moderator for the impact of entrepreneurial attitude, perceived entrepreneurial feasibility and desirability on sustainable entrepreneurship intentions among universities' students. Based on Upper Echelon Theory (UET), the essential characteristics like entrepreneurial passion impact on the success of any business [30]. Since it is based on UET, this study argues that the passion could facilitate the individual's entrepreneurial attitude towards sustainability, as well as the perceived entrepreneurial feasibility and desirability of sustainable entrepreneurship intentions among students.

During their research on entrepreneurial intention, [31,32] suggested the Entrepreneurial Potential Model (EPM). In their study, [31] mentioned two significant constructs including perceived desirability and perceived feasibility. Krueger and Brazeal proposed the EPM model by integrating the Entrepreneurial Event Model (EEM) and the Theory of Planned Behaviour (TPB) in which few concepts were overlapping. Findings indicate that the credibility depends on the perception of desirability and on understanding of feasibility of the venture opportunity that leads towards the behaviour which also depends on the person's potential who wants to start the venture [19]. The researchers such as [33] mentioned that EPM interacts with the two significant models, i.e., EEM and TPB. This study is based on the modified model of Entrepreneurial Potential Model [31] due to addition of two relevant constructs namely entrepreneurial attitude and entrepreneurial passion. EPM conceptualizes that the individuals can create entrepreneurial ventures based on their ability and potential to start a business which are explained by three main constructs namely perceived feasibility, perceived desirability, and propensity to act. However, in this study, the impacts of entrepreneurial attitude, perceived feasibility and perceived desirability have been analysed on sustainable entrepreneurship intention under the moderating influence of entrepreneurial passion. Based on modified EPM, we have proposed our hypothesized model as shown in Figure 1.

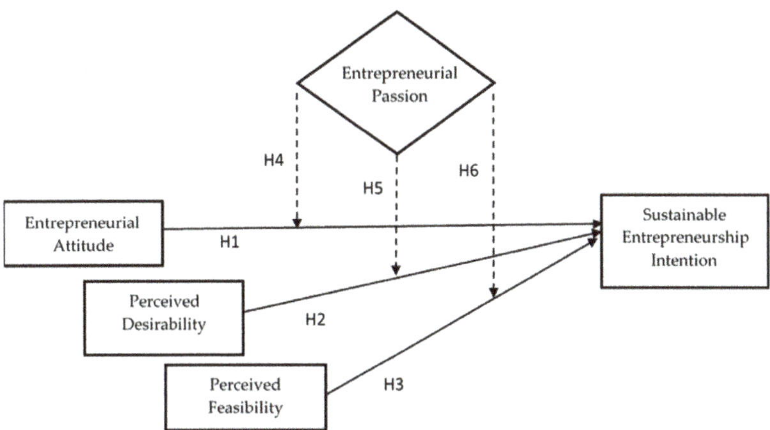

Figure 1. Proposed Research Model.

3. Development of Hypotheses

3.1. The Impact of Entrepreneurial Attitude on Sustainable Entrepreneurship Intentions

Sustainable entrepreneurship is in contrast with commercial entrepreneurship in order to focus on merging different types of orientations including, social, economic and environmental [13]. There are two substitute ways to sustainability-oriented entrepreneurial practice, that are assisted by a supportive operational environment and are created as a reaction to an unsupportive environment. Studies revealed that the first way refers to sustainability-oriented including sustainability-oriented entrepreneurial ideas, emphasising on perceiving business and value formation and social support. On the other hand, the second way on a high level of entrepreneurial intention toward sustainability, excluding sustainability-oriented entrepreneurial concepts and not perceiving social and contextual support [34]. Reference [35] argued that attitudes are formed by value priorities, shape intentions and the following behaviour, therefore, studies regarding entrepreneurial intention in the context of sustainable entrepreneurship must include attitude toward sustainability [36]. According to [37], entrepreneurial attitudes assess the character to achieve the specific targets, therefore, they are different from traits. Additionally, entrepreneurial attitudes also impact the individual intentions and behaviour simultaneously [33]. The entrepreneurial attitudes have a significant role in developing intentions and has become the determinant factor in forming behaviour [38,39]. Many existing studies have found the positive and significant influence of entrepreneurial attitude on entrepreneurial intentions under various contexts [38,40]. Therefore, the following hypothesis is developed in this regard:

H1. *Entrepreneurial attitude positively impacts the sustainable entrepreneurship intentions among universities' students with and without dual/triple award degree programs.*

3.2. The Impact of Perceived Desirability on Sustainable Entrepreneurship Intention

Existing empirical studies have found positive and significant impact of perceived desirability towards entrepreneurial intentions [24,41]. This is because the individuals are more intended to become entrepreneurs if they believe that being an entrepreneur is more desirable to them than working for others [42]. The desire of individuals to become entrepreneur or to be self-employed provides a positive drive to become an entrepreneur [41]. Individuals would prefer to become an entrepreneur when they believe that the benefits and rewards of entrepreneurship outweigh the advantages of work because of the fact that the expected rewards depend on the individual's evaluation of entrepreneurship and desirability to become an entrepreneur [41]. Thus, based on existing studies, this study develops the following hypothesis:

H2. *Perceived desirability positively impacts the sustainable entrepreneurship intentions among universities' students with and without dual/triple award degree programs.*

3.3. The Impact of Perceived Feasibility on Sustainable Entrepreneurship Intentions

Additionally, people who are more concerned with sustainable development and preserving nature also tend to act according to their values [43–45]. While, it has been identified that under more entrepreneurial self-efficacy and more industry resource-scarcity, people do not follow to their pro-environmental standards when assessing environmental destruction caused by grabbing opportunities [46]. Thus, partiality for environmental and social value creation composed with a positive opinion of entrepreneurship as a career choice could be positively referred to sustainability-oriented entrepreneurial intentions. Social problems are often considered to be much challenging, which suggests that the chances of effectively solving them are perceived to be low or even non-existent [47]. Consequently, perceived entrepreneurial feasibility has been anticipated to be positively associated with entrepreneurial intentions [48–50]. Based on the empirical evidences provided by the entrepreneurship literature, it is proposed that perceived feasibility influence the sustainable entrepreneurial intentions. Thus, the following hypothesis is developed.

H3. *Perceived feasibility positively impacts the sustainable entrepreneurship intentions among universities' students with and without dual/triple award degree programs.*

3.4. Moderating Impact of Entrepreneurial Passion

Numerous researches have shown that entrepreneurial passion contributes as one of essential factors in the creation processes of new venture [51–53]. Reference [54] referred passion with the 'fire of desire' that acts as a fuel for the entrepreneurs' efforts and creativity and enables them to face all the difficulties they encounter [53]. Entrepreneurial passion is related to the positive attitudes and feelings for activities that are important for the individual's self-identity [55,56]. Passion has been regarded as the "heart of entrepreneurship" that is also a key element of entrepreneurial behavior action towards the business creation [57]. Existing literature has widely proven that entrepreneurial passion has a significant role in entrepreneurial intention [23,55,58–61]. Furthermore, some researchers have also identified that entrepreneurial passion improves motivational factors and develops the positive feelings in high turbulent business environment with restraint resources [3,60].

Reference [61] concluded that entrepreneurial passion motivates individuals to identify opportunities for innovations and thus, develops intention to create a new business. Likewise, other researchers have also found the positive and significant impact of entrepreneurial passion on entrepreneurial intention under various contexts [51,52,58,62,63].

Reference [54] have described three types of entrepreneurial passion relevant to many characteristics of entrepreneurial activities. The first type of passion indicates the inventor identity [52,54] which is regarding the involvement of the entrepreneur in identifying, inventing, and then exploring new opportunities. This type of passion indicates the funder identity [52,54] which is regarding the involvement of the entrepreneur in entrepreneurial process of creating a business venture and related commercializing and exploiting activities [55]. The third type of passion indicates developer identity [52,54] which is regarding the entrepreneur's involvement in the nurturing, forecasting, progress, and growth activities of the venture after its establishment [55]. These entrepreneurial passions relevant with three different types of role identities impact the entrepreneurial intention [44]. Thus, individuals with their higher level of entrepreneurial passion are most likely to create a business and execute their passion into action [55]. Thus, the study suggests the following hypotheses:

H4. *Entrepreneurial passion positively moderates the impact of entrepreneurial attitude on sustainable entrepreneurship intention among universities' students, i.e., the positive impact of entrepreneurial attitude on sustainable entrepreneurship intentions will be more when the entrepreneurial passion is high.*

H5. *Entrepreneurial passion positively moderates the impact of perceived desirability on sustainable entrepreneurship intention among universities' students, i.e., the positive impact of perceived desirability on sustainable entrepreneurship intentions will be more when the entrepreneurial passion is high.*

H6. *Entrepreneurial passion positively moderates the impact of perceived feasibility on sustainable entrepreneurship intention among universities' students, i.e., the positive impact of perceived feasibility on sustainable entrepreneurship intentions will be more when the entrepreneurial passion is high.*

4. Methodology

We searched the data bases from Google Scholar, Emerald, Springer, Sage, Elsevier, Taylor and Francis, Academy of Management (AOM) Journals, and Wiley Online Library by using the combinations of various keywords such as "Entrepreneurial Attitude and Entrepreneurship Intentions among Students", "Perceived Desirability and Entrepreneurship Intentions among Students", "Perceived Feasibility and Entrepreneurship Intentions among Students", "Entrepreneurial Passion and Entrepreneurship Intentions among Students", and "Sustainable Entrepreneurship Intentions among Students". The observation period was last 11 years. We screened all the revenant studies and did review of only those studies that could qualify the two criteria. First, those studies that were mostly published in academic journals, excluding other sources such as trade publications, country reports or magazines. Second, we used empirical studies that mostly used the sample of undergraduate students.

Furthermore, the students were selected from management and business related programmes to collect data for this study. This study involves a quantitative study and data were collected using standard survey questionnaire from 600 undergraduate students of Kuala Lumpur and Selangor. Only 542 questionnaires were useable for data analysis which consisted 271 students in Group B with dual/triple award degree programmes who took entrepreneurship module and other 271 students in Group A who also took entrepreneurship module but their programmes did not offer dual/triple award degree programmes. Non-probability sampling techniques including snowball, convenience, and quota samplings were used to get target respondents. According to WarpPLS Software, the minimum sample size required for current model is 160 for the inverse square root with power level of 0.80 with significance level of 0.05 and 146 for the gamma exponential method [64]. Since researcher was able to collect data from 542 students (271 for Group A and 271 for Group B) which exceeded the minimum required sample size. All the constructs were measured using the items that were adapted from existing literature. For instance, Entrepreneurial Attitude (EA) and Perceived Desirability (PD) were measured with 3 items each adapted from [29], Perceived Feasibility (PF) was measured with 4 items adapted from [65], Sustainable Entrepreneurship Intention (SEI) and Entrepreneurial Passion (PASS) were measured with 5 and 4 items adapted from [36,55] respectively. All the constructs and their measures are presented in Table 1 as below:

Table 1. Constructs with Items and Source.

Items of Constructs	Source
Entrepreneurial Attitude (EA) EA1. Social impact (poverty reduction, employment, and increasing equality) that the venture would have. EA2. Environmental impact (e.g., use of natural resources, protecting biodiversity, and energy type) that the venture could have EA3. I'm determined to create a sustainable firm in the future.	[29]
Perceived Desirability (PD) PD1. A career as entrepreneur is interesting to me. PD2. If I have opportunities, capital, and abilities, I will start a new firm. PD3. Being an entrepreneur will give a large amount of satisfaction for me.	[29]
Perceived Feasibility (PF) PF1. I can control the creation process of a new firm. PF2. I know necessary practical details to start a firm. PF3. I know how to develop an entrepreneurial project. PF4. If I tried to start a new firm, I would have a high probability of succeeding.	[65]
Entrepreneurial Passion (PASS) PASS1. It is exciting to figure out new ways to solve unmet market needs that can be commercialized. PASS2. Searching for new ideas for products/services to offer is enjoyable to me. PASS3. I am motivated to figure out how to make existing products/services better. PASS4. Scanning the environment for new opportunities really excites me.	[55]
Sustainable Entrepreneurship Intentions (SEI) SE1. I prefer to be a sustainable entrepreneur rather than to be an employee of a company. SE2. My professional goal is to become a sustainable entrepreneur. SE3. I will make every effort to start and run my own sustainable firm. SE4. I am determined to create a new sustainable firm in the future. SE5. I have very seriously thought about in starting a sustainable firm.	[36]

Table 2 represents the demographic information about respondents. In both groups A and B, female respondents are more than male respondents. The majority respondents were Malays and Chinese having diploma and high school degree and belonged to the age group of 22–23. Group A students belonged to private as well as public universities located at Kuala Lumpur while Group B students belonged to only private universities located at Selangor.

Table 2. Descriptive Statistics.

Demographics	Categories	Group A (Without Dual/Triple Award Degree Programmes)		Group B (With Dual/Triple Award Degree Programmes)	
		Frequency	Percent	Frequency	Percent
Gender	Male	132	48.7	123	45.4
	Female	139	51.3	148	54.6
Age	18–19	93	34.3	68	25.1
	20–21	49	18.1	90	33.2
	22–23	129	47.6	113	41.7
Race/Ethnicity	Malay	167	61.6	154	56.8
	Chinese	71	26.2	79	29.2
	Indian	21	7.7	18	6.6
	Other	12	4.4	20	7.4
Highest education completed	Certificate	37	13.7	39	14.4
	Diploma	124	45.8	117	43.2
	High School	110	40.6	115	42.4
Location of your University	Kuala Lumpur	271	100	0	0
	Selangor	0	0	271	100
Your University Sector	Malaysian Private University	139	51.3	271	100
	Malaysian Public University	132	48.7	0	0

5. Data Analysis and Results

The current study used the WarpPLS software version 7.0 to test the proposed framework [66]. While performing analysis on WarpPLS, there are few requirements which are needed to be fulfilled to ensure that the instrument is reliable [66]. To evaluate the goodness of research model fit, several indicators were checked including: Average path coefficient (APC), Average R-squared (ARS), Average adjusted R-squared (AARS), Average block VIF (AVIF), Average full collinearity VIF (AFVIF), Tenenhaus GoF (GoF), Sympson's paradox ratio (SPR), R-squared contribution ratio (RSCR), Statistical suppression ratio (SSR) and Nonlinear bivariate causality direction ratio (NLBCDR) [67].

Table 3 indicates the evaluation of goodness of fit of this research based on APC value 0.132 with $p < 0.001$, ARS value 0.375 with $p < 0.001$ and AARS value 0.368 with $p < 0.001$. The AVIF value of 2.089 is ideally ≤ 5 and AFVIF values of 1.999 ideally ≤ 5 which means neither vertical nor lateral multicollinearity occurs in the research model. The GoF value is found 0.540 which is greater than 0.36 which means that the fit of the model is very good. Also, the SPR, RSCR, SSR and NLBCDR values meet their threshold criteria as shown in Table 3. This means that the predictors are not found to be mutually correlated in the research model and there is no collinearity problem between the predictors and the criterion as well.

Table 3. Model fit and quality indices.

No.	Model Fit and Quality Indices	Criteria Fit	Results	Remarks
1	Average path coefficient (APC)	$p < 0.001$	0.132	Good
2	Average R-squared (ARS)	$p < 0.001$	0.375	Good
3	Average adjusted R-squared (AARS)	$p < 0.001$	0.368	Good
4	Average block VIF (AVIF)	acceptable if ≤ 5, ideally ≤ 3.3	2.089	ideally
5	Average full collinearity VIF (AFVIF)	acceptable if ≤ 5, ideally ≤ 3.3	1.999	ideally
6	Tenenhaus GoF (GoF)	small ≥ 0.1, medium ≥ 0.25, large ≥ 0.36	0.540	large
7	Sympson's paradox ratio (SPR)	acceptable if ≥ 0.7, ideally = 1	1.000	ideally
8	R-squared contribution ratio (RSCR)	acceptable if ≥ 0.9, ideally = 1	0.958	Good
9	Statistical suppression ratio (SSR)	acceptable if ≥ 0.7	1.000	ideally
10	Nonlinear bivariate causality direction ratio (NLBCDR)	acceptable if ≥ 0.7	0.900	ideally

6. Multiple Group Invariant Assessment

To conduct the multiple group analysis (MGA) test, the current study divided the respondents into two groups based on university affiliation (Group A: without dual/triple award degree programmes and Group B: with dual/triple award degree programmes) as suggested by [68]. It is critical to establish the measurement invariance before conducting MGA. After that, the researchers confirmed that any differences in model ratings parameters between subgroups are not caused by content differences or perceived differences in the description of the steps that make up the model for both groups. It should be noted that rating error may increase when measurement imbalances can be established; It should be noted that measurement error can be inflated when measurement invariance is not established; this can lead to biased results [69]. Table 4 shows absolute latent coefficients for loadings and their *p* values greater than 0.05, which means that no significant difference occurred between groups due to factor loadings. After, establishing the partial measurement invariance, the MGA was made to compare the coefficients of the two groups to predict the purpose of smartwatch adoption.

The measurement model (outer model) is initially evaluated, which indicates the construct reliability and validity of Entrepreneurial Attitude (EA), Perceived Desirability (PD), Perceived Feasibility (PF), Entrepreneurial Passion (PASS) and Sustainable Entrepreneurship Intentions (SEI) variables that are measured as reflectively. To evaluate the outer model, the three criteria were used including the construct reliability, convergent validity, and discriminant validity (Fornell-Larcker criterion and Heterotrait-Monotrait (HTMT)) [70]. While evaluating the reliability and validity of the model, Cronbach alpha(α), composite reliability (CR), and average variance extracted (AVE) were checked. In general, value of outer loadings needs to be greater than 0.70 [71]. Those items whose outer loadings fall in the range of 0.40–0.70 should be removed only if deleting them increases α, CR or AVE values [72]. Hence, composite reliability is appropriate measure of reliability and varies from 0 to 1. Values above 0.70 are recommended as threshold [72]. The threshold level of AVE is 0.50 or above according to criteria [73]. Table 4 shows convergent validity and reliability of the model.

Discriminant validity is used to ensure that each concept of latent variable is different from other variables standards [73]. In Fornell-Larcker criteria, the comparison is done between square root value of AVE and the correlation coefficient of each construct. For a construct to have discriminant validity, square root value of AVE of a construct needs to be greater than the correlation coefficients of other constructs [74]. In Table 5, it can be seen that the root AVE value for each variable is higher than the AVE value for the other variables. This shows that the prerequisites for the discriminant validity test have been met. Thus, the instrument used in this study has met the requirements of the validity test.

The Heterotrait-Monotrait (HTMT) ratio indicates the average of correlation of the indicators among different constructs and the average of the correlation of indicators of the related construct. According to [71], models with constructs that are conceptually similar have threshold level of 0.90 while those constructs that are unrelated to each other have threshold value of 0.85 or below. From Table 6, it can be observed that not a single value is greater than 0.85. Hence, discriminant validity is established.

Table 4. Measurement Model and Invariance Analysis.

Variables/Items	Group A (Without Dual/Triple Award Degree Programmes)				Group B (With Dual/Triple Award Degree Programmes)				Invariance Analysis		
	Factor Loadings	α	CR	AVE	Factor Loadings	α	CR	AVE	Absolute Loadings	S. E	*p* Value
Entrepreneurial Attitude (EA)		0.687	0.829	0.624		0.734	0.763	0.524			
EA1. Social impact (poverty reduction, employment, and increasing equality) that the venture would have.	0.884				0.818				0.013	0.043	0.767
EA2. Environmental impact (e.g., use of natural resources, protecting biodiversity, and energy type) that the venture could have	0.847				0.741				0.001	0.043	0.981
EA3. I'm determined to create a sustainable firm in the future.	0.610				0.781				0.022	0.043	0.615
Perceived Desirability (PD)		0.909	0.943	0.846		0.737	0.806	0.585			
PD1. A career as entrepreneur is interesting to me.	0.919				0.834				0.054	0.043	0.202
PD2. If I have opportunities, capital, and abilities, I will start a new firm.	0.920				0.616				0.071	0.043	0.096
PD3. Being an entrepreneur will give a large amount of satisfaction for me.	0.921				0.826				0.014	0.043	0.747
Perceived Feasibility (PF)		0.792	0.866	0.619		0.701	0.726	0.502			
PF1. I can control the creation process of a new firm.	0.810				0.766				0.005	0.043	0.906
PF2. I know necessary practical details to start a firm.	0.681				0.656				0.007	0.043	0.541
PF3. I know how to develop an entrepreneurial project.	0.841				0.761				0.034	0.043	0.433
PF4. If I tried to start a new firm, I would have a high probability of succeeding.	0.806				0.737				0.013	0.043	0.708
Entrepreneurial Passion (PASS)		0.895	0.928	0.763		0.701	0.724	0.510			
PASS1. It is exciting to figure out new ways to solve unmet market needs that can be commercialized.	0.888				0.788				0.081	0.043	0.057
PASS2. Searching for new ideas for products/services to offer is enjoyable to me.	0.911				0.795				0.062	0.043	0.144
PASS3. I am motivated to figure out how to make existing products/services better.	0.895				0.692				0.017	0.043	0.691
PASS4. Scanning the environment for new opportunities really excites me.	0.795				0.713				0.030	0.043	0.486
Sustainable Entrepreneurship Intentions (SEI)		0.784	0.854	0.541		0.700	0.807	0.531			
SE1. I prefer to be a sustainable entrepreneur rather than to be an employee of a company.	0.574				0.799				0.078	0.043	0.067
SE2. My professional goal is to become a sustainable entrepreneur.	0.786				0.676				0.069	0.043	0.105
SE3. I will make every effort to start and run my own sustainable firm.	0.782				0.739				0.031	0.043	0.470
SE4. I am determined to create a new sustainable firm in the future.	0.796				0.654				0.031	0.043	0.473
SE5. I have very seriously thought about in starting a sustainable firm.	0.717				0.685				0.005	0.043	0.903

Abbreviation: Cronbach alpha (α), composite reliability (CR) and average variance extracted (AVE), Standard Error (S.E).

Table 5. Discriminant validity coefficients.

Constructs	EA	PF	PASS	SEI	PD
EA	**0.760**				
PF	0.727	**0.749**			
PASS	0.227	0.159	**0.778**		
SEI	0.538	0.546	0.067	**0.826**	
PD	0.505	0.544	0.115	0.535	**0.889**

Notes: The Items displayed in boldface represents the square roots of the AVE.: Entrepreneurial Attitude (EA), Perceived Desirability (PD), Perceived Feasibility (PF), and Sustainable Entrepreneurship Intentions (SEI).

Table 6. Heterotrait-Monotrait (HTMT) ratio.

	EA	PF	PASS	SEI	PD
EA					
PF	0.355				
PASS	0.336	0.211			
SEI	0.676	0.666	0.117		
PD	0.645	0.675	0.148	0.610	

Abbreviation: Entrepreneurial Attitude (EA), Perceived Desirability (PD), Perceived Feasibility (PF), Entrepreneurial Passion (PASS) and Sustainable Entrepreneurship Intentions (SEI).

7. Results Structural Model

After examining the measurement model, the structural model is assessed for the values of R^2, Q^2, f^2, and significance of relationships. R^2 for endogenous latent variable is assessed in order to find the amount of variance explained by all constructs [75]. Though a satisfactory value of R^2 depends upon the setting of study. According to [76], the value of 0.26, 0.13, and 0.09 express high, moderate and low amount of variance respectively. Table 7 shows the R^2 value of sustainable entrepreneurship intention of both groups i.e., Group A and Group B. The EA, PF and PD represent only 19.3% variance in sustainable entrepreneurship intention in Group A (without dual/triple degree awards) and 44.7% in Group B (with dual/triple degree awards). Furthermore, a cross-validated redundancy measure (Q^2) was applied to quantify the estimate significance of the research model [71]. There was support for sufficient estimates' significance of the direct effect model because Table 7 shows that the value of Q^2 is greater than zero in both Group A = 0.200 and Group B = 0.350. Therefore, it can be considered as a satisfactory predictive relevance of the model.

Table 7. Coefficient of Determination in the PLS method.

Groups	Construct	R Square	R Square Adjusted	Q^2
Universities A	Sustainable Entrepreneurship Intentions	0.193	0.175	0.200
Universities B	Sustainable Entrepreneurship Intentions	0.447	0.434	0.350

Reference [71] describe (f^2) estimations between 0.02, 0.15 and 0.35 as having small, medium, and large effects respectively. Thus, following [76] rule, the impacts' sizes of these exogenous construct on endogenous construct can be reflected as small, medium and large, respectively as shown in Table 8. Moreover, we calculated the *p*-values for the one-tailed test to interpret the significance of the coefficients. The Figure 2 shows that the EA has significant effect on SEI in Group A ($\beta = 0.322$, $p < 0.05$) also Group B ($\beta = 0.110$, $p < 0.05$). Thus, the H1 is supported for Group A and Group B. But the result for group pairs analysis is non-significant ($\beta = 0.065$, $p > 0.05$). H2 is not supported for Group A because the direct effect of PD on SEI is non-significant ($\beta = 0.055$, $p > 0.05$) but H2 is supported for Group B where its impact is significant ($\beta = 0.316$, $p < 0.05$) and group pairs result is significant ($\beta = 0.226$, $p < 0.05$). Lastly, the direct impact of PF on SEI is non-significant in Group A ($\beta = 0.043$, $p > 0.05$) but is significant in Group B ($\beta = 0.274$, $p < 0.05$). Thus, H3 is supported for Group B but is not supported for Group A. The group pairs result is significant too for this relationship as well (β = 0.137, $p < 0.05$). The results of group pairs imply that there is a significant difference in the path coefficients of Group A and Group B for the relationships of PD and PF with SEI, whereas, no significant difference is found for the path coefficients of relationship of EA and SEI between these groups.

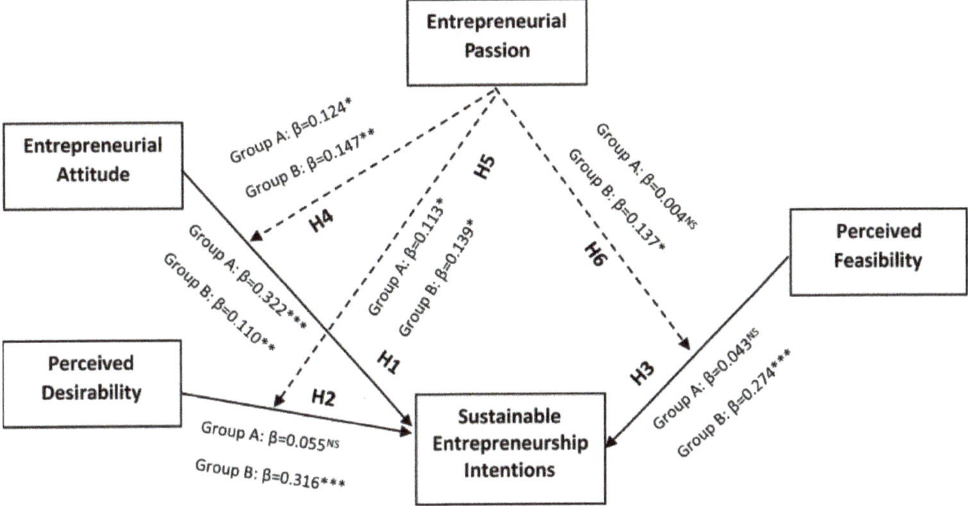

Figure 2. Groups path results. Note: * $p < 0.05$, ** $p < 0.01$, *** $p < 0.001$, and Not Supported (NS).

Since this study has also hypothesized the positive moderating impact of Entrepreneurial Passion (PASS) for the impacts of EA, PD, and PF on SEI. As described in Table 8, the moderating effect of PASS between EA and SEI is positive and significant in both Group A (β = 0.124, $p < 0.05$) and Group B (β = 0.147, $p < 0.01$). Thus, H4 is supported for both groups. Likewise, the moderating effect of PASS between PD and SEI is significant in both Group A (β = 0.113, $p < 0.05$) and in Group B (β = 0.139, $p < 0.05$). Thus, H5 is also supported for both groups.

Finally, the moderating effect of PASS between PF and SEI is non-significant in Group A (β = 0.004, $p > 0.05$) but is significant in Group B (β = 0.137, $p < 0.05$). Thus, H6 is supported only for Group B but not for Group A. Furthermore, the results of group pair were non-significant for the moderating impact of PASS between EA and SEI (β = 0.053, $p > 0.05$). However, the result of group pair was significant for the moderating impact of PASS between PD and SEI (β = 0.076, $p < 0.05$) and moderating impact of PASS between PF and SEI (β = 0.116, $p < 0.01$). Thus, the group pair results reveal that there is significant difference in the path coefficients for the PASS*PD and SEI as well as and for PASS*PF and SEI between Group A and Group B. However, no significant difference was found for the path coefficients of PASS*EA and SEI between both groups. Figure 2 and Table 8 show the results regarding the significance of all direct and moderating relationships of the hypothesized research model.

Table 8. Path coefficients.

Hypotheses	Relationship between Constructs	Group A (Without Dual/Triple Award Degree Programmes)					Group B (Without Dual/Triple Award Degree Programmes)					Results for Group Pair			
	Direct Effect	**β**	**S. E**	**f^2**	**p**	**Remarks**	**β**	**S. E**	**f^2**	**p**	**Remarks**	**β**	**S. E**	**p**	**Remarks**
H1	EA→SEI	0.322 ***	0.058	0.126	<0.001	S	0.110 **	0.051	0.104	0.007	S	0.065	0.043	0.064	NS
H2	PD→SEI	0.055	0.060	0.002	0.181	NS	0.316 ***	0.058	0.149	<0.001	S	0.226 ***	0.042	<0.001	S
H3	PF→SEI	0.043	0.060	0.014	0.237	NS	0.274 ***	0.058	0.137	<0.001	S	0.137 ***	0.042	<0.001	S
	Moderating Effect														
H4	PASS * EA	0.124 *	0.060	0.035	0.019	S	0.147 **	0.059	0.051	0.007	S	0.053	0.043	0.105	NS
H5	PASS * PD	0.113 *	0.060	0.015	0.029	S	0.139 *	0.059	0.052	0.010	S	0.076 *	0.043	0.036	S
H6	PASS * PF	0.004	0.061	0.001	0.477	NS	0.137 *	0.059	0.054	0.011	S	0.116 **	0.042	0.003	S

Abbreviations: Entrepreneurial Attitude (EA), Perceived Desirability (PD), Perceived Feasibility (PF), Entrepreneurial Passion (PASS) and Sustainable Entrepreneurship Intentions (SEI), Standard Error (S. E), Effect Size (f^2), Supported (S), Not Supported (NS). * $p < 0.05$, ** $p < 0.01$, *** $p < 0.001$.

8. Discussion

This study has used the EPM model of Krueger and Brazeal with some modifications to examine the sustainable entrepreneurship intentions among undergraduate students. Three constructs including entrepreneurial attitude, perceived desirability, and perceived feasibility have been taken as independent variables to examine the sustainable entrepreneurial intention among students under the moderating influence of entrepreneurial passion who have studied entrepreneurship subject at undergraduate level in Malaysian universities. This study consisted the comparison of two groups of students. Group A students were those students who took entrepreneurship module in local Malaysian universities without dual/triple award degree programmes. And Group B students were those students who took entrepreneurship module in local Malaysian universities with dual/triple award degree programmes.

The main aim of this study was to compare the impact of entrepreneurial attitude, perceived desirability, and perceived feasibility under the moderating impact of entrepreneurial passion on undergraduate students' sustainable entrepreneurship intention with and without dual/triple award programmes. The findings of this study reveal a significant impact of entrepreneurial attitude, perceived desirability, and perceived feasibility on sustainable entrepreneurship intention among undergraduate students of Group B whose universities are offering dual/triple award degree programmes. The positive and significant impact of entrepreneurial attitude on sustainable entrepreneurship intention are consistent with findings of some other studies that have also found similar results under various contexts [38,40]. Likewise, the results regarding the positive and significant influence of perceived desirability and perceived feasibility on sustainable entrepreneurship intention are also congruent with similar type of existing studies' results i.e., perceived desirability [24,41] and perceived feasibility [48–50] in different contexts.

Furthermore, the positive and significant moderating impact of entrepreneurial passion was also found for the influence of entrepreneurial attitude, perceived desirability, and perceived feasibility on sustainable entrepreneurship intention among undergraduate students of Group B. The results show that students of entrepreneurship module with dual/triple award degree programmes develop positive entrepreneurial attitudes, perceive more desirable and capable of starting an entrepreneurial business and their passion improves their entrepreneurial attitude, perceive desirability, and feasibility in starting sustainable entrepreneurial business as well. Therefore, all the six hypotheses were supported for Group B students. However, for Group A students, only H1 which regards the impact of entrepreneurial attitude on sustainable entrepreneurship intention was supported among the direct hypotheses. H5 and H6 were also supported regarding the positive moderating influence of entrepreneurial passion for the impact of perceived desirability and perceived feasibility respectively on sustainable entrepreneurship intention among Group A students, whereas, H2 and H3 regarding the direct positive impact of perceived desirability and perceived feasibility respectively, were not supported for Group A students. These findings reveal that students with entrepreneurship modules from universities without dual/triple award programmes perceived less desirability and capability to start sustainable entrepreneurship businesses and show less entrepreneurial passion as compared to students with entrepreneurship modules from universities with dual/triple award degree programmes. The multi group analysis also reveal significant differences among students of Group A and Group B regarding the impact of perceived desirability and perceived feasibility on sustainable entrepreneurship intentions. Likewise, significant differences were also found for the moderating impact of entrepreneurial passion on perceived desirability and perceived feasibility on sustainable entrepreneurship intention among students of Group A and Group B as well.

Additionally, the findings show that entrepreneurial attitude, perceived feasibility, and perceived desirability explain and influence most of the sustainable entrepreneurship intentions of Group B students (e.g, R^2 value = 0.447) whose degree programmes are dual or triple awarded as compared to Group A students without dual or triple award degree programmes (e.g, R^2 value = 0.193). Thus, the findings of this study clearly reveal that only entrepreneurship education is not very affective in promoting the sustainable entrepreneurship intention among undergraduate students but the partnership of local universities with other overseas' universities is equally important in this regard. Thus, to improve the sustainable entrepreneurship intentions among students, the universities should develop partnerships with other universities of developed countries. Due to the partnership of local universities with other overseas' universities, the quality of entrepreneurship education can be enhanced and more resources could be provided to assist students in their learning about entrepreneurship. Due to dual/triple award degree programmes, the students may feel more motivation and confidence in their abilities to start their own businesses and desire for sustainable entrepreneurial businesses. Their entrepreneurial attitude increases too and they feel more passionate in starting their own businesses. Whereas, based on the findings of this study, the students in local universities without dual/triple award degree programmes have entrepreneurial attitude towards sustainable entrepreneurship intentions, however, they don't perceive desirability and feasibility for it. In other words, they are not willing to be self-employed in their own business and feel less abilities for starting their own businesses. The reason could be that the entrepreneurship modules taught in local universities could only develop their entrepreneurial attitudes but could not develop their confidence and passion to start their own business. The local universities should also develop more partnerships with other overseas universities to give more exposure to their students for the new idea generation process. The quality of existing entrepreneurship modules could be improved and more learning resources could be accessed for students to learn about entrepreneurship due to partnership with other universities of overseas. Likewise, university-industry partnership could be another important factor that could impact on sustainable entrepreneurship intention among students. Local universities can use their contacts as well as industry networks of their partnership universities to create more internship opportunities for university students to improve their knowledge and understanding regarding sustainable entrepreneurial businesses.

9. Study Limitations and Future Recommendations

Despite the practical implications of this study, there are some limitations as well. For instance, the data were collected from students of only two cities namely Kuala Lumpur and Selangor in Malaysia in a cross-sectional setting. The future studies can include sample of students from other universities of Malaysia and longitudinal approach could be used in carrying out the research. The future studies should investigate the impact of academic entrepreneurship in motivating students towards sustainable entrepreneurship businesses by using qualitative/quantitative or mix methodologies. The future researchers should also examine the cultural conditions of the region to propose a model for an academic entrepreneur for significant contribution in the literature related to sustainable entrepreneurship intentions among students. Future researchers are suggested to conduct similar type of studies in other countries to contribute in the international literature. The comparative studies on the topic of sustainable entrepreneurship intentions among students of developing and developed countries could make a significant contribution in the existing literature. Likewise, future researchers are also suggested to compare the academic entrepreneurship courses in Malaysia and those of overseas universities to see their impact on sustainable entrepreneurship intentions among undergraduate students. As mentioned in above discussion that university-industry partnership could be another potential contributing factor in developing the sustainable entrepreneurship intention among students. Thus, future research is recommended to investigate the influence of university-industry partnership on sustainable entrepreneurship intentions among university students. Moreover, the future researchers can conduct the interview of alumni of universities who took entrepreneurship module at their undergraduate level and are successfully operating their businesses. The effectiveness of entrepreneurship education and impact of universities' dual/triple award degree programmes could be also explored through qualitative or mix-methodology research towards sustainable entrepreneurship intention.

10. Conclusions

It is essential for the students to grasp the depth understanding of entrepreneurship module and venture creation, equip the specific skills to implement new business ideas, and should develop propensity to act towards sustainable entrepreneurship intentions. The high-quality entrepreneurship modules can improve the entrepreneurial ability of students towards venture creation. The entrepreneurship curriculum in universities that have partnerships with overseas' universities, is creating the significant influence on the mindset of the students in Malaysia and enhancing their implementation behaviour for new business idea. The presence of high risk-taking skill, practical experience in incubators, more awareness regarding the government policies, increased engagement with entrepreneurial network is promoting entrepreneurial behaviour among students with dual/triple award degree programmes. Hence, it is strongly recommended that the universities should develop partnerships with overseas universities to improve quality of entrepreneurship curriculum and should provide more practical experience in incubation centers, and access to entrepreneurial networks that will boost the entrepreneurship thinking process of the students. Universities should provide more knowledge regarding government policies to increase the perceive desirability of students towards sustainable entrepreneurship intentions. The entrepreneurship curriculum should be designed with partner universities' experts to develop essential entrepreneurial skills among undergraduate students towards their sustainable entrepreneurial intention.

Author Contributions: Conceptualization, S.T.; methodology, S.T.; software, S.A.H.; validation, S.A.H. and S.T.; formal analysis, S.A.H.; investigation, S.T.; resources, S.T.; data curation, S.T.; writing—original draft preparation, S.T., writing—review and editing, S.T., and S.A.H.; visualization, S.T., and S.A.H.; supervision, S.T., and S.A.H.; project administration, S.T., and S.A.H.; funding acquisition, S.T. All authors have read and agreed to the published version of the manuscript.

Funding: This research was funded under the Research Funds; Project Codes #GRTIN-RRO-83-2020 and #GRTIN-IRG-65-2021 sponsored by Sunway University.

Institutional Review Board Statement: The study was approved by the Sunway University Research Ethics Committee of Sunway University (protocol code: SUREC 2020/009 and 27/02/2020).

Informed Consent Statement: Informed consent was obtained from all subjects involved in the study.

Data Availability Statement: The datasets used and/or analyzed during the current study are available from the corresponding author on reasonable request.

Acknowledgments: The authors thank all the students who participated in this study.

Conflicts of Interest: The authors declare no conflict of interest.

References

1. Ahmad, R.; Mahli, F.; Tehseen, S.; Qureshi, Z.H.; Jannat, U.M.U. Entrepreneurial intentions among university's students in Malaysia. In *2019 13th International Conference on Mathematics, Actuarial Science, Computer Science and Statistics (MACS)*; IEEE: New York, NY, USA, 2019; pp. 1–7. [CrossRef]
2. Sun, H.; Lo, C.T.; Liang, B.; Wong, Y.L.B. The impact of entrepreneurial education on entrepreneurial intention of engineering students in Hong Kong. *Manag. Decis.* **2017**, *56*, 1371–1393. [CrossRef]
3. Moses, C.; Olokundun, A.M.; Akinbode, M.; Agboola, M.G.; Inelo, F. Entrepreneurship education and entrepreneurial intentions: The moderating role of passion. *Soc. Sci.* **2016**, *11*, 645–653.
4. Hahn, D.; Minola, T.; Bosio, G.; Cassia, L. The impact of entrepreneurship education on university students' entrepreneurial skills: A family embeddedness perspective. *Small Bus. Econ.* **2020**, *55*, 257–282. [CrossRef]
5. Barringer, B.R.; Ireland, R.D. *Entrepreneurship: Successfully Launching New Ventures*, 6th ed.; Pearson Prentice-Hall: Upper Saddle River, NJ, USA, 2019.
6. Tomy, S.; Pardede, E. An entrepreneurial intention model focussing on higher education. *Int. J. Entrep. Behav. Res.* **2020**. [CrossRef]
7. Sardeshmukh, S.R.; Smith-Nelson, R.M. Educating for an entrepreneurial career: Developing opportunity-recognition ability. *Aust. J. Career Dev.* **2011**, *20*, 47–55. [CrossRef]

8. Hofer, A.-R.; Potter, J.; Redford, D.; Stolt, J. Promoting Successful Graduate Entrepreneurship at the University of Applied Sciences Schmalkalden, Germany. In *OECD Local Economic and Employment Development Programme*; OECD: Paris, France, 2013. [CrossRef]
9. Karlsson, T.; Moberg, K. Improving perceived entrepreneurial abilities through education: Exploratory testing of an entrepreneurial self-efficacy scale in a pre-post setting. *Int. J. Manag. Educ.* **2013**, *11*, 1–11. [CrossRef]
10. Chang, J.; Rieple, A. Assessing students' entrepreneurial skills development in live projects. *J. Small Bus. Enterp. Dev.* **2013**, *20*, 225–241. [CrossRef]
11. Shepherd, D.A.; Patzelt, H. The new field of sustainable entrepreneurship: Studying entrepreneurial action linking "what is to be sustained" with "what is to be developed". *Entrep. Theory Pract.* **2011**, *35*, 137–163. [CrossRef]
12. Schaltegger, S.; Wagner, M. Sustainable entrepreneurship and sustainability innovation: Categories and interactions. *Bus. Strategy Environ.* **2011**, *20*, 222–237. [CrossRef]
13. Hockerts, K.; Wüstenhagen, R. Greening Goliaths versus emerging Davids—Theorizing about the role of incumbents and new entrants in sustainable entrepreneurship. *J. Bus. Ventur.* **2010**, *25*, 481–492. [CrossRef]
14. Zahra, S.A.; Gedajlovic, E.; Neubaum, D.O.; Shulman, J.M. A typology of social entrepreneurs: Motives, search processes and ethical challenges. *J. Bus. Ventur.* **2009**, *24*, 519–532. [CrossRef]
15. Fayolle, A.; Liñán, F.; Moriano, J.A. Beyond entrepreneurial intentions: Values and motivations in entrepreneurship. *Int. Entrep. Manag. J.* **2014**, *10*, 679–689. [CrossRef]
16. Cohen, B.; Winn, M.I. Market imperfections, opportunity and sustainable entrepreneurship. *J. Bus. Ventur.* **2017**, *22*, 29–49. [CrossRef]
17. Vuorio, A. Opportunity-Specific Entrepreneurial Intentions in Sustainable Entrepreneurship. Ph.D. Thesis, Acta Universitatis Lappeenrantaensis, Lappeenranta, Finland, 2018.
18. Iakovleva, T.; Kolvereid, L. An integrated model of entrepreneurial intentions. *Int. J. Bus. Glob.* **2009**, *3*, 66–80. [CrossRef]
19. Tan, L.P.; Pham, L.X.; Bui, T.T. Personality Traits and Social Entrepreneurial Intention: The Mediating Effect of Perceived Desirability and Perceived Feasibility. *J. Entrep.* **2020**. [CrossRef]
20. Al-Jubari, I.; Hassan, A.; Liñán, F. Entrepreneurial intention among University students in Malaysia: Integrating self-determination theory and the theory of planned behavior. *Int. Entrep. Manag. J.* **2018**, 1–20. [CrossRef]
21. Ahmad, M.A. Entrepreneurial Intention Among Postgraduate Students of UUM. Master's Thesis, Universiti Utara, Kuala Lumpur, Malaysia, 2015.
22. Amanamah, R.B.; Acheampong, A.; Owusu, E.K. An exploratory study of entrepreneurial intention among university students in Ghana. *Int. J. Sci. Technol. Res.* **2018**, *7*, 140–148.
23. Biraglia, A.; Kadile, V. The role of entrepreneurial passion and creativity in developing entrepreneurial intentions: Insights from American homebrewers. *J. Small Bus. Manag.* **2017**, *55*, 170–188. [CrossRef]
24. Esfandiar, K.; Sharifi-Tehrani, M.; Pratt, S.; Altinay, L. Understanding entrepreneurial intentions: A developed integrated structural model approach. *J. Bus. Res.* **2019**, *94*, 172–182. [CrossRef]
25. Israr, M.; Saleem, M. Entrepreneurial intentions among university students in Italy. *J. Glob. Entrep. Res.* **2018**, *8*, 1–14. [CrossRef]
26. Lechuga Sancho, M.P.; Martín-Navarro, A.; Ramos-Rodríguez, A.R. Will they end up doing what they like? the moderating role of the attitude towards entrepreneurship in the formation of entrepreneurial intentions. *Stud. High. Educ.* **2020**, *45*, 416–433. [CrossRef]
27. Ramayah, T.; Rahman, S.A.; Taghizadeh, S.K. Modelling green entrepreneurial intention among university students using the entrepreneurial event and cultural values theory. *Int. J. Entrep. Ventur.* **2019**, *11*, 394–412. [CrossRef]
28. Liñán, F.; Fayolle, A. A systematic literature review on entrepreneurial intentions: Citation, thematic analyses, and research agenda. *Int. Entrep. Manag. J.* **2015**, *11*, 907–933. [CrossRef]
29. Vuorio, A.M.; Puumalainen, K.; Fellnhofer, K. Drivers of entrepreneurial intentions in sustainable entrepreneurship. *Int. J. Entrep. Behav. Res.* **2018**, *24*, 359–381. [CrossRef]
30. Hambrick, D.C.; Mason, P.A. Upper echelons: The organization as a reflection of its top managers. *Acad. Manag. Rev.* **1984**, *9*, 193–206. [CrossRef]
31. Krueger, N.F., Jr.; Brazeal, D.V. Entrepreneurial potential and potential entrepreneurs. *Entrep. Theory Pract.* **1994**, *18*, 91–104. [CrossRef]
32. Moghavvemi, S.; Salleh, N.A.M.; Abessi, M. Determinants of IT-related innovation acceptance and use behavior: Theoretical integration of unified theory of acceptance and use of technology and entrepreneurial potential model. *Soc. Technol.* **2013**, *3*, 243–260. [CrossRef]
33. Ajzen, I. The theory of planned behavior. *Organ. Behav. Hum. Decis. Process.* **1991**, *50*, 179–211. [CrossRef]
34. Muñoz, P.; Dimov, D. The call of the whole in understanding the development of sustainable ventures. *J. Bus. Ventur.* **2015**, *30*, 632–654. [CrossRef]
35. Fischer, R.; Schwartz, S. Whence differences in value priorities? Individual, cultural, or artifactual sources. *J. Cross Cult. Psychol.* **2011**, *42*, 1127–1144. [CrossRef]
36. Munir, H.; Jianfeng, C.; Ramzan, S. Personality traits and theory of planned behavior comparison of entrepreneurial intentions between an emerging economy and a developing country. *Int. J. Entrep. Behav. Res.* **2019**, *25*, 554–580. [CrossRef]
37. Ajzen, I. *Attitudes, Personality, and Behaviour*; Open University Press: Milton Keynes, UK, 1988.

38. Mahfud, T.; Triyono, M.B.; Sudira, P.; Mulyani, Y. The influence of social capital and entrepreneurial attitude orientation on entrepreneurial intentions: The mediating role of psychological capital. *Eur. Res. Manag. Bus. Econ.* **2020**, *26*, 33–39. [CrossRef]
39. Nguyen, A.T.; Do, T.H.H.; Vu, T.B.T.; Dang, K.A.; Nguyen, H.L. Factors affecting entrepreneurial intentions among youths in Vietnam. *Child. Youth Serv. Rev.* **2019**, *99*, 186–193. [CrossRef]
40. Jena, R.K. Measuring the impact of business management Student's attitude towards entrepreneurship education on entrepreneurial intention: A case study. *Comput. Hum. Behav.* **2020**, *107*, 106275. [CrossRef]
41. Yousaf, U.; Shamim, A.; Siddiqui, H.; Raina, M. Studying the influence of entrepreneurial attributes, subjective norms and perceived desirability on entrepreneurial intentions. *J. Entrep. Emerg. Econ.* **2015**. [CrossRef]
42. Levesque, M.; Shepherd, D.A.; Douglas, E.J. Employment or self-employment: A dynamic utility-maximizing model. *J. Bus. Ventur.* **2002**, *17*, 189–210. [CrossRef]
43. Wagner, M. Ventures for the public good and entrepreneurial intentions: An empirical analysis of sustainability orientation as a determining factor. *J. Small Bus. Entrep.* **2012**, *25*, 519–531. [CrossRef]
44. Criado-Gomis, A.; Cervera-Taulet, A.; Iniesta-Bonillo, M.A. Sustainable entrepreneurial orientation: A business strategic approach for sustainable development. *Sustainability* **2017**, *9*, 1667. [CrossRef]
45. Bruyere, B.; Rappe, S. Identifying the motivations of environmental volunteers. *J. Environ. Plan. Manag.* **2007**, *50*, 503–516. [CrossRef]
46. Shepherd, D.A.; Patzelt, H.; Baron, R.A. I care about nature, but . . . : Disengaging values in assessing opportunities that cause harm. *Acad. Manag. J.* **2013**, *56*, 1251–1273. [CrossRef]
47. Seelos, C.; Mair, J. Social entrepreneurship: Creating new business models to serve the poor. *Bus. Horiz.* **2005**, *48*, 241–246. [CrossRef]
48. Zhang, M.; Fang, Z. Research on the Influence Mechanism of Multi-level Information Ecological Environment on Entrepreneurial Intention. In Proceedings of the International Conference on Education and Cognition, Behavior, Neuroscience (ICECBN2018), Zhengzhou, China, 23–25 November 2018. [CrossRef]
49. Smith, I.H.; Woodworth, W.P. Developing social entrepreneurs and social innovators: A social identity and self-efficacy approach. *Acad. Manag. Learn. Educ.* **2012**, *11*, 390–407. [CrossRef]
50. Mair, J.; Noboa, E. Social Entrepreneurship: How intentions to create a social venture are formed. In *Social Entrepreneurship*; Palgrave Macmillan: London, UK, 2006; pp. 121–135. [CrossRef]
51. Syed, I.; Butler, J.C.; Smith, R.M.; Cao, X. From entrepreneurial passion to entrepreneurial intentions: The role of entrepreneurial passion, innovativeness, and curiosity in driving entrepreneurial intentions. *Personal. Individ. Differ.* **2020**, *157*, 109758. [CrossRef]
52. Neneh, B.N. Entrepreneurial passion and entrepreneurial intention: The role of social support and entrepreneurial self-efficacy. *Stud. Higher Educ.* **2020**, 1–17. [CrossRef]
53. Cardon, M.S.; Kirk, P.C. Entrepreneurial Passion as Mediator of the Self-Efficacy to Persistence Relationship. *Entrep. Theory Pract.* **2015**, *39*, 1027–1050. [CrossRef]
54. Cardon, M.S.; Wincent, J.; Singh, J.; Drnovsek, M. The Nature and Experience of Entrepreneurial Passion. *Acad. Manag. Rev.* **2009**, *34*, 511–532. [CrossRef]
55. Li, C.; Murad, M.; Shahzad, F.; Khan, M.A.S.; Ashraf, S.F.; Dogbe, C.S.K. Entrepreneurial passion to entrepreneurial behavior: Role of entrepreneurial alertness, entrepreneurial self-efficacy and proactive personality. *Front. Psychol.* **2020**, *11*, 1–19. [CrossRef] [PubMed]
56. Huyghe, A.; Knockaert, M.; Obschonka, M. Unraveling the "passion orchestra" in academia. *J. Bus. Ventur.* **2016**, *31*, 344–364. [CrossRef]
57. Santos, S.C.; Cardon, M.S. What's love got to do with it? Team entrepreneurial passion and performance in new venture teams. *Entrep. Theory Pract.* **2019**, *43*, 475–504. [CrossRef]
58. Karimi, S. The role of entrepreneurial passion in the formation of students' entrepreneurial intentions. *Appl. Econ.* **2020**, *52*, 331–344. [CrossRef]
59. Schenkel, M.T.; Farmer, S.; Maslyn, J.M. Process improvement in SMEs: The impact of harmonious passion for entrepreneurship, employee creative self-efficacy, and time spent innovating. *J. Small Bus. Strategy* **2019**, *29*, 71–84.
60. Türk, S.; Zapkau, F.B.; Schwens, C. Prior entrepreneurial exposure and the emergence of entrepreneurial passion: The moderating role of learning orientation. *J. Small Bus. Manag.* **2020**, *58*, 225–258. [CrossRef]
61. Cardon, M.S.; Glauser, M.; Murnieks, C.Y. Passion for what? Expanding the domains of entrepreneurial passion. *J. Bus. Ventur. Insights* **2017**, *8*, 24–32. [CrossRef]
62. Hockerts, K. Determinants of social entrepreneurial intentions. *Entrep. Theory Prac.* **2017**, *41*, 105–130. [CrossRef]
63. Campos, H.M. Impact of entrepreneurial passion on entrepreneurial orientation with the mediating role of entrepreneurial alertness for technology-based firms in Mexico. *J. Small Bus. Enterp. Dev.* **2017**, *24*, 353–374. [CrossRef]
64. Kock, N.; Hadaya, P. Minimum sample size estimation in PLS-SEM: The inverse square root and gamma-exponential methods. *Inf. Syst. J.* **2018**, *28*, 227–261. [CrossRef]
65. Masrury, M.J. *The Impact of Perceived Feasibility and Perceived Desirability on Entrepreneurial Intention among Undergraduate Students in Universitas Muhammadiyah Surakarta*; Universitas Muhammadiyah: Surakarta, Indonesia, 2016.
66. Kock, N. Factor-based structural equation modeling with WarpPLS. *Aust. Mark. J.* **2019**, *27*, 57–63. [CrossRef]

67. Sholihin, M.; Ratmono, D. *Analisis SEM-PLS Dengan WarpPLS 7.0: Untuk Hubungan Nonlinier Dalam Penelitian Sosial dan Bisnis*; Penerbit Andi: Surabaya, Indonesia, 2013.
68. Kock, N. *WarpPLS User Manual: Version 6.0*; ScriptWarp Systems: Laredo, TX, USA, 2017.
69. Hair, J.F.; Sarstedt, M.; Ringle, C.M.; Gudergan, S.P. *Advanced Issues in Partial Least Squares Structural Equation Modeling*, 3rd ed.; Sage: Thousand Oaks, CA, USA, 2018.
70. Hair, J.F.; Risher, J.J.; Sarstedt, M.; Ringle, C.M. When to use and how to report the results of PLS-SEM. *Eur. Bus. Rev.* **2019**, *31*, 2–24. [CrossRef]
71. Hair, J.F.; Hult, G.T.M.; Ringle, C.M.; Sarstedt, M. *A Primer on Partial Least Squares Structural Equation Modeling (PLS-SEM)*, 2nd ed.; Sage: Thousand Oaks, CA, USA, 2017.
72. Hair, J.F., Jr.; Sarstedt, M.; Matthews, L.M.; Ringle, C.M. Identifying and treating unobserved heterogeneity with FIMIX-PLS: Part I–method. *Eur. Bus. Rev.* **2016**. [CrossRef]
73. Fornell, C.; Larcker, D.F. Evaluating structural equation models with unobservable variables and measurement error. *J. Mark. Res.* **1981**, *18*, 39–50. [CrossRef]
74. Ringle, C.M.; Wende, S.; Becker, J.M. *SmartPLS 3*; SmartPLS GmbH: Boenningstedt, Germany, 2015.
75. Ramayah, T.; Cheah, J.; Chuah, F.; Ting, H.; Memon, M.A. *Partial Least Squares Structural Equation Modeling (PLS-SEM) Using smartPLS 3.0*; Pearson: Kuala Lumpur, Malaysia, 2018.
76. Cohen, J. *Statistical Power Analysis for the Behavioural Sciences*; Lawrence Earlbaum Associates: Hillside, NJ, USA, 1988; pp. 19–74.

MDPI

Article

Can Talent Management Improve Training, Sustainability and Excellence in the Labor Market?

Francisco J. Ferreiro-Seoane [1,*], Vanessa Miguéns-Refojo [2] and Yago Atrio-Lema [3]

[1] Department of Applied Economics, University of Santiago de Compostela, 15782 La Coruña, Spain
[2] Accounting Department, Faculty of Economic Sciences, Autonomous University of Madrid, 28049 Madrid, Spain; vanessa.miguens@uam.es
[3] Applied Financial Valuation Group, University of Santiago de Compostela, 15782 La Coruña, Spain; yago.atrio@rai.usc.es
* Correspondence: franciscojesus.ferreiro@usc.es; Tel.: +34-629-88-19-09

Abstract: The objective of this article is to analyze the characteristics of the most attractive companies in the labor market, which each year maintained their position in the ranking published by the Spanish business magazine *Actualidad Económica* (*AE*) for the period 2013–2020. The research study will focus on permanence in ranking, global valuation, and training. To do this, control variables were added: business management gender, geo-cultural areas, regional areas, economic activity, size and stock market membership. This is a quantitative work, where statistics such as partial correlations, Pearson coefficients and independent sample means were used with the Levene test; in modeling, multiple regressions of ordinary least squares (OLS) and panel data were used. It is concluded that the permanence in the ranking significantly increases the total value and training, which leads companies to excellence, along with the fact that they are in the capital of the country and that they focus on the commerce, professional, scientific and technical, and finance and insurance sectors. On the other hand, assessment of training is explained by employee valuation, the work environment and talent management. On the contrary, factors such as the gender variable in the business direction, nationality, size and stock market membership do not significantly influence the overall valuation.

Keywords: human capital; labor market; training; talent management; corporate governance

Citation: Ferreiro-Seoane, F.J.; Miguéns-Refojo, V.; Atrio-Lema, Y. Can Talent Management Improve Training, Sustainability and Excellence in the Labor Market? *Sustainability* **2021**, *13*, 6645. https://doi.org/10.3390/su13126645

Academic Editors: Jaana Seikkula-Leino, Mats Westerberg, Priti Verma and Maria Salomaa

Received: 26 April 2021
Accepted: 7 June 2021
Published: 10 June 2021

Publisher's Note: MDPI stays neutral with regard to jurisdictional claims in published maps and institutional affiliations.

1. Introduction

Successive economic crises around the world have resulted in job losses for millions of workers, many of whom will struggle to re-enter the labor market. The present COVID-19 crisis, which is impacting at the health, social and economic levels, has only aggravated the situation in the labor market, with an unemployment rate of 16% in the last quarter of 2020 in Spain. Therefore, if having a stable job in the current situation is a great asset, it is even more valuable if it is performed in one of the 100 most valued organizations in the labor market, sustaining itself throughout the period of analysis (2013–2020) in the ranking published by the business magazine *Actualidad Económica*, hereinafter *AE*, even during the COVID-19 pandemic.

The objective of this article is to study the factors that can influence the valuation obtained both in training, and the total valuation of the companies that persist every year (2013–2020) in the ranking of the 100 best companies to work for in Spain. Sustainability in the ranking will be shown to be synonymous with excellence and attributes that may explain this fact will be also analyzed, such as: the gender in business management, nationality, geo-cultural areas, regional areas, economic activity, size and stock market membership.

2. Theoretical Framework

Markets are evolving towards globalization, in which the importance of information technologies increases, which allows business competitiveness to increase [1]. This phenomenon has been reinforced with the COVID-19 pandemic, as telework spreads in companies and society at large. In this context, the knowledge economy and business training have become more relevant thanks to information and communication technologies (ICT), so that companies have more tools to compete in a globalized market [2]. The intangible assets of companies, including the training and skills of workers, knowledge and technological integration, and knowledge about the functioning of the market and business management in a global economy [3], increase productivity and the efficient management of resources. Investment in business training is essential to maintaining the competitiveness of companies [4] and human capital is a great asset that, through investment in business training, increases the productive capacity of the organization [5].

Conversely, there is an existential crisis in business education driven by the conflict between social and financial objectives. A paradigm shift in business education requires leaders to learn how to incorporate new competencies. It leads companies to continuous innovation and highly sustainable performance [6]. The misalignment between the education system and companies hinders the job market [7]. Business education should contribute to companies' members generating social value and demonstrating sustainable performance [8,9].

The existence of rankings that categorize companies according to job performance, such as Fortune 100 in the USA or Great Place to Work in Europe, adds corporate value contributing to generating an image of a good organization [10]. There are, in this regard, studies that refer to human resources rankings, including that of the Spanish magazine *Actualidad Económica* [11].

It is worth noting the importance of innovation for companies, markets and society [12], since it is fundamental to long-term profitability and sustainability [13]. Innovation leads to commercial and financial success, but society demands that innovation be carried out in a responsible and ethical manner [14]. If change and doing things differently was already important in Schumpeter's time, the market is now evolving into one based on innovation and ICT [15], it being essential for most companies to adopt this approach. Business digitalization consists of the implementation of digital tools and technologies as well as data, which together can make business processes more efficient and effective [16]. Digitalization is also improving the sustainability of companies [17].

A study on the integration of responsible innovation (RI), a concept integral to companies' practice, obtained results that link responsible innovation (RI) practices in the context of corporate social responsibility (CSR), sustainability and ethics [18]. In order to better adapt to change, organizations must have some essential attributes, and nowadays the buzzwords have become sustainability, digitization, resilience and agility [19].

A few years after Schumpeter's theories were published in the 1960s, the role of education and how profitable investments in human capital were in increasing the productivity and motivation of professionals [20] were highlighted. Thus, economies that base their productive model on low-value-added activities, using low-skilled labor, become more vulnerable in periods of recession (such as that being lived through due to COVID-19), destroying many jobs, as they are the easiest and cheapest to destroy [21]. The absence of skilled human capital, training, and knowledge harms the economic development of an organization [22], and this prevents sustainable development over time of companies and economies. High professional qualification entails the need to manage talent, provide constant business training, provide higher remuneration, and engage in permanent recruitment. All this contributes to improving the working environment and an increasing identification of the employee with the organization, which results in greater business productivity [23].

It also highlights the importance of people's overall ability (work ethic, assimilation of experience, natural intelligence, commitment, etc.), which means that people tend to extract higher performance from training [5]. Talent management is related to training and companies, which in order to thrive need to develop a talent recruitment strategy through investment in incentives and training programs [24]. Commitment to leadership, as well as the autonomy, competence, and relationship between employees, is a good strategy of attracting and retaining talent positively related to labor commitment [25]. In addition, job satisfaction largely determines organizational success, as dissatisfaction has negative effects on productivity [26].

The basic competitive advantage of companies lies in the level of training and management of human talent or human resources [27]. In the same line, it is established that the systemic interrelationship between knowledge, competencies, innovation, and competitiveness is a tool for the management of human talent by skills, which allows organizations to increase their productivity and competitiveness [28]. Companies that develop strategies capable of attracting talent, with sustained training over time, will promote innovation, productivity and competitiveness in the market [29]. The satisfaction of professionals increases with human resources policies that promote talent management and training [30]. The conclusion therefore drawn is that training is a key element for attracting and retaining talent, as investment in training is a resource that benefits both businesses and workers [31]. A responsible company is one that allows professional development according to the worker's needs [32]. Training helps to improve the skills of professionals by increasing their intentions to remain in the company, productivity and the position of the company in the market [33,34]. Adopting a global approach to talent management can create long-term sustainable organizational success [35].

On the other hand, competitiveness lies in adapting and regenerating the assets of training, knowledge, and competencies, as well as developing and strengthening the organizational capacities that translate this knowledge into effective actions [36]. Research on talent management shows that management strategies promote companies' transformation and growth processes and increase their competitiveness in the global market [29]. Education is the preparation by and for life, the purpose of which is to prepare the person both within and outside the working environment [37].

Financial capital ceases to be the most important resource and gives its place to knowledge, as applying knowledge profitably is more important than money [38]. People, from the caretaker services to senior management, bring organizations to life with their dynamism, creativity, and rationality [39]. This implies that all people need training to carry out their tasks efficiently so as to be able to contribute to the development of the company. In the Information Age, employment has shifted from the industrial sector to the services sector and manual labor has been replaced by intellectual work, which marks the path of the post-industrialization era, based on knowledge and the tertiary sector [40]. The trend of the labor market is innovation and technology, the globalization of markets, the virtual economy and an emphasis on services and knowledge sectors [41]. Companies, in order to compete sustainably over time, will have to boost the knowledge economy, which implies good talent management and continued business training [42].

3. Materials and Hypothesis

The following hypotheses have been raised to achieve the objectives of this research.

Hypothesis 1. *"Adequate talent-management improves training in the company".*

The talent management of a company has a positive impact on the productivity and motivation of professionals, and the more talent there is in an organization, the more training is demanded to maintain the level of excellence, as can be seen in the works of [24,35,43].

Hypothesis 2. *"For training efforts to take effect, an adequate working environment is needed".*

A bad working environment can cause irreparable damage to businesses. For talent management and training to generate added value, a good working environment [44] is required. Satisfied workers increase the value of companies [45].

Hypothesis 3. *"High levels of corporate social responsibility (CSR) influence a company's sustainability in the ranking"*.

Companies with a good corporate reputation have a sustainable competitive advantage [46]. CSR is seen as an important element that influences the good perception of companies and their professional attractiveness [47].

Table 1 describes the variables used by *AE*, which make up the six pillars used in the preparation of the annual ranking, based on a questionnaire containing 100 questions. Next to each variable is the maximum score that each company can obtain for it. The questionnaire was completed out by human resources experts from more than 500 companies. Companies that can participated and qualified for the ranking of the 100 best companies to work for in Spain had to meet the following requirements: operating in Spain, having more than five years of operation and having a workforce of more than 100 employees.

Table 1. Variables used by *AE*.

Variable	Description	Maximum Score	% o/Total
Talent Management	Projection, performance and unwanted rotation	240	24%
Remuneration	Combination of fixed, variable wage, remuneration in kind and social benefits	225	22.5%
Work Environment	Working hours, telework, family reconciliation and working conditions	205	20.5%
Corporate Social Responsibility (CSR)	Social and volunteer policies involving staff	50	5%
Training	Educational Investment received by the employee	220	22%
Employees Perception	Valuation of the professionals of the company in which they work	60	6%
TOTAL	Sum of scores	1000	100%

Source: Own elaboration.

The variable under study (training) had a maximum value of 220 points, similar to the other two characteristics that are most important in determining the total valuation of the company, namely remuneration and talent management (a variable that relates to training). The total scores amount to 1000 points. In 2020, AE changed the scale to 1375 points, and these scores were adapted to the scale used in the first seven years (2013–2019), with a maximum of points being 1000 points.

Table 2 shows the objective control variables added for conducting the study. The first variable, the number of years that companies have remained in the ranking, focuses the analysis on the excellent companies, which are defined as the companies that are included in the ranking every year. The second variable ranks companies according to the gender of the company's management. No studies have been found that have linked management according to gender and management excellence through sustainability in a ranking of the companies which are best to work for. Geo-cultural areas bring companies together by country, classifying them in Anglo-Saxon; North-Central European, Mediterranean and Asian areas. The regional areas correspond to the locations of companies in the three most important areas of Spain (Madrid, the Mediterranean and Northern Spain). The following variables, such as the economic sector in which companies are engaged, are classified by the first digit of the National Classification of Economic Activities (CNAE) code. The size is measured according to the number of workers working in Spain. Finally, the variable of being listed on the stock market will be used.

Table 2. Control variables, added to the study, of *AE*'s ranking companies.

Variables	Justification
Number of years in the ranking	This is measured by the number of years that companies have remained in the ranking. Companies remaining in the ranking throughout the 8-year period are considered the most excellent from the point of view of the labor market.
Gender management	Classifies companies according to whether management is exercised by men or women.
Geo-cultural areas	They are grouped by countries and cultural areas, such as Anglo-Saxon, Central-North European or Mediterranean European and Asian areas.
Regional areas	Companies are grouped according to the tax domicile by regional areas that group regions (Autonomous Communities). Madrid, the Mediterranean and Northern Spain are identified
Economic activity	Analyses what the economic sectors of companies are, and, for this, they are classified according to the CNAE code and grouped by the first digit.
Size	Calculates the number of workers in the companies that work in Spain.
Trading in the stock market	Classifies companies according to whether they issue its shares on financial markets admitted to trading.

Source: Own elaboration.

Table 3 shows the 794 records corresponding to the ranking during the 8-year period of 2013–2020. Only 6% of companies managed to be in the ranking every year, and it is that group of companies that the research in this article will focus on. The objective is to analyze whether the permanence in the ranking influences a higher valuation obtained by companies in a significant way, as well as to analyze whether there are external variables that can influence the results obtained, such as gender in management, geo-cultural area, regional area, type of activity, size or stock market membership.

Table 3. Number of companies and records according to number of years in the ranking.

Number of Years in the Rankings	NO. Companies	%o/Total	NO. Records
1	115	40.64%	115
2	54	19.08%	108
3	34	12.01%	102
4	22	7.77%	88
5	15	5.30%	75
6	12	4.24%	72
7	14	4.95%	98
8	17	6.01%	136
Grand Total	283	100.00%	794

Source: Own elaboration from the data published in Economic News (2013–2020).

4. Quantitative Analysis

To achieve the objectives described above, different analysis techniques will be used: unifactorial variances with the Levene test, statistical analysis, as well as a model with different specifications of minimum ordinary least squares (OLS) and panel data, that will try to measure empirically whether there are certain dichotomous or numerical variables that have some kind of significant effect on the total valuation of companies in 2020, in the first type of model, or on what affects the valuation of training, throughout the period of the preparation of the ranking (2013–2020), in the second type of model.

4.1. Descriptive Statistical Analysis

Table 4 shows an upward trend in the valuation of all items as the number of years of companies remaining in the ranking increases. There is a 20.5% increase in total value and a 21.2% increase in training valuation reached by organizations that remain in the ranking for eight years compared to the ones which stay in the ranking for a single year. In turn, when compared to the potential value, training is the most valued item (85.5% over potential value), scoring above talent management, remuneration, and work environment. The conclusions obtained in a previous study that referred to the period 2013–2016 did not show the permanence in the ranking to be a factor that influenced the valuation of the items [48]. However, in this article, the conclusions differ, since it is appreciated that sustainability in the ranking influences total valuation and training.

Table 4. Average valuations of companies sorted by years in the ranking.

N⁰ Years in Ranking	Talent	Remuneration	Environment	CRS	Training	Employees	Total
1	165.3	141.7	140.7	37.2	155.3	54.2	694.4
2	171.0	155.4	150.2	38.7	165.7	55.0	736.1
3	172.6	155.9	154.5	40.1	165.2	55.9	744.3
4	176.9	162.8	152.6	40.4	178.6	53.5	764.5
5	185.4	166.7	163.0	42.5	180.3	57.2	795.1
6	184.4	156.7	155.6	43.1	183.0	58.4	781.3
7	179.5	169.5	161.5	44.3	177.8	59.3	791.8
8	190.1	183.6	172.0	43.6	188.1	59.2	836.7
TOTAL	177.9	162.1	156.5	41.2	173.9	56.6	768.0
% 8 years/1 year	115.0%	129.5%	122.3%	117.1%	121.2%	109.2%	120.5%
% 8 years/potential value	79.2%	81.6%	83.9%	87.2%	85.5%	98.6%	83.7%

Source: Own elaboration from *AE* data (2013–2020).

It can be seen in Table 5 that companies that have been in the ranking for eight years (excellent company) in relation to those that have been in the ranking for seven years obtain a higher and significant valuation in four of the six items, including training, in addition to total. Therefore, the mere fact of being in the rankings for one more year makes their ratings superior in a revealing way. This confirms that permanence in the ranking generates an increase in value and those companies listed every year in the ranking reach the highest level of valuation excellence, compared to those that are there for fewer years.

Table 5. Testing independent samples of average values classified according to permanence in the ranking (2013–2020).

Variables	Years in the Ranking	N	Media	F.	Sig.	Test Levene	Sig. (Bilateral)
Talent	8 years old	136	190.5	0.117	0.732	Equal variances are assumed	0.000
	7 years	98	179.47				
Remuneration	8 years old	136	183.56	12.594	0.000	No equal variances are assumed	0.000
	7 years	98	169.51				
Environment	8 years old	136	172.04	0.246	0.621	No equal variances are assumed	0.000
	7 years	98	161.45				
CSR	8 years old	136	43.61	1.386	0.240	Equal variances are assumed	0.334
	7 years	98	44.27				
Training	8 years old	136	188.13	0.242	0.623	Equal variances are assumed	0.000
	7 years	98	177.84				
Employee rating	8 years old	136	59.18	0.402	0.526	Equal variances are assumed	0.935
	7 years	98	59.29				
Total	8 years old	136	836.74	0.132	0.717	Equal variances are assumed	0.000
	7 years	98	791.83				

Source: Own elaboration from *AE* data (2013–2020).

Since excellent companies are those that are in the ranking every year, the following analyses will focus on this group of companies, studying whether the gender of the presidency or maximum management responsibility of the company influence the valuation within this group of excellence.

In Table 6, in all items, women-led companies achieve a higher valuation in a significant way: in working environment, in CSR and in total valuation.

Table 6. Testing independent samples of average values classified by gender for companies that remain in the ranking every year (2013–2020).

Variables	Gender	N	Media	F.	Sig.	Test Levene	Sig. (Bilateral)
Talent	man	120	189.91	2.611	0.108	Equal variances are assumed	0.706
	women	16	191.89				
Remuneration	man	120	182.78	3.357	0.069	Equal variances are assumed	0.193
	women	16	189.36				
Environment	man	120	170.41	7.302	0.008	No equal variances are assumed	0.000
	women	16	184.33				
CSR	man	120	43.19	0.530	0.468	Equal variances are assumed	0.010
	women	16	46.77				
Training	man	120	188.07	2.089	0.151	Equal variances are assumed	0.921
	women	16	188.61				
Employee rating	man	120	58.90	0.070	0.792	Equal variances are assumed	0.411
	women	16	61.25				
Total	man	120	833.35	3.266	0.073	Equal variances are assumed	0.045
	women	16	862.19				

Source: Own elaboration from *AE* data (2013–2020).

The following section discusses the possible influence of the nationality and geo-cultural area variables on the assessment of training and the total score. In an earlier study, the authors found no influence of nationality on human resources policies in Malaysian companies [49]. On the contrary, another work presented a list of systematic differences in human resources management in multinational enterprises depending on the country of origin [50]. Some more recent empirical research supported findings in the same direction [51,52].

Companies from 19 countries are present in the ranking during the period 2013–2020, but this number is reduced to less than half (seven countries) for those that are included in the ranking every year (Table 7). The most prevalent are those from Mediterranean Europe, which account for 58.8%, and Anglo-Saxon countries, who account for 29.4%, thus increasing their participation over the total population that have been included for just a few years. By contrast, those in Central-North Europe have a reduced presence in the most excellent companies list, and Asian companies are not even represented.

Another noteworthy fact is that the highest rated companies in total are those in Mediterranean Europe (839.1), followed by the Anglo-Saxon companies (833.9) and those in Northern-Central Europe (831.8), although these differences are not noticeable. While focusing attention on the valuation of the training, the same order is found. However, when considering the total sample, the most valued companies are the Anglo-Saxon companies, with significant differences from those from Mediterranean countries with a bilateral sig of (0.037) [53]. The explanation for this is that the companies that remain in the ranking are the most outstanding, with no differences between them depending on their nationality, their total value, or training.

Table 7. Valuations items sorted by international areas of companies that are held in the ranking every year.

Areas/ Countries	N° Records	%	Talent	Remuneration	Environment	CSR	Training	Employees	Total
Saxon	40	29.4%	189.3	182.1	172.7	43.1	186.5	60.2	833.9
USA	24	17.6%	194.8	178.9	175.3	43.9	198.2	59.5	850.7
UK	16	11.8%	180.9	186.9	168.8	41.9	169.0	61.2	808.7
North Central Europe	16	11.8%	204.8	175.3	170.0	43.1	181.3	56.6	831.8
Germany	8	5.9%	202.7	175.0	153.8	41.2	175.9	58.1	806.8
Netherlands	8	5.9%	206.9	175.5	186.3	45.0	186.6	55.1	856.8
Mediterranean Europe	80	58.8%	187.7	185.9	172.1	44.0	190.3	59.2	839.1
Spain	56	41.2%	186.5	187.3	178.4	44.5	188.9	59.4	845.1
France	16	11.8%	192.5	173.1	155.4	43.5	198.9	58.4	821.7
Italy	8	5.9%	186.0	202.2	161.5	40.8	182.6	59.6	832.6
TOTAL	136	100.0%	190.1	183.6	172.0	43.6	188.1	59.2	836.7

Source: Own elaboration from *AE* data (2013–2020).

By making the comparison according to specific items, it can be seen (Table 8) that companies in North-Central Europe have significantly higher values in talent management, highlighting companies in both the Netherlands and Germany. In all other items, the differences do not reach statistical relevance.

Table 8. Testing independent samples of significant average values by international areas.

Variable	Areas	N	Media	F.	Sig.	Test Levene	Sig. (Bilateral)
Talent	Saxon	40	189.26	4.831	0.032	No equal variances are assumed	0.001
	North Central Europe	16	204.78				
Talent	Mediterranean Europe	80	187.66	4.740	0.032	No equal variances are assumed	0.000
	North Central Europe	16	204.78				

Source: Own elaboration from *AE* data (2013–2020).

There are certain differences between human resources practices in different regions, detected by comparative research using econometric techniques [54–58]. Another study, which focused on companies that were in the ranking at some point during the period 2013–2016, did not reflect the fact that the regions with the highest GDP per capita in Spain had significantly better data, compared to other areas of the territory [59].

Table 9 shows that companies appearing on *AE*'s ranking every year (2013–2020) are mainly located in Madrid (78.8%), the country's capital, achieving the highest valuation in talent management and training. Considering that only one company is in the north and that part of its operational headquarters is in Madrid, this increases the total value and the relative importance of the capital of Spain, being the reference place for the most outstanding companies.

Table 9. Valuations of items sorted by geographical areas of companies included in the ranking every year.

Autonomous -Regions Grouping	N° Records	%	Talent	Remuneration	Environment	CSR	Training	Employees	Total
Madrid	107	78.7%	193.3	182.0	170.2	43.4	188.7	59.3	837.0
Mediterranean	21	15.4%	176.1	190.0	176.7	43.6	187.0	57.5	830.8
North	8	5.9%	184.9	187.8	183.9	46.4	183.8	61.4	848.2
TOTAL	136	100.0%	190.1	183.6	172.0	43.6	188.1	59.2	836.7

Source: Own elaboration from *AE* data (2013–2020).

Some authors include, among the contextual factors influential in human resources practices, the characteristics of the sector of activity [60]. The sector can also be categorized in several ways: services, industry, construction [55]. A previous study considering all companies concluded that the energy, financial and professional, scientific and technical sectors had higher values in a relevant manner compared to the remaining 16 economic sectors in which the companies operated [53].

In Table 10, seven economic sectors can be seen, with the most valuable regarding the total score and in the variable training being the commerce sector (876.2/193.4), forming 5.9% of the companies in the ranking, the professional, scientific and technical sector, with a valuation of (862.6/196.8), making up 23.5% of companies, and the financial and insurance sector, with a valuation of (839.9/187).

Table 10. Valuations items of companies classified by activities.

Description CNAE	N° Records	% 0/Total	Talent	Remuneration	Environment Agency	CSR	Training	Employees	Total
Commerce	8	5.9%	198.8	190.9	184.8	47.1	193.4	61.1	876.2
construction	8	5.9%	189,4	161.5	134.7	43.0	197.4	55.4	781.3
Financial and insurance companies	48	35.3%	183.2	192.9	172.9	44.7	187.0	59.2	839.9
Manufacturing industry	16	11.8%	179.3	183.8	164.5	40.3	172.4	61.3	801.7
Information and communication	8	5.9%	202.7	175.0	153.8	41.2	175.9	58.1	806.8
Professionals, scientific and technical	32	23.5%	202.8	174.5	184.1	44.4	196.8	59.6	862.6
Supply of energy	16	11.8%	186.3	184.9	174.3	41.9	188.7	57.5	833.7
TOTAL	136	100.0%	190.1	183.6	172.0	43.6	188.1	59.2	836.7

Source: Own elaboration from *AE* data (2013–2020).

Size is one of the most potentially influential factors in human resources practices [56], but there does not appear to be a consensus on the positive or negative signs of its effects. On the one hand, there can be a noticeable positive impact of small/medium size on employee behaviors (commitment or job satisfaction) and a negative impact of the same size on an operational performance indicator (absence and sick leave) [61]. On the other hand, there can be an association between increasing the size of businesses and formalizing human resources practices [62]. One study suggests that small businesses have several advantages, such as flexibility of roles, a close entrepreneur–worker relationship, among others [63]. On the other hand, another study stated that small businesses lack the resources to adopt progressive human resources management practices [49]. Small businesses have several advantages, such as a flexibility of roles, and a close employer–worker relationship, [63]. Ref. [59] showed in a clear and meaningful way that large organizations have better training ratios than small ones.

It can be seen in Table 11 that, as the number of years spent in the ranking increases, so do company size, total rating and training rating.

Table 11. N° workers and valuation according to years spent in the ranking.

N° of Years in the Rankings	N° Records	%	Average Workers	Total Rating	Training Rating
1	115	14.48%	3,211	694.4	155.3
2	108	13.60%	2,369	736.1	165.7
3	102	12.85%	4,093	744.3	165.2
4	88	11.08%	2,127	764.5	178.6
5	75	9.45%	4,370	795.1	180.3
6	72	9.07%	5,097	781.3	183.0
7	98	12.34%	6,249	791.8	177.8
8	136	17.13%	6,484	836.7	188.1
TOTAL	794	100.00%	4,306	768.0	173.9

Source: Own elaboration from *AE* data (2013–2020).

Table 12 calculated Pearson's correlation between the variables training, total score, workers and years in the ranking, there being a positive and significant relationship among them all.

Table 12. Correlations between training, number of employees and total valuation between companies in the ranking.

	Descriptive	Training	Total	Workers	Ranking Years
Training	Pearson correlation	1	0.621 **	0.092**	0.395 **
	Sig. (bilateral)		0.000	0.009	0.000
	n	794	794	794	794
Total	Pearson correlation	0.621**	1	0.081*	0.310 **
	Sig. (Bilateral)	0.000		0.022	0.000
	n	794	794	794	794
Workers	Pearson correlation	0.092 **	0.081 *	1	0.182 **
	Sig. (Bilateral)	0.009	0.022		0.000
	n	794	794	794	794
Years in the Ranking	Pearson correlation	0.395 **	0.310 **	0.182 **	1
	Sig. (Bilateral)	0.000	0.000	0.000	
	n	794	794	794	794

Source: Own elaboration from *AE* data (2013–2020). **. Correlation is significant at level 0.01 (bilateral). * Correlation is significant at level 0.05 (bilateral).

Companies listed on the stock market bear a higher degree of demand, as they are controlled by the National Securities Market Commission, by shareholders and by the market itself. Unlike those that are not publicly traded, a higher valuation of such companies could be expected. A previous study concluded that publicly traded companies achieve significantly better results in training management [59]. On the other hand, when considering the total sample of the companies that are listed in the ranking on occasion for the period 2013–2020, 63% of the companies in the total sample are listed on the stock market and have a valuation of 782.1***, higher and of significant difference from the non-listed stock market. However, when compared to the most excellent companies, striking differences occur. On the one hand, the relative weight of companies listed on the stock market increases (76.5%), and those listed in ibex-35 (29.4%). However, in terms of total valuation and training, it is those that are not publicly traded that reach the highest value, although the % decreases (Table 13). Therefore, the few companies that are not listed on the stock market and that remain in the rankings every year are very well valued and achieve high values in training, corresponding to companies in the world of consulting, professional and technical advice.

Table 13. Companies that remain for 8 years according to stock market membership (2013–2020).

	N° Records	%	Total	Training
Listed	104	76.5%	829.3	186.6
Ibex35	40	29.4%	832.2	186.5
No Ibex	64	47.1%	827.4	186.6
Not Listed	32	23.5%	861.0	193.1
TOTAL	136	100.0%	836.7	188.1

Source: Own elaboration from *AE* data (2013–2020).

4.2. Regression Analysis

This section establishes global models that seek to measure empirically whether there are certain dichotomous or numerical variables that have a significant effect on the total valuation of the companies in 2020. For this purpose, a multiple regression model (OLS) is used, since it is a technique that allows us to explain the relationship between the total valuation and the independent variables [64]. We choose the linear specification from the former estimation, since it fits better than other specifications that are more complex to interpret. The following equation would be a standard equation [65] for the estimation of a multiple regression analysis, where Xs would be explanatory variables, including intercept, while the ε would be the error term:

$$Total\ Valuation_i = \beta_i * X_i + \varepsilon_i$$

In the first specification (Table 14), there is a positive and significant value of the dichotomous variable that identifies companies that spent 8 years in the ranking (Excellent company), compared to those that spent less time, and the valuation of these companies is 69,988 points higher than the others. In the second specification, it can be seen how every passing year for companies in the ranking causes their total value to increase by 11,248 units. These results are as expected, given the rejection of most bilateral sigma in Table 5 as well as confirming the analysis of the data in Table 4. This would be the only noteworthy difference between the three specifications, so only the specification that we consider to best explain the relationship will be discussed, which would be the first.

Table 14. OLS method of the total variable relative to the control variables.

Dependent Variable	Total Valuation	
Specification	Specification 1	Specification 2
Variables	Coefficient (Std. Error)	Coefficient (Std. Error)
Constant	766.826 (22.395) ***	724.673 (22.945) ***
Male presidency	−0.998 (19.241)	4.080 (19.386)
Central-North European	6.074 (18.232)	0.908 (18.197)
Mediterranean	5.836 (14.670)	−0.656 (14.784)
Number of employees	−16.214 (13.284)	−7.383 (13.082)
Stock market listing	16.066 (12.484)	5.828 (13.384)
Excellent company	69.988 (15.794) ***	-
Number of years being an excellent company	-	11.248 (2.616) ***
Mediterranean region	−9.656 (16.278)	−7.282 (16.421)
Northern region	−41.071 (23.533) *	−33.500 (23.665)
Other regions	−60.059 (26.379) **	−53.567 (26.762) **
Finance and insurance (K)	57.220 (14.759) ***	54.466 (14.968) ***
Commerce (G)	34.944 (18.448) *	40.083 (18.697) **
Professional, scientific and technical (M).	31.078 (16.928) ***	27.562 (17.177)
Adjusted R-squared	0.343	0.336
N	100 (t = 1, i = 100)	100 (t = 1, i = 100)
F-statistic	5.311	5.177

Source: Own elaboration from *AE* 2020 data. Sig.*** = 0.01, ** = 0.05, * = 0.

A difference is found in the significance of the presidency's gender variable among the average tests in Table 6 and linear regression. This is because in a multiple linear regression, it is possible to control by more than one variable at the same time, making the explanatory power of the variable gender of the presidency disappear. Although Table 6 shows that certain differences in valuation (total, working environment and CSR) were due to the gender of the presidency, this variable is not relevant in the overall model.

As noted by the descriptive statistics in Table 7, the origin of the company or geo-cultural area do not affect the total valuation of the company, since it does not have a different valuation from that of the Anglo-Saxon companies, meaning that it should be omitted to avoid falling into the trap of the fictitious variables [65].

Unlike the geo-cultural area of the company (Table 9), the Spanish region in which the company is headquartered significantly affects the valuation, with all regions having a lower valuation than that omitted, which would be Madrid, while the regions with the least valued companies would be those that are part of other regions.

As shown in Table 10, the sector to which the companies belong significantly affects the total valuation of the companies, with the financial sector having the most positive effect, followed by commerce and professional, scientific and technical companies, relative to the omitted variable that would be made up of all companies that are not part of any of these three sectors.

Although Table 12 shows a weak positive correlation between total valuation and worker numbers, this is not manifested in the overall model, with no significant relationship between number of employees and this variable. This could be related to the fact that larger companies are those that remained in the ranking for the longest, as seen in Table 11.

In line with Table 13, the membership of companies in the stock market positively, but not significantly, affects the valuation of companies. This may be due to the presence of professional, scientific and technical companies which are highly valued but not publicly traded.

To try to explain the relationships between the valuation of the training of the 17 companies that spent 8 years in the ranking of the most valued companies, a model of panel data is used; that is, a combination of temporary data with cross-section data, which would be each of the companies, while the temporal variable would be the year.

The regression estimation with panel data has certain advantages over linear models, such as allowing one to control individual heterogeneity, as well as introducing a greater number of degrees of freedom and, in general, reducing multicollinearity [64]. The individual heterogeneity can be assumed to be random or fixed [66]; the choice between the two types of heterogeneity depends on the nature of the data. As data come from companies chosen by a valuation process, the choice of fixed effects on the cross-section seems more advisable [67]. A fixed-effects dashboard data model, where αi is a non-random and company-specific element, would be written as follows:

$$Valuation\ of\ Training_{it} = \beta_i * X_{it} + \alpha_i + \mu_{it}$$

Based on the partial correlations of Table 15, together with selection criteria, we develop three specifications, which examine the relationships between the individual valuations of each of the characteristics of the companies, upon the valuation of training.

Table 15. Partial correlations of companies' business valuations that are ranked every year.

	Talent	CSR	Remuneration	Training	Employees	Environment
Talent	1	0.158	0.079	0.352	0.041	0.135
CSR	0.158	1	0.167	0.283	0.065	0.37
Remuneration	0.079	0.167	1	0.014	0.192	0.168
Training	0.352	0.283	0.014	1	0.366	0.245
Employees	0,.041	0.065	0.192	0.367	1	0.034
Environment	0.135	0.37	0.168	0.245	0.034	1

Source: Own elaboration from *AE* data (2013–2020).

The first specification in Table 16 includes three explanatory variables: employees, environment and talent assessment [25,27,29]. The other two specifications include the CSR and remuneration variables, with the aim of evaluating different alternatives, although the best regression would be that of specification 1, which will be the one we will discuss. Since all the variables, both explanatory and those to be explained, are strictly positive, a logarithmic transformation can be made to them, to interpret the coefficients as elasticities. It is concluded that the variable that would be most positively related to the assessment of training would be the valuation of Employees, this being significant. A 1% increase in employee valuation would increase training valuation by 0.25%. The other variables would also be positive and significant.

Research reveals that H1 and H2 are confirmed. However, the results of the analysis in the Levene tests show that, in relation to H3, the difference is not significant, nor does it appear to influence the valuation of the company's training, as can be seen in Table 16.

Table 16. Panel data model, with training being the dependent variable for the period 2013–2020.

Dependent Variable	**Valuation of Training**	**2013–2020**	
Specification	Specification 1	Specification 2	Specification 3
Variables	Coefficient (Std. Error)	Coefficient (Std. Error)	Coefficient (Std. Error)
Constant	72.463 (23.474) ***	66.847 (25.025) ***	75.355 (24.694) ***
Val. Employees	0.740 (0.116) ***	0.738 (0.116) ***	0.749 (0.119) ***
Val. Environment	0.198 (0.090) **	0.186 (0.092) **	0.203 (0.091) **
Val. Talent	0.198 (0.081) **	0.197 (0.082) **	0.207 (0.085) **
Val. CSR	-	0.186 (0.282)	-
Val. Remuneration	-	-	−0.003 (0.081)
Adjusted R-squared	0.529	0.527	0.526
N	136 (t = 8, i = 17)	136 (t = 8, i = 17)	136 (t = 8, i = 17)
F-statistic	8.980	8.511	8.476

Source: Own elaboration from AE data 2013–2020. Sig.*** = 0.01, ** = 0.05.

5. Conclusions

Being present in the ranking of the most valued companies in the labor market developed by *AE* is exceedingly difficult, and only 6% of the companies remained in the ranking for every year during the period 2013–2020, as can be seen in Table 3.

Companies that hold their position in the ranking longer achieve more value in all items (Table 4). The training variable reaches the highest value relative to its potential (85.5%) among most representative variables (talent management, remuneration and work environment). By comparing the data of companies that are in the ranking over the eight-year period and those in the ranking for only seven, it can be observed that the result is higher in most cases and specifically in training, with a bilateral significance (0.000), as shown in Table 5. The OLS regression model (Table 14) also shows the relevance of permanence. This proves that sustainability in the ranking of the most valued companies in the labor market corresponds to value excellence, as well as attaching great importance to training.

Although it is appreciated that women-led companies that appear in the ranking every year achieve higher valuation in all items, including training, significantly in some cases (Table 6), the global regression model (Table 14) does not allow us to conclude that gender is a relevant variable.

Internationally, companies from 19 different countries can be observed for the period 2013–2020, this value being reduced to seven nations (36.8%) when referring to those that are included in the ranking every year. By focusing on the most excellent companies (Table 7), it is noticed that the value of talent management is higher with a sig 0.001 in favor of companies in Central Europe (Table 8). The global regression model (Table 14) confirms that nationality does not influence the results for total score or training.

The companies that are in the ranking every year operate in seven economic sectors, namely the commerce, professional, scientific and technical, and financial and insurance sectors, these being the most valued in general as well as in training (Tables 10 and 14).

The most resilient and valued companies are mainly located in Madrid (Table 9), and their relevance can be seen in the regression model (Table 14).

Positive and significant correlations between sustainability in ranking, size, training and total valuation are seen through Pearson's coefficient (Tables 11 and 12). Yet, when submitting it to the overall model (Table 14), it cannot be concluded that size influences the overall outcome in the most excellent companies.

Overall, 76.5% of companies are listed on the stock market and 29.4% belong to IBEX35. By contrast, the highest valuations, both for total score and training, are achieved by companies that do not trade on the stock market, although the differences are not significant according to the regression model (Tables 13 and 14).

In the panel data model (Table 16), it is appreciated that training as a dependent variable is significantly explained with employee valuation, work environment and talent management. We observe a global feedback effect, where investment in human capital improves the overall valuation of the company and its permanence in the ranking, given that maintaining high standards in training requires a high level of talent management, attracting and keeping the most excellent workers. On the other hand, it is not seen that CSR or remuneration influence the values acquired by training in the most excellent companies to work for.

It can be established that remaining in the ranking significantly increases total valuation and training, which leads to business excellence. These companies are in the capital of the country, and focus on the commerce; professional, scientific and technical; and financial and insurance sectors. It is also appreciated that the value of training is explained by employee valuation, work environment and talent management. This last result is consistent with human capital theories, where investment in this resource by firms and workers improves productivity and motivation, due to the identification that the skilled workers feel with the firm. On the contrary, the variable gender in the business direction does not influence the total valuation, nationality, size, or stock market membership.

In summary, for the most outstanding companies in the country, where the salary of their workers is already higher than the national average, improvements in pay do not necessarily lead to business excellence. However, guiding the organization's efforts to make the working day more flexible, to welcome teleworking or to improve talent management, will perpetuate the good management of the company by maximizing the retention of their most qualified personnel, guaranteeing their permanence in the ranking of the best companies to work for.

It should be mentioned as a limitation of this research that the recommendations provided are limited as to the level of disaggregation of the valuations offered by *AE* magazine. Future research should introduce new variables, expand the number of years, and extend the analysis to other countries in order to be able to perform comparative international analyses.

Author Contributions: Conceptualization, F.J.F.-S. and V.M.-R.; methodology, F.J.F.-S. and Y.A.-L.; software, F.J.F.-S. and Y.A.-L.; validation, F.J.F.-S. and Y.A.-L.; formal analysis, F.J.F.-S. and Y.A.-L.; investigation, F.J.F.-S., V.M.-R.; resources, F.J.F.-S., V.M.-R. and Y.A.-L.; data curation, F.J.F.-S. and Y.A.-L.; writing—original draft preparation, F.J.F.-S. and Y.A.-L.; writing—review and editing, F.J.F.-S., V.M.-R. and Y.A.-L.; visualization, F.J.F.-S., V.M.-R. and Y.A.-L.; supervision, F.J.F.-S.; project administration, F.J.F.-S. All authors have read and agreed to the published version of the manuscript.

Funding: This research received no external funding.

Institutional Review Board Statement: Not applicable.

Informed Consent Statement: Not applicable.

Data Availability Statement: The main data such as Talent Management; Remuneration; Work Environment; Corporate Social Responsibility (CSR); Training; Employees Perception; TOTAL; were obtained from: The 100 best companies to work for in 2013. Revista de Actualidad Económica (RAE), 2735, 33-41. The 100 best companies to work for in 2014. Revista de Actualidad Económica (RAE), 2747, 18-24. The 100 best companies to work for in 2015. Revista de Actualidad Económica (RAE), 2758, 18-24. The 100 best companies to work for in 2016. Revista de Actualidad Económica (RAE), 2770, 18-24. The 100 best companies to work for in 2017. Revista de Actualidad Económica (RAE), 2783, 20-26. http://www.expansion.com/actualidadeconomica/analisis/2017/09/06/59b01a4f468aebe25a8b45e0.html. The 100 best companies to work for 2018. Revista de Actualidad Económica (RAE), 2795, 20-26. http://www.expansion.com/actualidadeconomica/analisis/2018/09/27/5baca46ee5fdea307c8b463b.html. The 100 best companies to work for 2019. Revista de Actualidad Económica (RAE), 2807, 20-26. https://www.elmundo.es/economia/actualidad-economica/2019/09/23/5d83c2e0fdddff4b198b463a.html. The 100 best companies to work for 2020. Actualidad Económica Magazine (RAE), september 2020. https://www.elmundo.es/economia/actualidadeconomica/2020/09/26/5f6cbe80fc6c83c2498b45a9.html. Gender management; Geo-cultural areas; Regional areas; Economic activity; number of workers; Trading in the stock market, they were obtained directly from the web pages of each company.

Conflicts of Interest: The authors declare no conflict of interest.

References

1. Dewett, T.; Jones, G.R. The role of information technology in the organization: A review, model, and assessment. *J. Manag.* **2001**, *27*, 313–346. [CrossRef]
2. Peñaloza, M. Tecnología e innovación factores claves para la competitividad. *Actual. Contab. FACES* **2007**, *10*, 82–94. Available online: https://www.redalyc.org/pdf/257/25701508.pdf (accessed on 10 April 2020).
3. Rosegger, G. *The Economics of Production and Innovation: An Industrial Perspective*, 3rd ed.; Butterworth-Heinemann: Oxford, UK, 1996.
4. Joyce, K.E. Lessons for employers from Fortune's "100 best". *Bus. Horiz.* **2003**, *46*, 77–84. [CrossRef]
5. Eide, E.R.; Showalter, M.H. Human Capital. In *International Encyclopedia of Education*, 3rd ed.; Peterson, P., Baker, E., McGaw, B., Eds.; Elsevier: Oxford, UK, 2010; pp. 27–32. Available online: https://www.scopus.com/record/display.uri?eid=2-s2.0-84869851395&origin=inward&txGid=4e1d92ebb47843d5fb3755308344a726 (accessed on 18 February 2021).
6. Rocha, R.; Pinheiro, P. Business Education: Filling the Gaps in the Leader´s Awareness Concerning Organizational Phronesis. *Sustainability* **2021**, *13*, 2274. [CrossRef]
7. Alam, G.M.; Roslan, S. Contribution of prejudiced clustering education system in developing horizontal and vertical mismatch in job market: Quality of employees in banking sector. *Bus. Process Manag. J.* **2020**. [CrossRef]
8. Hoffman, A.J. Business education as if people and the planet really matter. *Strateg. Organ.* **2020**. [CrossRef]
9. Calleja, R.; Melé, D. Valero's "Enterprise Politics": A model of humanistic management and corporate governance. *J. Manag. Dev.* **2017**, *36*, 644–659. [CrossRef]
10. Carvalho, A.; Areal, N. Great places to work: Resilience in times of crisis. *Hum. Resour. Manag.* **2016**, *55*, 479–498. [CrossRef]
11. Guinot, J.; Chiva, R.; Mallén, F. Altruismo y capacidad de aprendizaje organizativo: Un estudio en las empresas mejor valoradas por los trabajadores en España. *Universia Bus. Rev.* **2015**, *45*, 92–109. Available online: http://repositori.uji.es/xmlui/bitstream/handle/10234/153475/chiva%202016.pdf?sequence-1&isAllowed-y (accessed on 5 January 2020).
12. Schumpeter, J. *Capitalismo, Socialismo, y Democracia*; Ediciones Orbis: Barcelona, Spain, 1983.
13. Fassin, Y. Innovation and ethics ethical considerations in the innovation business. *J. Bus. Ethic* **2000**, *27*, 193–203. Available online: https://link.springer.com/article/10.1023/A:1006427106307 (accessed on 18 February 2021).
14. Stilgoe, J.; Owen, R.; Macnaghten, P. Developing a framework for responsible innovation. *Res. Policy* **2013**, *42*, 1568–1580. Available online: https://www.sciencedirect.com/science/article/pii/S0048733313000930?via%3Dihub (accessed on 9 June 2021).
15. Rueda, A. *El sector TIC en Alerta y en Una Posición Clave ante la Posible Desaceleración de la Economía Española*; Vass Research: Madrid, Spain, 2019. Available online: https://www.vass.es/vass-research/ (accessed on 5 January 2020).
16. Parviainen, P.; Tihinen, M.; Kääriäinen, J.; Teppola, S. Tackling the digitalization challenge: How to benefit from digitalization in practice. *Int. J. Inf. Syst. Proj. Manag.* **2017**, *5*, 63–77.
17. Marinai, L.; David, R.; Di Sarra, A.; Escorcia, A.; Akhtar, M.M.J.; Al-Jefri, A.; Al-Hendasi, A. Digital Transformation of Production Governance and Assurance Process for Improving Production Efficiency. In Proceedings of the Abu Dhabi International Petroleum Exhibition & Conference; Society of Petroleum Engineers (SPE), Abu Dhabi, United Arab Emirates, 9 November 2020. [CrossRef]
18. Gurzawska, A. Responsible Innovation in Business: Perceptions, Evaluation Practices and Lessons Learnt. *Sustainability* **2021**, *13*, 1826. [CrossRef]
19. Miceli, A.; Birgit, H.; Pia, M.; Sotti, F.; Settembre-Blundo, D. Thriving, Not Just Surviving in Changing Times: How Sustainability, Agility and Digitalization Intertwine with Organizational Resilience. *Sustainability* **2021**, *13*, 2052. [CrossRef]
20. Shultz, T.W. *The Economic Value of Education*; Columbia University Press: New York, NY, USA, 1963.
21. Felgueroso, F.; Garcia-Perez, J.I.; Jansen, M. La contratación temporal en España: Nuevas tendencias, nuevos retos. *Pap. Econ. Española* **2018**, *156*, 47–61. Available online: https://dialnet.unirioja.es/servlet/articulo?codigo=6518996 (accessed on 5 January 2020).
22. Del Campo, M.O.; Salcines, J.V. El valor económico de la educación a través del pensamiento económico del siglo XX. *Rev. La Educ. Super. (ANUIES)* **2008**, *37*, 45–61. Available online: http://www.scielo.org.mx/scielo.php?pid=S0185-27602008000300004&script=sci_arttext (accessed on 5 January 2021).
23. Motellón, E. *Temporalidad. La Contratación Temporal en España*; Universitat Oberta de Catalunya: Barcelona, Spain, 2007.
24. Gandossy, R.; Kao, T. Talent wars: Out of mind, out of practice. *Hum. Resour. Plan.* **2004**, *27*, 15–19. Available online: https://go.galegroup.com/ps/anonymous?id=GALE%7CA126653495&sid=googleScholar&v=2.1&it=r&linkaccess=abs&issn=01998986&p=AONE&sw=w (accessed on 26 October 2020).
25. Rahmadani, V.G.; Schaufeli, W.B.; Ivanova, T.Y.; Osin, E.N. Basic psychological need satisfaction mediates the relationship between engaging leadership and work engagement: A cross-national study. *Hum. Resour. Dev. Q.* **2019**, 1–19. [CrossRef]
26. Weapons, Y.; Plains, M.; Traverso, P. *Human Talent Management and New Labor Scenarios*; ECOTEC University: Samborondón, Ecuador, 2017.
27. Fleitas, S. La gestión del talento humano y del conocimiento. *Rev. Latinoam. Psicol.* **2013**, *45*, 157–160. Available online: http://www.scielo.org.co/scielo.php?script=sci_arttext&pid=S0120-05342013000100012 (accessed on 5 January 2021).

28. Giraldo, A.M.; Jaramillo, M.; Castillo, M.B. Formación del talento humano: Factor estratégico para el desarrollo de la productividad y la competitividad sostenibles en las organizaciones. *Rev. Guillermo Ockham* **2006**, *4*, 43–81. [CrossRef]
29. Molina, R.I.R.; Zúñiga, H.C.C.; Alfaro, K.P.V. Gestión del talento humano como estrategia organizacional en las Pequeñas y Medianas Empresas. *CICAG* **2018**, *16*, 20–42.
30. Hur, H.; Maurer, J.A.; Hawley, J. The role of education, occupational match on job satisfaction in the behavioral and social science workforce. *Hum. Resour. Dev. Q.* **2019**, 407–435. [CrossRef] [PubMed]
31. Alandes, J. *Training and Talent Management*; Area RH.com: Lara, Venezuela, 2002. Available online: http://www.ucla.edu.ve/dac/Departamentos/AdmRecHumNV/Material/UNIDAD%20VI%20Capacitaci%F3n%20y%20development%20of%20RRHH/La%20formaci%F3n%20y%20la%20gestion%20del%20talento%20human.doc (accessed on 18 February 2020).
32. Pazmiño, W.; Fiallos, H.V. New trends in training and development: Tomorrow's careers gain prominence based on talent management by competencies. *J. Bus. Entrep. Stud.* **2017**, *1*, 19–25. [CrossRef]
33. Snyder, C.R.; López, S.J. *Handbook of Positive Psychology*; Oxford University Press: New York, NY, USA, 2002.
34. Armstrong, M. *A Handbook of Human Resource Management Practice*; Kogan Page: London, UK, 2006.
35. Ashton, C.; Morton, L. Managing talent for competitive advantage. *Strateg. HR Rev.* **2005**, *4*, 28–31. [CrossRef]
36. Lladós, J. *La Geografía de la Innovación*; FUOC: Barcelona, Spain, 2019.
37. Chiavenato, I. *Recursos Humanos*; Mc Graw Hill: Mexico City, Mexico, 2010.
38. Chiavenato, I. *Gestión del Talento Humano*; McGraw/Interamericana Editores S.A.: Mexico City, Mexico, 2009.
39. Alles, M. *Comportamiento Organizacional*; Ed Granica: Buenos Aires, Argentina, 2007.
40. Nova, P.; Beneyto, P. Teorías de la organización del trabajo y del capital humano. In *Trabajo y Empresa*; Beneyto, P.J., de la Torre, I., Nova, P., Eds.; Tirant Lo Blanch: Madrid, Spain, 2014; pp. 13–49.
41. Plains, M. El Desarrollo de los Sistemas de Producción y su influencia en las relaciones laborales y el rol del trabajador. *Econ. Desarro.* **2016**, *157*, 130–146.
42. Santos, W.P.; Ventimilla, H.V.F. Nuevas tendencias en formación y desarrollo: Las carreras del mañana ganan protagonismo en base a la gestión del talento por competencias. *J. Bus. Entrep. Stud.* **2017**, *1*, 19–25. [CrossRef]
43. DiRomualdo, T.; Joyce, S.; Bression, N. *Key Findings from Hackett's Performance Study on Talent Management Maturity*; Hackett Group: Palo Alto, CA, USA, 2009.
44. Davies, G.; Chun, R.; Vinhas da Silva, R.; Roper, S. *Corporate Reputation and Competitiveness*; Routledge: London, UK; New York, NY, USA, 2003.
45. Helm, S. One Reputation or Many? Comparing Stakeholders' Perceptions of Corporate Reputation. Corporate Communications. *Int. J.* **2007**, *12*, 238–254. [CrossRef]
46. Capraro, A.J.; Srivastava, R.K. Has the influence of financial performance on reputation measures been overstated? *Corp. Reput. Rev.* **1997**, *1*, 86–92. [CrossRef]
47. Arikan, E.; Kantur, C.; Maden, C.; Telci, E.E. Investigating the mediating role of corporate reputation on the relationship between corporate social responsibility and multiple stakeholder outcomes. *Qual. Quant.* **2016**, *50*, 129–149. [CrossRef]
48. Ferreiro-Seoane, F.J. Multianálisis de las organizaciones más valoradas para el desempeño laboral en España. *Rev. Métodos Cuantitativos Para La Econ. Empresa* **2020**, *29*, 57–78. [CrossRef]
49. Ibrahim, H.I.; Shah, K.A.M. Effects of organizational characteristics factors on the implementation of strategic human resource practices: Evidence from Malaysian manufacturing firms. *Econ. Ser. Manag.* **2013**, *16*, 5–24. Available online: http://management.ase.ro/reveconomia/2013-1/1.pdf (accessed on 9 June 2020).
50. Ferner, A. Country of origin effects and HRM in multinational companies. *Hum. Resour. Manag. J.* **1997**, *7*, 19–37. [CrossRef]
51. Liu, W. The cross-national transfer of HRM practices in MNCs: An integrative research model. *Int. J. Manpow.* **2004**, *25*, 500–517. [CrossRef]
52. Guthrie, J.P.; Liu, W.; Flood, P.C.; MacCurtain, S. *High Performance Work Systems, Workforce Productivity, and Innovation: A Comparison of MNCs and Indigenous Firms (WP 04-08)*; DCU Business School: Dublin, Ireland, 2008. Available online: Doras.dcu.ie/2421/1/wp0408.pdf (accessed on 9 June 2020).
53. Ferreiro-Seoane, F.J.; Miguéns-Refojo, V.; Campo-Villares, M.O. The most attractive sectors of the labour market and their international dimension. *Glob. Compet. Gov.* **2021**, *15*, 102–113. [CrossRef]
54. Amossé, T.; Bryson, A.; Forth, J.; Petit, H. Managing and working in Britain and France: An introduction. In *Comparative Workplace Employment Relations, An Analysis of Practice in Britain and France*; Amossé, T., Bryson, A., Forth, J., Petit, H., Eds.; Palgrave MacMillan: London, UK, 2016; pp. 1–26. [CrossRef]
55. Conway, N.; Deakin, S.; Komzelmann, S.; Petit, H.; Rebérioux, A.; Wilkinson, F. The influence of stock market listing on human resource management: Evidence for France and Britain. *Br. J. Ind. Relat.* **2008**, *46*, 632–673. [CrossRef]
56. Fields, D.; Chan, A.; Akhtar, S. Organizational context and human resource management strategy: A structural equation analysis of Hong Kong firms. *Int. J. Hum. Resour. Manag.* **2002**, *11*, 264–277. [CrossRef]
57. Grimshaw, D.; Rubery, J. Economics and HRM. In *The Oxford Handbook of Human Resource Management*; Boxall, P., Purcell, J., Wright, P., Eds.; Oxford University Press: New York, NY, USA, 2007; pp. 68–87. [CrossRef]
58. Paawee, J.; Boselie, P. HRM and societal embeddedness. In *The Oxford Handbook of Human Resource Management*; Boxall, P., Purcell, J., Wright, P., Eds.; Oxford University Press: New York, NY, USA, 2007; pp. 166–184. [CrossRef]

59. Ferreiro-Seoane, F.J.; Del-Campo-Villares, M.O.; Camino-Santos, M. La formación y la gestión del talento en las empresas más valoradas en recursos humanos en España. *Contad. Adm.* **2019**, *64*, 1–21. [CrossRef]
60. Jackson, S.; Schuler, R.S. Understanding human resource management in the context of organizations and their environments. *Annu. Rev. Psychol.* **1995**, *46*, 237–264. [CrossRef]
61. Kortekaas, F. HRM, Organizational Performance and the Role of Firm Size. Master's Thesis, Erasmus School of Economics, Rotterdam, The Netherlands, 2007. [CrossRef]
62. Kok, J.; Uhlaner, L.M. Organization context and human resource management in the small firm. *Small Bus. Econ.* **2001**, *17*, 273–291. [CrossRef]
63. Morgan, J. *The Future of Work: Attract New Talent, Build Better Leaders, and Create a Competitive Organization*; Wiley: Hoboken, NJ, USA, 2014.
64. Wooldridge, J.M. *Econometric Analysis of Cross Section and Panel Data*; MIT Press: Cambridge, MA, USA, 2010.
65. Dougherty, C. *Introduction to Econometrics*, 5th ed.; Oxford University Press: Oxford, UK, 2016.
66. Nyström, K. The institutions of economic freedom and entrepreneurship: Evidence from panel data. *Public Choice* **2008**, *136*, 269–282. [CrossRef]
67. Baltagi, B.H.; Bresson, G.; Pirotte, A. Fixed effects, random effects or Hausman-Taylor? A pretest estimator. *Econ. Lett.* **2003**, *79*, 361–369. [CrossRef]

Article

Key Elements and Their Roles in Entrepreneurship Education Ecosystem: Comparative Review and Suggestions for Sustainability

Haibin Liu [1,*], Sadan Kulturel-Konak [2] and Abdullah Konak [3]

[1] Institute of Employment and Entrepreneurship Education, Northeast Normal University, Changchun 130024, China
[2] Management Information Systems, Penn State Berks, Reading, PA 19610, USA; sxk70@psu.edu
[3] Information Sciences and Technology, Penn State Berks, Reading, PA 19610, USA; auk3@psu.edu
* Correspondence: liuhb@nenu.edu.cn

Abstract: This paper examines two core issues of the university-based entrepreneurship education ecosystem by explicating the key elements of the ecosystem and their roles, and the development process and sustainable construction strategy of the ecosystem. Thirty stakeholders of ecosystems from the US universities were interviewed, and the transcripts of these interviews were coded through a three-phase process, including open, axial, selective coding, and were analyzed based on the grounded theory. It was found that (i) the key elements of the university-based entrepreneurship education ecosystem consist of six units (colleges and universities, learners, educators, government, industry, and community) acting as initiators and seven factors (entrepreneurship curriculum, entrepreneurial activities and practices, organizational structure, resources, leadership vision, core faculty, and operating mechanism) acting as the intermediaries; (ii) These key elements constitute three independent functional subsystems, namely, teaching and innovation, support, and operation that are interconnected by the universities; (iii) The development process of a university-based entrepreneurship education ecosystem involves seven steps as preparation, germination, growth, equilibrium, stagnation, recession, and collapse; (iv) For sustainability, suggestions on a solid foundation, continuous investment, and constant monitoring are provided to university administrators and policymakers to advance higher education's contribution to social and economic development.

Keywords: entrepreneurship education; university-based ecosystem; key elements; development process; sustainable construction strategy

Citation: Liu, H.; Kulturel-Konak, S.; Konak, A. Key Elements and Their Roles in Entrepreneurship Education Ecosystem: Comparative Review and Suggestions for Sustainability. *Sustainability* **2021**, *13*, 10648. https://doi.org/10.3390/su131910648

Academic Editors: Jaana Seikkula-Leino, Mats Westerberg, Priti Verma and Maria Salomaa

Received: 30 August 2021
Accepted: 23 September 2021
Published: 25 September 2021

1. Introduction

The importance of entrepreneurship has been widely recognized as one of the driving forces of economic and social development [1,2]. University-based entrepreneurship education is a key link in the establishment of new enterprises, cultivation of innovative talents, scientific, and technological innovation [3,4]. Consequently, entrepreneurship education has become a buzzword in national strategies of various countries and regions, such as the Small Business Innovation Research program of the US [5], the European Education Area of the European Union [6,7], and the Innovation-driven development strategy of China [8]. For institutions of higher education, classroom teaching is the main channel to provide education services for the learners. Therefore, for a long time, scholars and educators have taken entrepreneurship education curriculum and programs as the main field of research [9,10] to reveal the causal conclusions related to entrepreneurship education and improve its impact [11,12]. As far as the output of the whole university is concerned, entrepreneurship courses and programs are the core components of entrepreneurship education, but they are not the whole story. The other components ignored by researchers to some degree may also play a vital role in achieving entrepreneurship-related outcomes.

Based on the recognition of this view, recent studies [13,14] have shown that scholars have begun to re-examine the entrepreneurship education in higher education institutions from a system theory perspective. More specifically, ecosystems have become a new research perspective for entrepreneurship education.

Typically, entrepreneurship education ecosystem research requires a systematic approach to explore and study entrepreneurship education issues. Compared with previous studies on entrepreneurship education, the research perspective of the entrepreneurship education ecosystem is considered as a truly systematic and interdisciplinary research method [13,15,16]. The term 'ecosystem' comes from biology [17] and has been adopted by many fields as a research perspective or method, including entrepreneurship education. Overall, an entrepreneurship education ecosystem research has two characteristics: case-based research and inductive research. In entrepreneurship education ecosystem research, scholars usually choose one or more university-based entrepreneurship education ecosystems as research objects (cases) [18,19] and then drive conclusions from the outcomes and attributes of the selected universities through a process of scientific induction [18,20]. Although the research on entrepreneurship education ecosystem has many advantages in research systematization [21], relationship research [22], and input–output comparison [23], the two characteristics mentioned above have shortcomings in terms of adaptability and portability, especially when administrators are looking for examples and guidance that are suitable for their universities to establish or enhance their entrepreneurship education ecosystems for sustainability. The reason for this dilemma is that different administrators and educators face different and complex university environments in terms of historical development, staff and student backgrounds, resources and culture, etc., and their entrepreneurship education ecosystems may also be at different stages of development. Inspired by the above issues, the purpose of this study is to seek answers to the following two research questions through interviews and analysis processes:

QI. What are the key elements of a university-based entrepreneurship education ecosystem?

QII. Based on the role, importance, and relationship of the constituent key elements, what strategies could be followed to build a sustainable university-based entrepreneurship education ecosystem?

The contribution of this study is to provide a dynamic and interactive view of the university-based entrepreneurship education ecosystem lifecycle process, rather than a static snapshot of the cases presented in previous studies at a certain stage. Hence, the conclusions of this research are drawn based on the views of 30 stakeholders from many different institutions. The study also focuses on the key elements that are generally applicable to the best practices of starting and growing an entrepreneurship education ecosystem for sustainability. The finding of this study may guide the executives and educators of higher education institutions in building their entrepreneurship education ecosystem to strengthen, modify, or rethink the role of different elements in the ecosystem and maintain their entrepreneurship education ecosystem in a healthy and sustainable operating status.

2. Literature Review

2.1. Entrepreneurial Ecosystem

Tansley [24] first coined the term "ecosystem" to recognize the integration of the biotic community and its physical environment as a fundamental unit of ecology within a hierarchy of physical systems that span the range from atom to universe. Since the 1950s, the idea of human–ecological systems as an integrated biosocial approach to understanding communities, urban areas, and environments has been on the rise [18]. In 1982, Nelson and Winter [25] brought the concept of ecosystems to evolutionary economic theory in their work, since economics has always been about the systems that explain differential output and outcomes [26]. As the strategic literature framework of the business ecosystems was constantly improved by Moore [27] and later Iansiti and Levien [28], the concept of an ecosystem was applied to more fields, including knowledge ecosystems [17], entrepreneurial ecosystems [29], organizational ecosystems [30], and so on. In general, the ecosystem is a biotic community encompassing the physical environment and all living entities, and all the complex interactions that facilitate its co-existence, co-dependence, and co-evolution [26]. Therefore, the advantage of ecosystem theory lies in the study of the individual within its overall environment [31]. This systematic view can consider both the individual and the environment [32], so it has been widely developed and applied in recent years, including in the field of the entrepreneurial ecosystem.

The entrepreneurial ecosystem utilizes a 'systemic view of entrepreneurship' [31] or 'ecosystem research approach' [26] to look at issues related to entrepreneurship. The concept of the entrepreneurial ecosystem has diversely been used in the literature, some of which definitions are concise and rich in general, such as 'an interconnected group of actors in a local geographic community committed to sustainable development through the support and facilitation of new sustainable ventures' [29], or 'a set of interdependent actors and factors coordinated in such a way that they enable productive entrepreneurship within a particular territory' [33]. Although many studies have shown a correlation between the number of new firms and economic growth, the earlier methods failed to produce a convincing causal relationship between them [34]. The entrepreneurial ecosystem approach is a better way to deal with these complex problems, which the neoclassical economic approach may not appropriately address [35]. The emerging research on the entrepreneurial ecosystem involves evolutionary, socially interactive, and non-linear approaches, and the interactive nature of entrepreneurial processes is in contrast with previous literature that treats it as unitary, atomistic, and individualistic [36,37]. Another feature of the entrepreneurial ecosystem approach is to emphasize the key role of entrepreneurs in innovation [38], while also focusing on several other key actors, including large firms, universities, financial firms, and public organizations that support new and growing firms [39,40]. With the unique perspective and approaches of the entrepreneurial ecosystem recognized by scholars, the number of relevant researches has gradually increased, especially since 2010. Malecki [41] and Cavallo, et al. [42] performed a meta-analysis of publications on the entrepreneurial ecosystem, and the results showed that the entrepreneurial ecosystem has quickly emerged as a promising research area in entrepreneurship. However, Isenberg [43] also pointed out that the popular ecosystem theory of the natural sciences should be applied prudently in the field of entrepreneurship since the ecosystem metaphor does not completely fit entrepreneurship in terms of the creation, the centralized control, the geography, the intention, and the entrepreneur-centrality.

Research on the entrepreneurial ecosystem has two major lineages: the strategy literature and the regional development literature [26]. Although the origins are the same, there are three differences between the two lineages. First, the regional development literature focuses more on the boundaries of ecosystems, while the strategy literature generally ignores regions or assumes a global context [44]. Second, the conclusion of the regional development literature is to explain regional performance or inter-regional performance differences, while the strategy literature focuses on the value creation and value capture by individual firms (units) [45]. Third, the strategy literature assumes that some units play a key or leading role in the ecosystem [46], and these units are called 'lead firm' [47], 'keystone' [48], or 'ecosystem captain' [49], etc. In contrast, the regional development literature tends to decentralize, which emphasizes the mutual promotion of units within the ecosystem [33].

According to the classification of the research literature on ecosystem theory by Jacobides, et al. [50], the literature on the entrepreneurial ecosystem can be divided into three categories: environmental, innovative, and framework research. Environmental research focuses on individual firms or start-ups, and ecosystems are viewed as the structural environment in which new ventures are launched. The research goals of this type of research are related to discovering the interaction mechanism between the company and the ecosystem, such as monitoring or responding to the business environment [51]. Innovative research focuses on a focal innovation and the set of components (upstream) and complements (downstream) that support it in the ecosystem [52]. The emphasis of this type of research is on understanding how interdependent units in an ecosystem interact, such as knowledge sharing [53], collaborative investment [54], product complementarity [55], etc., to create and commercialize innovation. Framework research focuses on the structure, operation, and development of entrepreneurial ecosystems. Case analysis is a prominent methodology in this kind of research. For example, Silicon Valley [23,56], Boulder [57], Stanford [58], and Oxford [59,60] are all studied as a case of an entrepreneurial ecosystem framework. In addition to the three types of research described above, there are also some systematic and comprehensive studies of the entrepreneurial ecosystem in the form of academic monographs [61,62]. Regardless of the type of research, almost all cited work above considers entrepreneurship education or universities as playing an important role in the entrepreneurial ecosystem [18,63].

2.2. *Entrepreneurship Education Ecosystem*

With the popularity and development of research on entrepreneurial ecosystems, the important role of entrepreneurship education and universities in the ecosystem has been increasingly recognized. For instance, with the implementation of the Bayh-Dole Act in the US, which allowed inventors and universities to retain the ownership of intellectual property arising from federally-funded research [64], the research from the perspective of an entrepreneurial ecosystem has started paying more attention to the university [16]. As Rice, et al. [65] note, universities are at the hub of economic development around the world, providing infrastructure, resources, and means to develop entrepreneurial communities. They argue that entrepreneurial ecosystems evolve and expand through the specialization of knowledge and innovation. Therefore, the concept of the entrepreneurship education ecosystem was proposed.

The entrepreneurship education ecosystem refers to those dynamic systems of integrated networks and associations aligned to entrepreneurship education programs [66]. Entrepreneurship education programs are various concrete manifestations of entrepreneurship education, whereby entrepreneurship skills, knowledge, guidance, assistance, and information are disseminated to various stakeholders [67]. In a monograph, Rice, Fetters and Greene [65] systematically analyzed the entrepreneurial ecosystems of six universities and summarized success factors. This research perspective marks the starting point for the research on the entrepreneurship education ecosystem, which decomposes university and entrepreneurship education as an ecosystem rather than as an integral unit within the ecosystem. Since 2010, the research on the entrepreneurship education ecosystem has gradually increased, especially in the past three years as summarized in Table 1.

Table 1. Number of the research results on entrepreneurship education ecosystem.

Years	2010–2015	2016	2017	2018	2019	2020	2021
	10	5	9	12	22	24	27

In line with the entrepreneurial ecosystem approach, the research on the entrepreneurship education ecosystem can also be divided into two categories. One is to view the entrepreneurship education ecosystem as a subsystem of the entrepreneurial ecosystem. For example, Regele and Neck [68] considered entrepreneurship education as a nested sub-ecosystem within the broader entrepreneurship ecosystem, discussed entrepreneurship education across three distinct levels-K12, higher education, and vocational training, and proposed that entrepreneurship education must develop such coherence by building a network of education programs that fit together in a coordinated way. Another kind of entrepreneurship education ecosystem research is university-based. Such research studies generally point out that universities are at the hub of economic development around the World, providing infrastructure, resources, and means to develop entrepreneurial communities. It is argued that a university-based entrepreneurial ecosystem may evolve and expand through the specialization of knowledge and innovation. An example of such research was given by Brush [18], who explored the concept of an entrepreneurship education ecosystem and adopted the dimensions of infrastructure, stakeholders, and resources to characterize the internal entrepreneurship education ecosystem based on the research of Rice, Fetters and Greene [65]. Although these two types of studies both cover the entrepreneurship education ecosystem and universities in terms of content, few achievements in the current literature specifically take the university-based entrepreneurship education ecosystem as the research object. The university-based entrepreneurship education ecosystem plays an important role in the whole entrepreneurial ecosystem, which is the intersection of these two kinds of studies. Therefore, the research on the university-based entrepreneurship education ecosystem has important significance to the literature contribution, which is also the starting point of this study.

2.3. Composition of the Entrepreneurship Education Ecosystem

Similar to the entrepreneurial ecosystem research, the entrepreneurship education ecosystem is dominated by case studies or comparative studies between cases, while the composition is still the focus and foundation. Composition research studies the framework of the functional units in the entrepreneurship education ecosystem and the interactions among these units. Examples of relatively simple frameworks are the three-dimensional ecosystem framework of entrepreneurship curriculum, entrepreneurship co-curricular, and entrepreneurship research proposed by Brush [18]. Ferrandiz, et al. [69] expanded Brush's framework to seventeen constituent components. Based on Brush's framework, Wraae and Thomsen [14] also explored the interaction among stakeholders, including educators, students, institutions, external organizations, and communities. Examples of complex frameworks are the research of De Jager, et al. [70] and Mogollón, et al. [71] on the entrepreneurship education ecosystem in universities. In addition, the research on the interaction among twelve stakeholder groups in the entrepreneurship education ecosystem framework of Bischoff, et al. [72] is also representative, and the relationship among stakeholders was divided into nine relationship types according to the cross combination of coordination and strength. The representative studies on the composition of the entrepreneurship education ecosystem in the literature are presented in Table 2.

Table 2. Representative composition of entrepreneurship education ecosystem in the literature.

Authors	Article (Book Section)'s Title	Composition of Entrepreneurship Education Ecosystem
Brush [18]	Exploring the Concept of an Entrepreneurship Education Ecosystem	Curriculum, co-curricular activities, research, culture, stakeholders, infrastructure, and resources.
De Jager, Mthembu, Ngowi and Chipunza [70]	Towards an Innovation and Entrepreneurship Ecosystem: A Case Study of the Central University of Technology, Free State	The academic program, research and innovation, idea generator, incubator program, innovation services, students, staff, entrepreneurial activity, local community, business and industry support.
Bischoff, Volkmann and Audretsch [72]	Stakeholder Collaboration in Entrepreneurship Education: An Analysis of the Entrepreneurial Ecosystems of European Higher Educational Institutions	Entrepreneurs, companies, financial institutions, support service providers, incubators and accelerators, student organizations, alumni, higher educational institutions, science and technology parks, governmental organizations, non-governmental organizations, and other organizations.
Mogollón, Portillo, Escobedo and Pérez [71]	The Approach of the Entrepreneur Microecosystem for University Entrepreneurial Education: Model M2E EMFITUR	Model of value contribution, objectives, stakeholders, instruments, teaching methodologies, teacher, progress tests, processes and procedures, syllabus, resources, common errors, skills-competencies, milestones (the rhythm of the course), connection-progress, brakes-barriers, programs, promotive strategy, evaluation of the model.
Wraae and Thomsen [14]	Introducing a New Framework for Understanding Learning in an Entrepreneurship Education Ecosystem	Educator, student, institution, external organization, and community.

From these representative studies on the composition of the entrepreneurship education ecosystem, four characteristics could be summarized. First, although there are differences in the naming and classification of the composition, these elements are entrepreneurial stakeholders that are centered around the main functions of universities [73–75], among which education [76] and scientific research [23] are at the core of an ecosystem. Second, the boundaries of the university-based entrepreneurship education ecosystem are not determined by the 'walls' of the university [19] because institutions, organizations, industries, communities, and other external elements, which are independent of the university, are also included in the scope of the ecosystem [56,66,77]. Third, the elements that make up the entrepreneurship education ecosystem may be tangible such as incubators, teachers, and online entrepreneurship courses [78,79], or they may be intangible such as relationships, values, cultural traditions, and atmosphere [20]. Fourth, although terms such as dimensions [18], units [21,26], factors [72,79], levels [80], and items [22] could be found in the literature to describe the components that make up the entrepreneurship education ecosystem, the element [19,56,78] is the most common term because of its generality and compatibility. Although these case-based studies have made important contributions to the literature, the limitation is that the generalizability of the conclusions is applicable to all university-based entrepreneurial education ecosystems.

2.4. Construction Strategy of the Entrepreneurship Education Ecosystem

The relationship between theoretical research and practical application is mutually reinforcing, so it may provide suggestions and references for the practice of the entrepreneurship education ecosystem after summarizing theories and regularity in case studies. Therefore, the research on the construction strategy, or related to it, has become another important branch of the research on the entrepreneurship education ecosystem, although the number of such studies is still very limited. In addition, due to the limitation of work on the definition and scope of the entrepreneurship education ecosystem, the corresponding construction strategies also showed great dispersion.

The development of the entrepreneurship education ecosystem is an important perspective of construction strategy research. Brush [18] divided the entrepreneurship education ecosystem into domains (curricular, co-curricular, scholarship, and outreach) and dimensions (stakeholders, resources, culture, and infrastructure) in her research, and proposed a four-step construction strategy. (1) Assess: information gathering, benchmarking, discussions, and determine strengths and capabilities. (2) Engage: identify champions, develop definitions and concepts, and research groups. (3) Test ideas/experiment: pilot classes, workshops, and stakeholder events. (4) Build: courses and programs, metrics/monitoring, and practices for innovation and expansion. The entrepreneurship education ecosystem is built based on multiple universities in the industry or region is a research topic with practical significance [81]. The initial stage of the construction of the entire ecosystem was divided into three steps. (1) Develop models for ingraining entrepreneurship education into specific engineering and technology curricula and drive new course concepts into the policy action of the partnering universities. (2) Build a network for entrepreneurship education by implementing a pilot between the universities within their innovation and entrepreneurial ecosystems. (3) Share best practices for promoting hands-on entrepreneurial skills within accelerations, local hubs, technology platforms, and student-driven start-up activities.

Another kind of research on the construction strategy of the entrepreneurship education ecosystem is closely related to the composition because their starting point is the relationship between these constituent elements and the successful construction of the ecosystem. For example, Rice, Fetters and Greene [65] summarized success factors in their work as follows: (1) Senior leadership vision, engagement, and sponsorship; (2) strong programmatic and faculty leadership; (3) sustained commitment over a long period; (4) commitment of substantial financial resources; (5) commitment to continuing innovation in curriculum and programs; (6) an appropriate organizational infrastructure;

and (7) commitment to building the extended enterprise and achieving critical mass. After discussing three cases of entrepreneurship education ecosystem adopted by Cornell University, the University of Rochester, and Syracuse University, Antal, et al. [82] proposed four viewpoints for success: (1) Some administrative support from parts of the university at the start; (2) a focus on entrepreneurship education and programs for students; (3) faculty champions in schools and colleges across campus; (4) emotional and financial investment of alumni. Similar studies have been carried out by Lyons and Alshibani [83], Schmidt and Molkentin [84], and Belitski and Heron [21].

The perspectives of the development process and the ecosystem composition have their own advantages in the research of the construction strategy of the entrepreneurship education ecosystem. The strategy proposed from the perspective of the development process of the ecosystem is more prominent in its guiding significance to practice, while the strategy proposed from the perspective of the composition reflects the role and importance of different elements in the ecosystem.

However, there are still many meaningful issues that could be explored and discussed urgently to enrich the comprehensiveness of the research on the sustainable construction strategy of the entrepreneurship education ecosystem. For example, there is almost no research focus on the entire process of the development of the entrepreneurship education ecosystem and revealing the characteristics of each stage at the theoretical or case study level. Similar to small enterprises in different developmental stages having different institutional settings and business models, the entrepreneurship education ecosystem in various developmental states might contain different compositions and face different situations. In addition, just as the factors and causes of entrepreneurial success were advocated and pursued at the early stage of the entrepreneurship research, the study on the success of entrepreneurship education ecosystem construction has become mainstream, while the failure factors or cases are ignored and marginalized.

3. Methodology

3.1. Research Design

The research of the entrepreneurship education ecosystem is still in its infancy. Limited to the complexity and cost of research, this study follows the inductive approach commonly used in entrepreneurial ecosystem research. Specifically, this study used the interview method and analyzed the interview responses of stakeholders in entrepreneurship education ecosystems with qualitative analysis software, and then drew meaningful findings and conclusions for best practices.

The research was carried out in three stages: (1) The exploration stage, which included searching, sorting out, and analyzing references, defining research boundaries, focusing on research gap in the literature, and forming an interview outline; (2) The data collection stage through interviewing 30 entrepreneurship education ecosystem stakeholders selected based on the principle of purposeful sampling; (3) The data analysis stage, which involved transcribing the interview recordings into the text format and text analyses in the NVivo12 qualitative analysis software. Based on the grounded theory, the collated interview transcripts were sequentially coded with open coding, axial coding, and selective coding [85].

3.2. Sample Selection

Following the logic of qualitative research, the sampling strategy was purposeful and oriented towards achieving theoretical saturation [86]. Specifically, this study collected the contact information of participants at ten academic conferences/events of different sizes held in the US on the theme of entrepreneurship education or university-based innovation and entrepreneurship. Then, participants were asked whether they would like to take part in either in-person or video conferencing interviews of this study. Participants in these conferences were university-based entrepreneurship education ecosystem stakeholders who are familiar with, passionate, and looking forward to the improvement of the ecosystem.

Based on the premise of fully respecting the interviewees' consent and compliance with research ethics, this study comprehensively considered the distribution of factors such as gender, industry, position, work experience, and educational background. Finally, 30 positive respondents who expressed willingness to take part in our research from ten regions, nine universities or campuses, five other organizations were identified as interviewees in this study. The details of the relevant demographics of the interviewees are shown in Table 3.

Table 3. Interview sample details (n = 30).

Category			Frequency	Weight (%)
Gender	Male		16	53.33
	Female		14	46.67
Occupation *	University professor		15	50.00
	University administrator		9	30.00
	University support staff		8	26.67
	Enterprise mentor		5	16.67
	Entrepreneur		5	16.67
	Community/industry support staff		8	26.67
Experience	Less than 5 years		4	13.33
	5 to10 years		9	30.00
	More than 10 years		17	56.67
Educational Background	Highest Degree	Doctorate	12	40.00
		Master	10	33.33
		Bachelor	8	26.67
	Major	Engineering	6	20.00
		Business	8	26.67
		Natural Sciences	4	13.33
		Humanities and Social Science	6	20.00
		Education	3	10.00
		Art	1	3.33
		Health Science	2	6.67

* Some participants had multiple occupations.

3.3. Data Collection

The interviews were conducted either in-person or remotely via video conferencing by three interview team members, which were uniformly trained on the research objectives and interview skills before the interviews started. Interviews were conducted on relatively independent occasions at scheduled times, and each interview lasted 25–40 min. The interview was semi-structured and consisted of five main questions, including (1) the key constituent elements of the entrepreneurship education ecosystem, (2) the role and (3) importance of these key elements, (4) how to build the ecosystem, and (5) the best practices for sustainability of the entrepreneurship education ecosystem the interviewees had observed. Interviewees were appropriately questioned based on their responses to fixed questions. The entire recordings of the interview were transcribed into text, which was imported into the database built by NVivo12 after clarifying the interviewees' vague views and confirming the content.

3.4. Data Analysis

Each interview's text data was analyzed in NVivo12. According to the grounded theory, the analysis of the interview text and the coding process was also divided into three phases [85]. Firstly, during the open coding phase, the interview text was reviewed carefully and repeatedly. The meaningful local concepts in the interview text were identified and established as free nodes. In NVivo's process of text analysis, a node refers to a collection of references about a specific theme or case, and researchers gather the references by coding sources to a node. Secondly, in the axial coding phase, similar free nodes were classified and merged concerning the research purpose and interview outline. For example, most of the interviewees mentioned "enough places", "gathering space", "physical places", "workspace", "together space", and "office space" when talking about the key elements of the entrepreneurship education ecosystem, so a node named "space" was established in the "elements" node directory. Then, the characteristics of different classification nodes were mined and analyzed, and the nodes were reconstructed and renamed accordingly. For example, when interviewees talked about the importance of the constituent elements of the ecosystem, almost all the local concepts that they used involved the comparison of two different elements or the comparison of one element with other elements. Therefore, the free nodes describing the importance of the constituent elements were transformed into the relationship nodes. Finally, during the selective coding phase, the classified nodes were organized into a tree-like hierarchical structure based on logic, containment, and dependencies in the node directory.

Trustworthiness is considered a more appropriate criterion for evaluating qualitative studies [87]. Therefore, three team members independently coded and compared the coding repetition rate with each other to ensure the reliability of the qualitative analysis of the interview text. To avoid errors caused by different coders using different vocabularies for the same node, the three sets of coding tables were modified and clarified before performing the reliability test. Specifically, similar words and synonyms, such as universities and colleges, teaching methods and pedagogy, financial and economic support were negotiated and unified. After such corrections, a team was set as the basic group, and the other two teams were set as the comparison group. The code comparison tool of NVivo was used to measure the repetition rate of the coding tables of the base group and the comparison group. According to the average weighting of each interview text, the comparison results were 90.75% (Cohen's Kappa coefficient = 0.8219) and 86.31% (Cohen's Kappa coefficient = 0.7328), which were within acceptable ranges. Table 4 presents the number of nodes at all levels of the interview text after analysis and coding.

Table 4. Number of nodes at all levels after coding.

Node Directory	Node Type	Number of the Primary Node	Number of the Secondary Node	Number of the Tertiary Node
Elements	Themes	2	13	77
Development processes	Themes	3	7	29
Roles	Relationship	8	33	-
Importance	Relationship	4	21	-

3.4.1. Key Elements of the Entrepreneurship Education Ecosystem

Although this study attempted to summarize the common aspects in the composition of the university-based entrepreneurship education ecosystem, almost all interviewees expressed that the key elements of the ecosystem should or could be different in various environments, stages, and situations. For example, *"They (the elements) should be drawn depending on the specific situation of the university"* (Interviewee #2). However, in the process of encoding and analyzing the interview text, it was found that the differences expressed by the interviewees were actually due to variations in the characterization of the same or similar elements under different conditions. Therefore, it did not affect the summary of the key elements frequently mentioned by most interviewees after clarification. For example, "*entrepreneurship degrees prevalent in Australian universities*" (Interviewee #5), "*entrepreneurship education minor offered by many universities*" (Interviewee #11), "*a range of entrepreneurship education courses*" (Interviewee #21) were shown in the responses of different interviewees to describe the same key elements. They are just different forms of the entrepreneurship education curriculum, so the "entrepreneurship curriculum" as a key element of the entrepreneurship education ecosystem could not be denied. Similarly, various forms of institutions were mentioned by interviewees as constituent elements of the ecosystem, including "*entrepreneurship center or similar institutions are the core of the ecosystem*" (Interviewee #1), "*cooperation between entrepreneurship education institutions and supporting institutions is the key*" (Interviewee #3), "*entrepreneurship education institutions at both the university and college levels are necessary*" (Interviewee #17). The interviewees were presenting what they thought would be the most ideal framework for university institutions within the entrepreneurship education ecosystem. The significance of this divergent view is that organizational structure is an important part of the entrepreneurship education ecosystem. As a result, "organizational structure" is one of the key elements in the analysis of the presentation of the characteristics of the interviewees.

The analysis process sought common elements in the interview text but also aimed to preserve differences in the perspective of the interviewees, and these two contradictory objectives became the basis for the classification of key elements in the coding process and led to interesting revelations. Especially in the selective coding stage, the characteristics of the key elements presented two distinct classifications. One type of the key elements, encoded as units, was an institution, organization, or stakeholder group in the entrepreneurship education ecosystem, which was the source or target of behavior, resources, information, and interaction. Another type of the key elements, encoded as factors, was an intermediary that linked the units of the entrepreneurship education ecosystem together, or the conditions and environment associated with the units. Table 5 presents the results of the interview text analysis on the key elements of the entrepreneurship education ecosystem, which were composed of six units and seven factors. Among them, six units referred to colleges and universities, learners, educators, government, industry, and community. On the other hand, seven factors included entrepreneurship curriculum, entrepreneurial activities and practices, organizational structure, resources, leadership vision, core faculty, and operating mechanism. Based on this, the key elements of the entrepreneurship education ecosystem generated by NVivo Map Tool are shown in Figure 1.

Table 5. Key elements of the entrepreneurship education ecosystem.

Primary Node (Frequency)	Secondary Node	Frequency	Tertiary Node (Partial)
Units (412)	Colleges and universities	75	Entrepreneurship education idea, university orientation, mission, service …
	Learners	74	Students, graduates, potential entrepreneurs, entrepreneurial thinking, entrepreneurial intention, entrepreneurial ideas, entrepreneurial mindset, entrepreneurial competency, entrepreneurial skills, entrepreneurial enthusiasm …
	Educators	91	Teachers, staff and mentors, teacher competency, number of teachers, teacher team structure, teacher background, teaching method application, teacher initiative …
	Government	59	Government policies, policy implementation, supporting institutions, supporting projects (government), entrepreneurship education standards, and evaluation …
	Industry	44	Industry-university-research cooperation, support projects (enterprise), internship opportunities, enterprise, small business, company, firm …
	Community	69	Support projects (community), tolerance for failure, partners …

Table 5. *Cont.*

Primary Node (Frequency)	Secondary Node	Frequency	Tertiary Node (Partial)
Factors (565)	Entrepreneurship curriculum	81	Courses, programmer, lectures and workshops, course quantity, course credit, course effect, minor course, course content . . .
	Entrepreneurial activities and practices	105	Co-curricular activities, entrepreneurship competitions, entrepreneurial practice, entrepreneurial activities, entrepreneurship laboratory, entrepreneurship simulation . . .
	Organizational structure	89	Entrepreneurship education institution, incubator, entrepreneurship research institutions, entrepreneurship curriculum management institutions, entrepreneurship club or center, supporting institutions, scientific research institution . . .
	Resources	80	Financial resources, human resources, space, time, information, materials . . .
	Leadership vision	63	Clear and achievable goals, great and persistent concern, strong determination . . .
	Core faculty	66	Backbone faculty, key faculty, model faculty, champion faculty . . .
	Operating mechanism	81	Management system, rules, operating guide, platform, negotiation mechanism, innovation mechanism, problem-solving mechanism, technology transformation mechanism . . .

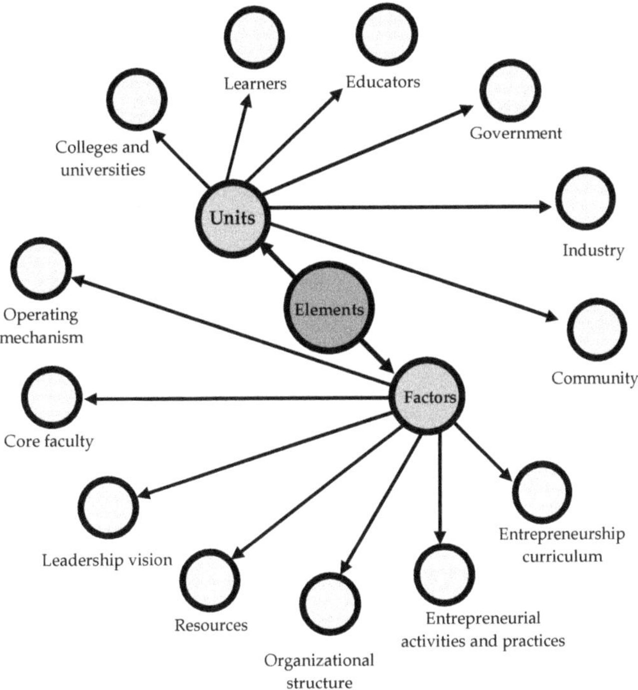

Figure 1. Map of key elements of the entrepreneurship education ecosystem.

To further clarify the relations among the key elements that make up the entrepreneurship education ecosystem, Figure 2 graphically presents the relationship nodes identified from the analysis of the roles and their importance along with the key elements. In the figure, the units and factors are depicted by two types of circles that are connected by arrows of different thicknesses according to the frequency of the secondary relationship nodes. The frequency values were calculated as the number of coded relationship nodes used to describe the relations among the elements in the interview text. The relations between any two elements had a maximum frequency of 31 and a minimum frequency of 0, and the frequencies were categorized into four levels from "Very Low" to "High", respectively represented by different types of arrows in Figure 2. The identified units and factors are grouped into three areas as the teaching and innovation, support, and operation subsystem based on their roles in the entrepreneurship education ecosystem, and the colleges and universities interconnect these main areas.

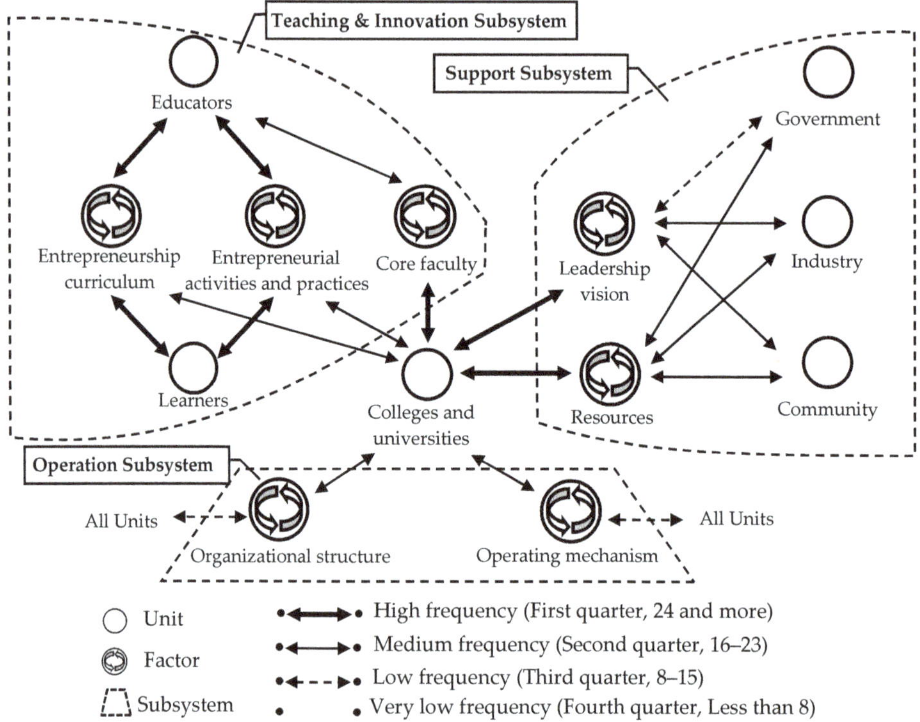

Figure 2. The relation among the key elements of the entrepreneurial education ecosystem.

3.4.2. Development Cycles of the Entrepreneurship Education Ecosystem

Biological ecosystems are dynamic, and the entrepreneurship education ecosystem is no exception. Some terms describing dynamicity, such as time, order, stages, steps, and process were identified in the analysis of the interview text in this study. Moreover, these terms did not appear independently in a vacuum but were accompanied by the description of the characteristics of a specific period of the entrepreneurship education ecosystem. Then, these terms associated with the corresponding descriptions were coded and classified according to the similarity of the descriptions of the characteristics. As one of the analysis results of the interview text, the development cycles of the entrepreneurship education ecosystem could be divided into three stages and seven steps, which correspond to the primary and secondary nodes, respectively (see Table 6).

Table 6. Development cycles of the entrepreneurship education ecosystem.

Primary Node (Frequency)	Secondary Node	Frequency	Tertiary Node (Partial)
Ascending Stage (121)	Preparation	59	Arrangements for, plan, purpose ...
	Germination	62	Start with, move out, the first move ...
Stabilization Stage (131)	Growth	38	Development, progress, expansion ...
	Equilibrium	22	Maintenance, self-sustaining, balance ...
	Stagnation	71	Crisis, difficulty, risk, bottleneck ...
Declining Stage (44)	Recession	28	Cut down, turnover, retrenchment ...
	Collapse	16	End, over, breakdown, disappear ...

3.4.3. Importance and Role of the Factors in Development Cycles

Ignoring the sudden changes to the ecosystem caused by extreme external forces, such as university mergers or natural disasters, the entrepreneurship education ecosystems are all in one of the seven steps of the development cycle found in this study. These seven steps were divided based on the characteristics and the relationships of the elements in the ecosystem. From a dialectical perspective, the elements of the entrepreneurship education ecosystem may play different roles and have varying importance in the seven steps of the development cycle. The role and importance of the unit-type elements are more intuitive (e.g., the role of educators and learners). Compared with the units, understanding the role and importance of the factor elements that link the units together is a more valuable and meaningful study [88]. Although one-third of the interviewees initially responded "equally" to the question about the importance of the factors of the ecosystem, they provided more specific answers while discussing the steps of the development cycle. Then, it was found that when interviewees described or discussed a certain step of the development cycle, some factors of the ecosystem were mentioned frequently. Moreover, the role and importance of certain factors were always associated with a specific step of the development cycle. To explore the role and importance of ecosystem factors in the different steps of the development cycles, this study used the coding tool of NVivo to conduct a cross-analysis of elements coding and development processes coding. The factors part of the secondary node of elements coding and the secondary node of development processes coding were cross-tabulated, as shown in Figure 3.

The numbers in the cross-analysis matrix are the frequency that corresponds to both the row and column encoding nodes in the interview text. Different factors have different frequencies in different steps of the development cycle. A higher frequency meant that the corresponding factors of the ecosystem are strongly related to a specific step of the cycle. This strong correlation is essentially an internal manifestation of the role and importance of the factor in a specific step, which provides an innovative research perspective for this study. Meanwhile, the frequency distribution of factors in different steps of the development cycles also exhibited certain patterns: Some factors were more frequent at the beginning steps; some factors were more frequent in the stabilization stage; the frequencies of some factors were evenly distributed throughout all steps of the development cycle. Based on the emerged patterns in the cross-analysis matrix, this study divided the factors of the entrepreneurship education ecosystem into three categories, namely: Launch factors, process factors, and maintenance factors.

Factors/Steps	Preparation	Germination	Growth	Equilibrium	Stagnation	Recession	Collapse
Entrepreneurship curriculum	9	9	6	2	5	1	-
Entrepreneurial activities and practices	5	11	9	6	5	3	1
Organizational structure	10	10	5	2	12	6	-
Resources	5	5	1	3	9	2	-
Leadership vision	7	5	2	1	10	5	1
Core faculty	9	6	2	1	11	5	-
Operating mechanism	5	10	5	2	9	2	-

High frequency The first 1/3, 9, or more; Medium frequency The middle 1/3, 5-8; Low frequency The last 1/3, 4, or less

Figure 3. Cross-matrix analysis of factors and steps.

4. Findings

4.1. Key Elements of Categories

The key elements of the university-based entrepreneurship education ecosystem consist of six units and seven factors, which constitute three independent functional subsystems.

4.1.1. Unit Category

In terms of the key elements of the unit category, the findings of this study were consistent with the research results of scholars on the ecosystem of entrepreneurship education. Learners and educators were mentioned as key elements repeatedly by almost all interviewees. In particular, government, industry, and community were identified as key elements. Consistent conclusions were determined by the inherent nature of entrepreneurship education [14,18,71,89] *because it is not the general education of universities, but the education about entrepreneurship* (Interviewee #9). Entrepreneurship education has a positive impact on innovation and entrepreneurship activities [90], which is an important driving force for economic and social development. Therefore, the government, industry, and community encouraged and invested various resources in entrepreneurship education with the expectation that *entrepreneurship education could export new enterprises, new jobs, new talents, new technologies*, etc. (Interviewee #19). When analyzing the composition of the entrepreneurship education ecosystem, *the ecosystem must be placed in a specific context that is determined by the policies of the country and state, the needs of the industry, and the situation of the local community where the university is located* (Interviewee #7). The government, industry, and community were identified as the key elements of the entrepreneurship education ecosystem because they are located in different parts of the ecosystem's relation chain [91,92]. Simultaneously, *entrepreneur teams in universities need to tap the entrepreneurship education ecosystem for capital, which always comes from outside organizations such as angel investors, companies, governments, and industries* (Interviewee #26). These inputs and expected outputs of the government, industry, and community also constitute the main components of the resource and information flow in the entrepreneurship education ecosystem.

The unique finding of this study was to identify the university as an element placed in the entrepreneurship education ecosystem, rather than considering specific colleges or different academic programs. *The centrality of universities to entrepreneurship education has never been challenged* (Interviewee #6). However, the research perspective of the university-based entrepreneurship education ecosystem usually focuses on individual academic programs [18,21,90,93], so that the university as an element of the ecosystem is ignored by most scholars. For the ecosystem level, the internal composition of the university presents some common characteristics in terms of interests, behaviors, and value trends. *This uniqueness of a university is the foundation and start point for building an entrepreneurship education ecosystem* (Interviewee #15). Therefore, it is necessary to consider the university as a whole in the ecosystem, especially when exploring the relationship between the university and the government, industry, and community. In addition, any element that is considered to be an internal element of the university, such as students or faculty, may not represent the university. *The boundary of the entrepreneurship education ecosystem is not the university campus, but the strength of the relationship between the candidate element and other elements of the ecosystem* (Interviewee #13). In other words, whether *entrepreneurship education activities occur inside or outside the university* (Interviewee #1), whether the learners are *current students or graduates* (Interviewee #21), and whether the educator is a *faculty member or part-time entrepreneur mentor* (Interviewee #24) is not the criterion for identifying an element that belongs to the entrepreneurship education ecosystem. Therefore, our findings suggest considering the entire university, rather than individual academic units, as an element in the entrepreneurship education ecosystem.

4.1.2. Factor Category

The key elements in the factor category could be divided into three types according to the characteristics of the roles that they play in the entrepreneurship education ecosystem. These three types are the intermediary, the condition, and the environment, which represent the main interaction and relationship among different units. *The vibrant ecosystem needs the interactions and the relationships to be mapped and managed* (Interviewee #23). The factors of intermediary type mainly refer to entrepreneurship curriculum and entrepreneurial activities and practice, which connect educators and learners. The scope of entrepreneurship curriculum encompasses *formal degree courses of the entrepreneurship* (Interviewee #30), *courses in entrepreneurship minors* (Interviewee #14), *entrepreneurship co-curricular* (Interviewee #6), and *lectures and workshops on entrepreneurship* (Interviewee #11), most of which take place in the classroom. The scope of entrepreneurial activities and practices is much broader, including *activities that take place outside the classroom* (Interviewee #5) for teaching and improving the results of entrepreneurship, as well as *technological innovation and entrepreneurship research in the form of "teaching in doing"* (Interviewee #19), most of which take place in laboratories or research institutions. The main function of entrepreneurship curriculum, as well as entrepreneurial activities and practices, are to *transfer entrepreneurial skills, spirit, knowledge, and information between educators and learners* (Interviewee #9), so these two factors are the main channel and method of entrepreneurship education. *These transmitted things may have existed in the past from knowledge and experience, or they may have been created during the transfer process* (Interviewee #13), which reflects on the uniqueness of entrepreneurship education for future-oriented and practice-oriented education.

The factors of condition type mainly refer to resources, leadership vision, and core faculty. This type of factor has a profound impact on the existence and development of the entrepreneurship education ecosystem and presents the status of the causal relationship between the units. Specifically, the resources factor composes of *human, material, financials, time, space, and information* (Interviewee #5) and links some related units in the entrepreneurship education ecosystem. The leadership vision presents a higher education institution's positioning, concept, and goals for entrepreneurship education, and *its role is to generate cohesion* (Interviewee #25) that brings other units in the entrepreneurship education ecosystem together and *gives stakeholders encouragement and confidence* (Interviewee #22). Core faculty refers to leaders, who are prominent figures among full-time and part-time members in the entrepreneurship education ecosystem. They were found to *be closely related to the creation, development, and crisis exclusion of the ecosystem* (Interviewee #14).

The factors of environment type mainly include the operating mechanism and organizational structure, which are the background and context of the elements of the ecosystem. At least half of the interviewees repeatedly mentioned the institutions and guidelines/policies, such as the entrepreneurship center, club, incubator, laboratory, management system, technology transformation, library support mechanism, and so on. Depending on whether they are tangible or not, the factors of environment type were divided into organizational structure and operating mechanism. It needs to be pointed out that both of them are most widely used to classify and even name the whole entrepreneurship education ecosystem because of their distinctive characteristics. For example, the "Triple Helix" and "Magnet-Model" frequently mentioned by interviewees originated from the description of institutions or mechanisms and have become the "brand" of the entrepreneurship education ecosystem for some universities, and they are widely spread.

4.2. Three Subsystems of Division

The teaching and innovation, support, and operation subsystems were identified by analyzing the relations among key elements of the university-based entrepreneurship education ecosystem. The teaching and innovation subsystem is the core of the ecosystem, including educators, core faculty, learners, entrepreneurship curriculum, and entrepreneurial activities/practices. The teaching and innovation subsystem relies *on the process of knowledge transfer to catalyze the birth of potential entrepreneurs* (Interviewee #12) and relies on universities' capacities in research to continue to *create new value through innovative technologies, business models, products, and services* (Interviewee #3). The support subsystem consists primarily of leadership vision, resources, government, industry, and community, and *it provides two-way, mutually beneficial support* (Interviewee #5) for the other two subsystems. It emphasizes "support" rather than "input" and "output" to avoid the boundary of the entrepreneurship education ecosystem being preconceived as the campus of the university. *The formation and adjustment of the Leadership vision are inextricably linked to the concept of entrepreneurship education in government, industry, and community* (Interviewee #20). With a high degree of adaptation between them, the resources needed for the entrepreneurship education ecosystem are drawn from the government, industry, and community and used by other subsystems. In contrast, entrepreneurs and new technologies, business models, products, and services are also the resources to the government, industry, and community, so the working mode of the support subsystem is two-way and reciprocal. The operation subsystem is mainly composed of the organization structure and operation mechanism, the function of which is to ensure the daily operation of the entire entrepreneurship education ecosystem.

In addition, the significance and contribution of the above findings of the relation among the elements of this study also provide a new research perspective for other entrepreneurship ecosystem studies by explaining the details of phenomena and events in the entrepreneurship education ecosystem. For example, the process of an entrepreneurial student team's obtaining start-up funds to develop their businesses within the entrepreneurship education ecosystem can be explained according to the framework of the key elements and relationships of the ecosystem discovered in this study as follows. An entrepreneurship education concept (leadership vision) of the university attracts the attention of an enterprise in an Industry. This enterprise sets up an entrepreneurship scholarship (resources) at the university to encourage students to start businesses and develop industry innovation. With the scholarship, the director (core faculty) of the university's entrepreneurship center launches an entrepreneurship competition (entrepreneurial activities and practice) according to the responsibilities and functions of the organization (organizational structure). Under the guidance and assistance of entrepreneurship teachers (educators), students (learners) who have passed the entrepreneurship training program (entrepreneurship curriculum) form entrepreneurship teams to participate in the competition. The entrepreneurship competition selects the best teams according to the relevant rules (operating mechanisms) and awards them with some scholarships. Most of the events and phenomena similar to this example could be presented in detail in the entrepreneurship education ecosystem.

4.3. *Stages of Development*

The development process of a university-based entrepreneurship education ecosystem involves three stages, which can be subdivided into seven steps.

In chronological order, the first stage was the ascending stage, including the two steps of preparation and germination. The second stage was the stabilization stage, including the three steps of growth, equilibrium, and stagnation. The last was the declining stage, including the two steps of recession and collapse. Among them, the two stages of ascending and declining were a one-way development process. However, the three steps of the stabilization stage showed typical cyclic characteristics. Based on these findings, the diagram of development cycles of the entrepreneurship education ecosystem was constructed, as shown in Figure 4.

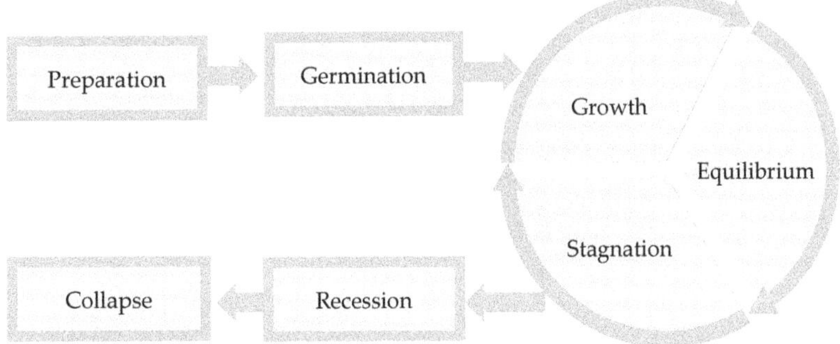

Figure 4. Development cycles of the entrepreneurship education ecosystem.

4.3.1. Ascending Stage

For the entrepreneurship education ecosystem, the ascending stage of the development cycle is the process of establishing the ecosystem from scratch. Depending on whether the plan is put into action, this initial stage of the ecosystem development cycle can be divided into two steps: preparation and germination. In the preparation step, the development blueprint of the entrepreneurship education ecosystem is proposed, which *clarifies the goals* (Interviewee #2) or the *development direction of the ecosystem* (Interviewee #25). This development blueprint is critical to ecosystem development because it *marked the difference between the entrepreneurship education ecosystem and the spontaneous entrepreneurship curriculum improvement or organizational collaboration* (Interviewee #9). Therefore, the blueprint must be panoramic, systematic, and framed, and *in most cases was proposed by leadership* (Interviewee #16) or *in line with the leadership's vision* (Interviewee #22). The uniqueness of the germination step lies in the action. However, this kind of action is limited to the development of relatively independent elements within the ecosystem, and these early-developed elements rarely have regular contacts, interactions, and information exchanges with each other. *Once the elements were regularly linked by some mechanism* (such as interaction, information exchange, or mutual dependence), *which marked the formation of the entrepreneurship education ecosystem* (Interviewee #29), and then the development of the ecosystem entered the next stage.

4.3.2. Stabilization Stage

The stabilization stage is the core stage of the development cycles of the entrepreneurship education ecosystem. Moreover, the three steps of growth, equilibrium, and stagnation are included in this core stage. These three steps represent a sustainable cyclical sequence as shown in Figure 3. In the process of the growth step, the entrepreneurship education ecosystem presents a thriving situation. Specific manifestations include: *The increase of entrepreneurship courses* (Interviewee #6); *increased participation in entrepreneurship courses, practices, and activities* (Interviewee #17); *resources, including finance, were constantly enriched* (Interviewee #13); *normalization of multi-element cooperation* (Interviewee #10); *with universities as the center, ecosystem cooperation in industry and community continued to expand and strengthen* (Interviewee #15), etc. On the contrary, the stagnation step indicates that the development of the entrepreneurship education ecosystem is in crisis and cannot be sustained. This unsustainable state of the ecosystem is reflected as follows: *Entrepreneurship courses gradually failed to meet learners' needs in terms of content or timeliness* (Interviewee #27); *resources tend to be exhausted or unable to meet the needs of ecosystem development* (Interviewee #30); *industry or community cooperation was inefficient, stagnant or shrinking* (Interviewee #16); *the institutional framework or operational mechanism was not compatible with other elements* (Interviewee #24), etc. In particular, the crisis identified by the stagnation step does not refer to minor problems and contradictions that occur in the operation of the various elements of the ecosystem. For example, a case used in an entrepreneurship classroom did not receive positive feedback from learners, or occasionally the number of students choosing entrepreneurship courses has decreased. There are some minor problems and contradictions that often exist in a variety of daily teaching and ecosystem operations. The so-called crisis in the stagnation step means that the state or situation caused by an event has a greater impact on the entire entrepreneurship ecosystem. Related to a specific negative event is the distinctive feature of the stagnation step of the entrepreneurship education ecosystem development cycles. The typical examples of the stagnation step in the responses of the interviewees talking about the best practice of the ecosystem were the cases of Emlyon Business School and Babson College. The entrepreneurship education ecosystem of Emlyon Business School has experienced three crises, corresponding to the promotion of entrepreneurship education, the reform of teaching methods, and the development of inter-academic courses, which were called 'Three Waves' by Fayolle and Byrne [94]. Additionally, when summing the experience of the Babson College, Fetters, et al. [95] mentioned that the development of the ecosystem is both staged and wavy, and the valley represents the crisis of the ecosystem and major events. When opportunity-rich growth and challenging stagnation are in check and balance, the entrepreneurship education ecosystem is in equilibrium. The equilibrium step is a *dynamic and changeable state* (Interviewee #26), and *its duration can be long or short* (Interviewee #13). The state of most entrepreneurship education ecosystems, in reality, is in one of these three steps of the stabilization stage. Furthermore, a sustainable cycle was formed between these three steps: An entrepreneurship education ecosystem at the equilibrium step is in crisis due to unexpected events and has transitioned into the stagnation step. Then, if the internal or external forces of the ecosystem correctly respond to and resolve this crisis, the entrepreneurship education ecosystem in the stagnation step will enter the next growth step. Compared with the original state of the ecosystem, this new growth may gradually compensate for the damage or loss caused by the stagnation period, or it may progress toward a larger and healthier state based on the original. Regardless of the way of growth, the ecosystem development direction is to enter the next step of the equilibrium. As mentioned by the interviewees, *the development process of the entrepreneurship education ecosystem was very similar to the development of a new enterprise* (Interviewee #12). New enterprises must *constantly face new business development and market challenges* (Interviewee #30), and they were always in a state of *opportunity and crisis* (Interviewee #3). Therefore, in the stabilization stage of the entrepreneurship education ecosystem, the development process can be summarized as sustainable cyclical growth, as shown in Figure 5.

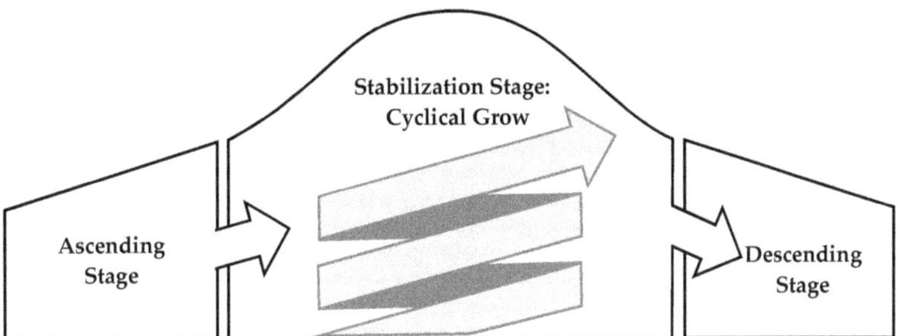

Figure 5. Sustainable cyclical growth in the development cycles of the entrepreneurship education ecosystem.

4.3.3. Descending Stage

The descending stage is the last stage of the development of the entrepreneurship education ecosystem. Interviewees rarely gave realistic examples of ecosystems in the descending stage, because successful and active ecosystems were often widely advertised rather than failed and unhealthy ones. In the context of the development process of the idealized entrepreneurship education ecosystem set in this study, interviewees gave negative responses to the development of the ecosystem that could not overcome the crisis step. Intuitively, the ecosystem is bound to go from bad to worse and to recession and collapse when its sustainability is affected. Therefore, the difficult problem facing the entrepreneurship education ecosystem in the crisis of the stagnation step is resolved as a sign, which determined whether the ecosystem continues to enter the development cycle or enters the descending stage. The two steps of recession and collapse constitute the declining stage. The recession step is characterized by the gradual shrinking or declining of elements and the reduction of interactions and information exchange between the elements in the ecosystem. For example, *the departure of core faculty and staff* (Interviewee #13); *the decrease in the number of teachers engaged in entrepreneurship courses* (Interviewee #16); *insufficient resources to support the daily operation of the incubator* (Interviewee #1); *the cooperation program between the university and the community or industry were canceled* (Interviewee #28), etc. When regular entrepreneurship education activities cannot be carried out in universities, the entrepreneurship education ecosystem enters the collapse step. The collapse step is the last one in the development process of the entrepreneurship education ecosystem, and it means the gradual end of the ecosystem.

4.4. *Roles of the Factors*

The factors play different functions in each stage of the development of a university-based entrepreneurship education ecosystem.

4.4.1. Launch Factors

The launch factors play a core role in the startup of the entrepreneurship education ecosystem by providing the foundation for the germination, growth, and expansion of the ecosystem. The launch factors mainly include entrepreneurship curriculum and entrepreneurial activities and practices, which could take place in or out of the classroom. Regardless of theory or practice, entrepreneurship education is at the foundation of the entrepreneurship education ecosystem. When a blueprint for building an ecosystem was planned, many factors could not be started at the same time due to the availability of resources, but entrepreneurship courses and practices were always the priority according to the interviewees. *If we were asked to drop these factors one by one according to the importance, there is no doubt that the last thing left must be the entrepreneurship course in the initial stage of the entrepreneurship education ecosystem* (Interviewee #11). Self-sustaining is an important feature of biological ecosystems, but it is often overlooked in other fields. Without considering self-sustaining, some researchers decomposed entrepreneurship curriculum and practices into more factors [14,18], thus forming a narrow sense of entrepreneurship education ecosystem. From a broad perspective, these studies were the best interpretations of the launch factors that play a core role in the ecosystem. The launch factors are very important in the early stage of the entrepreneurship education ecosystem, but it does not mean that they can be ignored at later stages. Downplayed or weakened launch factors in the stabilization stage may lead to stagnation of the development of the ecosystem due to unstable foundations. To be specific, in the early stage of ecosystem development, the role of the launch factors is as follows: *Provision of formal entrepreneurship curriculum system* (Interviewee #5); *regular entrepreneurship lecture series or workshop* (Interviewee #13); *daily guide and assistance potential entrepreneurs or entrepreneurial team practice activities* (Interviewee #20). In other development stages, the improvement and solidification of the launch factors are *the abundance and variety of courses* (Interviewee #19), *the cutting-edge and timeliness of course content* (Interviewee #7), and *the personalization and diversification of practical activities* (Interviewee #14). Essentially, the launch factors, which consist of entrepreneurship courses and practices, are the main way to achieve the goals of the entrepreneurship education ecosystem. Supported by the environment, conditions, and resources of the ecosystem, the launch factors realize the delivery and transformation of entrepreneurial-related knowledge through the interaction of Learners and Educators. Furthermore, the launch factors promote potential entrepreneurs and entrepreneurial teams and provide the foundation for the birth of new ideas, new solutions, and new enterprises. Therefore, metaphorically, the launch factors are the *chloroplast of synthetic entrepreneurial "Organic Matter"* (Interviewee #7) and *the "Engine" that keeps the ecosystem alive* (Interviewee #11).

4.4.2. Process Factors

The process factors are defined as factors that function continuously at various steps in the development cycle of the entrepreneurship education ecosystem, and they are primarily important to the ecosystem in a supportive manner. Resources and leadership vision constitute process factors, both of which respectively provide endless dynamics for the ecosystem at the realistic and spiritual levels. Resources from the government, industry, and community in addition to internal university resources provide the necessary human, economic, and material support in the construction and development of the entrepreneurship education ecosystem. Support and attitudes also lead to a more positive risk perception among potential entrepreneurs within the ecosystem [96]. Moreover, the availability of resources determines to some extent the length or efficiency of ecosystem construction and development. *Resources are vital to the ecosystem at all times, just as water and electricity are to factories* (Interviewee #8). The leadership vision works by encouraging, spurring, or motivating all human factors within the ecosystem, including teachers, students, faculty, managers, entrepreneurs, investors, and officials. Specifically, the role of the leadership vision is reflected by several interviewees: *The outline of the ecosystem blueprint* (Interviewee #19); *the formulation of phased goals* (Interviewee #30); *The potential allocation of benefits and resources* (Interviewee #4); *the stable and sustainable development direction* (Interviewee #13); *the promotion of cooperation inside and outside the university* (Interviewee #15); *the acquisition of the sense of accomplishment and mission of faculty and staff* (Interviewee #28). In addition, it has been found that the stagnation step of ecosystem development is often caused by changes in the state of process factors. Changes in resources and leadership vision are among the reasons that trigger the entrepreneurship education ecosystem to enter a crisis state, which is directly reflected in the high frequency of the corresponding data in the cross-matrix table. For example, *changes in the economic situation and markets would affect the resource input of industry to the ecosystem* (Interviewee #17), or *the replacement of university president may lead to subversive changes in leadership vision* (Interviewee #10). For sustainability, the stagnation step of the development cycle is a crossroad: if managed properly, it can make the ecosystem evolve and upgrade into the next level of the development cycle, or it can also make the ecosystem decline and shrink into the decline stage. In this regard, the process factors are the key to diagnosing problems in an ecosystem and taking measures, whether passively influenced by external factors of the ecosystem or actively initiating internal reform of the ecosystem.

4.4.3. Maintenance Factors

The maintenance factors are critical to the early stages and stagnation step of the entrepreneurship education ecosystem because their role is to get the ecosystem out of its unhealthy state and through the crisis. The maintenance factors include organizational structure, operating mechanism, and core faculty, which play the maintenance function for the ecosystem in different ways. The emergence of two high-frequency peaks in the early stages and the stagnation step of the ecosystem development cycle is the unique feature of maintenance factors through the analysis of the cross-matrix table. Organizational structure and operating mechanism are classified as environment types in the presentation of the composition of the entrepreneurship education ecosystem. However, unlike other natural environments, which are difficult to be transformed, organizational structures and operational mechanisms need to be adjusted promptly to facilitate the development of the ecosystem. *Organizational structure was like tangible computer hardware, and operating mechanism is like intangible software running on a computer . . . both of them must be adapted to exert maximum efficiency* (Interviewee #20). This adaptability is reflected in two aspects. The first is the internal matching of both the organizational structure and operating mechanism. In the face of numerous institutions and complex relationships within the ecosystem, the unclear division of responsibilities and extensive management regimes could inevitably lead to low efficiency and poor communication [79]. The other is that the organizational structure and operating mechanism match the development scale and stage of the entire entrepreneurship education ecosystem. The small-scale ecosystem in the early stage and the large-scale ecosystem in the equilibrium stage necessarily correspond to different organizational structures and operating mechanisms. Core faculty is also one of the maintenance factors, although it was not intended to be an element at the beginning of the study because it is more like a unit, essentially. The reason why core faculty members stand out from educators, leaders, administrators, or other primary and secondary nodes of coding as an independent ecosystem factor was that interviewees frequently mentioned it in the early stage and stagflation step of ecosystem development. Core faculty may be the president of a university, the administrator of an institution, the leader of a discipline, or even a socially versatile person. In addition, most of the core faculty were described by interviewees in terms of genius, talent, legend, leadership, excellence, distinction, preeminence, etc. Core faculty are the ones who can lead others to overcome difficulties and turn the situation when the entrepreneurship education ecosystem was established or in crisis. Unsurprisingly, when the interviewees listed and talked about specific cases of the development process of the entrepreneurship education ecosystem, these core faculty were one or more specific people with identified names. For example, *one of the reasons for Silicon Valley's success was the role of universities in promoting entrepreneurship and innovation. If an entrepreneurship education ecosystem was defined based on Stanford University, its early growth and expansion will undoubtedly be inseparable from his work and vision during Frederick Terman's tenure as the dean of the School of Engineering* (Interviewee #12). Another typical example is an entrepreneurship education ecosystem centered on the Massachusetts Institute of Technology (MIT), including the Route 128 Entrepreneurship Cluster. *Its development process had been up and down several times, but in its development process, it is necessary to mention a legendary figure who shoulders of industry, academia, and research, Professor Robert Langer of MIT. With him at the forefront, Kendall Square had developed into an innovation-intensive area for life sciences* (Interviewee #10). The development of the entrepreneurship education ecosystem is always linked to the names of a few core faculty, whether these names are known by a few or many people. Therefore, the behavior of core faculty and the adjustment of organizational structures and operating mechanisms have the same maintenance effect on the ecosystem. If the process factor is the underlying cause of the crisis of the entrepreneurship education ecosystem, then the maintenance factors may provide specific measures and key persons to resolve these crises.

5. Discussion

The growth of the university-based entrepreneurship education ecosystem could not be completed overnight. It requires a long time, resources, and continuous input from all stakeholders, as well as careful planning, continuous adjustment, and innovation in dealing with adverse situations, and may even require a little luck. The center of the university-based entrepreneurship education ecosystem is the university. Although almost all the stakeholders may contribute to the growth and development of the ecosystem from their respective roles, university administrators play an irreplaceable main role in coordinating and promoting the construction of the ecosystem due to their unique advantages in power, responsibility, and resources. Therefore, based on the analysis of the key elements of the entrepreneurship education ecosystem and the role and importance of these elements in different steps of the development cycle, the following strategies for constructing the entrepreneurship education ecosystem could be recommended for university administrators.

University administrators need to pay great attention and lay a solid foundation to accumulate, improve, and develop the identified launch factors. As the first step in building an entrepreneurship education ecosystem, the planning blueprint and specific actions begin with entrepreneurship courses and practices, whether it is entrepreneurship curriculum in the classroom or co-curricular entrepreneurial activities and practices outside the classroom. The launch factors are both the core and the foundation of the entire entrepreneurship education ecosystem. If significant progress could not be achieved in the launch factors at the early stage, these factors would inevitably restrict the development scale and lay hidden dangers for the future development of the ecosystem. Therefore, for the establishment of an idealized entrepreneurship education ecosystem, the following strategies for the launch factors are suggested. (1) The entrepreneurship curriculum, activities, and practices are taken as the first step to build an idealized entrepreneurship education ecosystem, and continuous development and improvement. (2) The development and improvement of entrepreneurship curriculum, activities, and practices should be both competency-driven and problem-driven, to ensure that the course content could meet learners' long-term expectations for entrepreneurship education. (3) After completing the initial construction of the entrepreneurship curriculum, activities, and practices, with the expansion of entrepreneurship education ecosystem scale, it is necessary to gradually expand and enrich in quantity, level, and form. (4) The pioneering and coverage of entrepreneurship curriculum, activities, and practices should be constantly maintained to keep the vitality and attraction of entrepreneurship education for all kinds of learners. (5) Learner-centered teaching methods need to be encouraged, explored, and innovated to highlight the interactivity and practicality of entrepreneurship education.

For the process factors, university administrators need to invest, expand, and innovate continuously based on inheritance. It is a long process to build a self-sustainable entrepreneurship education ecosystem. This extensive construction process requires constantly creating educational value with resources from within and outside the university, such as the government, industry, and community. In addition, this construction process is also the process of transforming the blueprint and goals in the leadership vision from intangible to reality. The leadership vision needs to be recognized by ecosystem stakeholders at various institutions and levels within the university, and it also needs to be consistent with the values of government, industry, and community. This recognition and consensus would create synergy between the internal and external resources of the university, thereby promoting the establishment of the entrepreneurship education ecosystem. The process factors of leadership vision and resource composition could adopt the following strategies in the construction of an ideal entrepreneurship education ecosystem. (1) The senior leaders of the university need to have a deeper understanding of entrepreneurship education and have lasting enthusiasm for the establishment of an entrepreneurship education ecosystem. (2) The university leaders actively participate in the construction of the ecosystem and incline to support entrepreneurship education in resource allocation are often the prerequisite for the sustainable and stable development of the entrepreneurship education ecosystem. (3) The brain trust or advisory board for the entrepreneurship education ecosystem is proposed to be established in university to provide intellectual support for senior leadership decisions. (4) The leadership vision should maintain certain relative stability from subversive changes caused by the switch of position candidates while fine-tuning the leadership vision in response to changing circumstances and ecosystem needs to be encouraged. (5) The output of the entrepreneurship education ecosystem, which could be in the form of new business creation, innovation, and economic growth, should be maximized and publicized to ensure contentious support from governments, industries, and communities. (6) The acquisition of resources is not the sole responsibility of the leader, and faculty and staff are encouraged to seek outside resources to collectively build the entrepreneurship education ecosystem rather than to act like outsiders.

The maintenance factors ensure the efficient operation of the entrepreneurship education ecosystem. The adjustment of maintenance factors seems to be more suitable for university senior leaders than other factors because the organizational structure, operating mechanisms, and core faculty are all within the scope of their position duties. University administrators need to constantly monitor the ecosystem to maintain its healthy development through timely interventions. It is unwise to pay more attention to maintenance factors only when the ecosystem is in the stagnation phase of the development cycle. A better approach for the sustainability of the ecosystem is to continuously innovate and regulate maintenance factors regularly, thereby preventing the ecosystem from entering stagnation. Therefore, the following strategies for building an ideal entrepreneurship education ecosystem could be inferred in terms of the maintenance factors. (1) Risk awareness is necessary for senior administrators in university, and it is suggested to introduce risk management into the whole process of constructing the entrepreneurship education ecosystem. (2) The reasonable organizational structure and efficient operating mechanism are very important for the development of the early entrepreneurship education ecosystem, and when the ecosystem encounters problems, they should also be considered as solutions and measures first. (3) The favorable conditions and environments within the university need to be created to initiatively cultivate localized core faculty, and the same effect could be achieved by actively introducing appropriate experts and talents from outside the university. (4) Based on common values and goals, creating an innovation culture and context with full participation in universities is a sign of the success of entrepreneurship education ecosystem, which is also a comprehensive embodiment of the organizational structure with a clear division of responsibilities, the operating mechanism with an unimpeded communication, and the harmonious and prosperous relationship between the core staff and team.

6. Conclusions

The entrepreneurial ecosystem is currently a focus of attention because of its role in promoting economic development and social progress. From the research of scholars or the practice of ecosystem development, higher education institutions have become the core of the entrepreneurial ecosystem because of their roles in education, research, and service to society [2–4]. Therefore, with the deepening of research, the dominant lineages of strategy and regional development of entrepreneurial ecosystem research tend to converge on a promising and key research topic, which is the university-based entrepreneurship education ecosystem. This research examines two core issues of the entrepreneurship education ecosystem: what elements make up the ecosystem and their role, and what is the development process and construction strategy of the ecosystem. Based on the interview and analysis of 30 stakeholders of the entrepreneurship education ecosystem, this study provides an ecosystem perspective for researchers and practitioners in the field of entrepreneurship education, through which entrepreneurship education could be placed in a specific environment for systematic implementation, analysis, and improvement. In addition, specific strategies were also provided for universities on how to promote the growth, development, and maintenance of the entrepreneurship education ecosystem from the perspective of the executives of higher education institutions. specifically, the following conclusions with theoretical significance and practical value were drawn.

The constituent elements of the university-based entrepreneurship education ecosystem could be divided into two categories as units and factors. Units refer to the institutions, organizations, or stakeholders in the ecosystem, including colleges and universities, learners, educators, government, industry, and community. These units are the sources or targets of behaviors, resources, information, and interactions. Factors are the intermediaries that link the units of the ecosystem together or the conditions and environment associated with the units, including entrepreneurship curriculum, entrepreneurial activities and practices, organizational structure, resources, leadership vision, core faculty, and operating mechanism. These key elements constitute three independent functional subsystems, namely, teaching and innovation, support, and operation with the universities interconnecting these subsystems.

The development of the entrepreneurship education ecosystem could be divided into three stages and seven steps. In chronological order, the first stage is ascending, including the two steps of preparation and germination. The second stage is stabilization, including the three steps of growth, equilibrium, and stagnation. The last is the declining stage, including the two steps of recession and collapse. In the stabilization stage between the ascending and declining stages, the development of an entrepreneurship education ecosystem can be summarized as a sustainable cyclical growth process. This development process is named the development cycle of the entrepreneurship education ecosystem, in which all the ecosystems of reality are located.

There are differences in the role and importance of the constituent factors of the ecosystem in each step of the development cycle of the entrepreneurship education ecosystem. According to their common aspects, the constituent factors are divided into three categories: launch, process, and maintenance factors. Launch factors mainly include entrepreneurship curriculum and entrepreneurial activities and practices, which play a core role in the startup of the development cycles and provide the foundation for the germination, growth, and expansion of the ecosystem. Resources and leadership vision constitute process factors, which function smoothly and continuously at various steps in the development cycle of the ecosystem in a supportive manner. Maintenance factors include organizational structure, operating mechanism, and core faculty, which are critical to the early stages and stagnation step of the development cycles because their role is to get the ecosystem out of its unhealthy state and through the crisis.

Based on the research results of the constituent elements, development cycles, role, and importance, three strategies for constructing an ideal university-based entrepreneurship education ecosystem are proposed from the perspective of university administrators. For the launch factors, university administrators need to pay great attention and lay a solid foundation for the accumulation, improvement, and development of the ecosystem. For the process factors, university administrators need to invest continuously and expand and innovate continuously based on inheritance. For the maintenance factors, university administrators need to be constantly alert and monitoring and maintain the healthy and sustainable development of the ecosystem through timely adjustment and reform.

One limitation of this study is the selection of the interview sample. The sample of 30 stakeholders could be improved to gain more weight. However, it is needless to say that interview-based research is very hard, as it needs a lot of time dedication not only by the researchers but also by the interviewees. In addition, the interviewees are all from the U.S., which also has some degree of impact on the achievement of the goal of the universality of the research conclusions. Another limitation is that the construction strategy of the entrepreneurship education ecosystem is mainly aimed at the executives of higher education institutions, while the perspectives of other stakeholders are ignored. As a further research avenue, the findings of this research by the induction method might be applied to the practice of case studies to further verify and improve.

Author Contributions: Conceptualization, H.L., S.K.-K. and A.K.; methodology, H.L., S.K.-K. and A.K.; data collection, H.L., S.K.-K. and A.K. All authors have read and agreed to the published version of the manuscript.

Funding: This research received no external funding.

Institutional Review Board Statement: The study was reviewed and determined at the Exempt level by the Pennsylvania State University Institutional Review Board (STUDY00013623 and 21 October 2019).

Informed Consent Statement: Informed consent was obtained from all subjects involved in the study.

Data Availability Statement: Not applicable.

Conflicts of Interest: The authors declare no conflict of interest.

References

1. Su, J.; Zhai, Q.; Landström, H. Entrepreneurship research in three regions-the USA, Europe and China. *Int. Entrep. Manag. J.* **2015**, *11*, 861–890. [CrossRef]
2. Lazzaro, E. Linking the creative economy with universities' entrepreneurship: A spillover approach. *Sustainability* **2021**, *13*, 1078. [CrossRef]
3. Portuguez Castro, M.; Ross Scheede, C.; Gómez Zermeño, M.G. The impact of higher education on entrepreneurship and the innovation ecosystem: A case study in Mexico. *Sustainability* **2019**, *11*, 5597. [CrossRef]
4. Bărbulescu, O.; Constantin, C.P. Sustainable growth approaches: Quadruple Helix approach for turning Brașov into a startup city. *Sustainability* **2019**, *11*, 6154. [CrossRef]
5. Siegel, D.S.; Wessner, C. Universities and the success of entrepreneurial ventures: Evidence from the small business innovation research program. *J. Technol. Transf.* **2012**, *37*, 404–415. [CrossRef]
6. Veiga, A.; Magalhães, A.; Amaral, A. Disentangling policy convergence within the European Higher Education Area. *Eur. Educ. Res. J.* **2019**, *18*, 3–18. [CrossRef]
7. Seikkula-Leino, J.; Salomaa, M.; Jónsdóttir, S.R.; McCallum, E.; Israel, H. EU policies driving entrepreneurial competences—Reflections from the case of EntreComp. *Sustainability* **2021**, *13*, 8178. [CrossRef]
8. Warnecke, T. Social entrepreneurship in China: Driving institutional change. *J. Econ. Issues* **2018**, *52*, 368–377. [CrossRef]
9. Henry, C.; Lewis, K. A review of entrepreneurship education research. *Educ. Train.* **2018**, *60*, 263–286. [CrossRef]
10. Fayolle, A.; Verzat, C.; Wapshott, R. In quest of legitimacy: The theoretical and methodological foundations of entrepreneurship education research. *Int. Small Bus. J.* **2016**, *34*, 895–904. [CrossRef]
11. Fellnhofer, K. Toward a taxonomy of entrepreneurship education research literature: A bibliometric mapping and visualization. *Educ. Res. Rev.* **2019**, *27*, 28–55. [CrossRef]
12. Wardana, L.W.; Narmaditya, B.S.; Wibowo, A.; Fitriana; Saraswati, T.T.; Indriani, R. Drivers of entrepreneurial intention among economics students in Indonesia. *Entrep. Bus. Econ. Rev.* **2021**, *9*, 61–74. [CrossRef]
13. Ács, Z.J.; Autio, E.; Szerb, L. National systems of entrepreneurship: Measurement issues and policy implications. *Res. Policy* **2014**, *43*, 476–494. [CrossRef]
14. Wraae, B.; Thomsen, J. Introducing a new framework for understanding learning in an entrepreneurship education ecosystem. *J. High. Educ. Theory Pract.* **2019**, *19*, 170–184. [CrossRef]
15. Kuratko, D.F.; Fisher, G.; Bloodgood, J.M.; Hornsby, J.S. The paradox of new venture legitimation within an entrepreneurial ecosystem. *Small Bus. Econ.* **2017**, *49*, 119–140. [CrossRef]
16. Ratten, V. Entrepreneurial education ecosystems. *J. Sci. Technol. Policy Manag.* **2019**, *10*, 857–860. [CrossRef]
17. Bahrami, H.; Evans, S. The research laboratory: Silicon Valley's knowledge ecosystem. In *Super-Flexibility for Knowledge Enterprises*; Springer: Berlin/Heidelberg, Germany, 2005; pp. 43–60.
18. Brush, C.G. Exploring the concept of an entrepreneurship education ecosystem. In *Innovative Pathways for University Entrepreneurship in the 21st Century*; Advances in the Study of Entrepreneurship, Innovation & Economic Growth; Emerald Group Publishing Limited: Bradford, UK, 2014; Volume 24, pp. 25–39.
19. Ribeiro, A.T.V.B.; Uechi, J.N.; Plonski, G.A. Building builders: Entrepreneurship education from an ecosystem perspective at MIT. *Triple Helix* **2018**, *5*, 1–20. [CrossRef]
20. Thomsen, B.; Muurlink, O.; Best, T. The political ecology of university-based social entrepreneurship ecosystems. *J. Enterp. Communities People Places Glob. Econ.* **2018**, *12*, 199–219. [CrossRef]
21. Belitski, M.; Heron, K. Expanding entrepreneurship education ecosystems. *J. Manag. Dev.* **2017**, *36*, 163–177. [CrossRef]
22. Spigel, B. The relational organization of entrepreneurial ecosystems. *Entrep. Theory Pract.* **2017**, *41*, 49–72. [CrossRef]

23. Pique, J.M.; Berbegal-Mirabent, J.; Etzkowitz, H. Triple helix and the evolution of ecosystems of innovation: The case of Silicon Valley. *Triple Helix* **2018**, *5*, 1–21. [CrossRef]
24. Tansley, A.G. The use and abuse of vegetational concepts and terms. *Ecology* **1935**, *16*, 284–307. [CrossRef]
25. Nelson, R.R.; Winter, S.G. *An Evolutionary Theory of Economic Change*; Belknap Press of Harvard University Press: Cambridge, MA, USA; London, UK, 1985.
26. Acs, Z.J.; Stam, E.; Audretsch, D.B.; O'Connor, A. The lineages of the entrepreneurial ecosystem approach. *Small Bus. Econ.* **2017**, *49*, 1–10. [CrossRef]
27. Moore, J. Predators and prey: A new ecology of competition. *Harv. Bus. Rev.* **1993**, *71*, 75–86.
28. Iansiti, M.; Levien, R. Strategy as Ecology. *Harv. Bus. Rev.* **2004**, *82*, 68–81.
29. Cohen, B. Sustainable valley entrepreneurial ecosystems. *Bus. Strategy Environ.* **2006**, *15*, 1–14. [CrossRef]
30. Mars, M.M.; Bronstein, J.L.; Lusch, R.F. The value of a metaphor: Organizations and ecosystems. *Organ. Dyn.* **2012**, *41*, 271–280. [CrossRef]
31. Song, A.K. The digital entrepreneurial ecosystem—A critique and reconfiguration. *Small Bus. Econ.* **2019**, *53*, 569–590. [CrossRef]
32. Doanh, D.C. The role of contextual factors on predicting entrepreneurial intention among Vietnamese students. *Entrep. Bus. Econ. Rev.* **2021**, *9*, 169–188. [CrossRef]
33. Stam, E. Entrepreneurial ecosystems and regional policy: A sympathetic critique. *Eur. Plan. Stud.* **2015**, *23*, 1759–1769. [CrossRef]
34. Glaeser, E.L.; Kerr, S.P.; Kerr, W.R. Entrepreneurship and urban growth: An empirical assessment with historical mines. *Rev. Econ. Stat.* **2015**, *97*, 498–520. [CrossRef]
35. McAdam, M.; Miller, K.; McAdam, R. Understanding Quadruple Helix relationships of university technology commercialisation: A micro-level approach. *Stud. High. Educ.* **2018**, *43*, 1058–1073. [CrossRef]
36. Cooke, P. The virtues of variety in regional innovation systems and entrepreneurial ecosystems. *J. Open Innov. Technol. Mark. Complex.* **2016**, *2*, 1–19. [CrossRef]
37. Adner, R. Ecosystem as structure: An actionable construct for strategy. *J. Manag.* **2017**, *43*, 39–58. [CrossRef]
38. Ylinenpää, H. Entrepreneurship and innovation systems: Towards a development of the ERIS/IRIS concept. *Eur. Plan. Stud.* **2009**, *17*, 1153–1170. [CrossRef]
39. Brown, R.; Mason, C. Looking inside the spiky bits: A critical review and conceptualisation of entrepreneurial ecosystems. *Small Bus. Econ.* **2017**, *49*, 11–30. [CrossRef]
40. Nguyen, T.T. The impact of access to finance and environmental factors on entrepreneurial intention: The mediator role of entrepreneurial behavioural control. *Entrep. Bus. Econ. Rev.* **2020**, *8*, 127–140. [CrossRef]
41. Malecki, E.J. Entrepreneurship and entrepreneurial ecosystems. *Geogr. Compass* **2018**, *12*, e12359. [CrossRef]
42. Cavallo, A.; Ghezzi, A.; Balocco, R. Entrepreneurial ecosystem research: Present debates and future directions. *Int. Entrep. Manag. J.* **2018**, *15*, 1291–1321. [CrossRef]
43. Isenberg, D.J. Applying the ecosystem metaphor to entrepreneurship: Uses and abuses. *Antitrust Bull.* **2016**, *61*, 564–573. [CrossRef]
44. Zahra, S.A.; Nambisan, S. Entrepreneurship in global innovation ecosystems. *AMS Rev.* **2011**, *1*, 4–17. [CrossRef]
45. Evans, D.S.; Schmalensee, R. *Matchmakers: The New Economics of Multisided Platforms*; Harvard Business Review Press: Boston, MA, USA, 2016.
46. Autio, E.; Thomas, L.D. Tilting the playing field: Towards an endogenous strategic action theory of ecosystem creation. In *World Scientific Reference on Innovation: Volume 3: Open Innovation, Ecosystems and Entrepreneurship: Issues and Perspectives*; World Scientific: Singapore, 2018; pp. 111–140.
47. Williamson, P.J.; Meyer, A.D. Ecosystem advantage: How to successfully harness the power of partners. *Calif. Manag. Rev.* **2012**, *55*, 24–46. [CrossRef]
48. Barnett, M.L. The keystone advantage: What the new dynamics of business ecosystems mean for strategy, innovation, and sustainability. *Acad. Manag. Perspect.* **2006**, *20*, 88–90. [CrossRef]
49. Price, E.L. The palgrave encyclopedia of strategic management. *Reference Reviews* **2014**, *28*, 22–23. [CrossRef]
50. Jacobides, M.G.; Cennamo, C.; Gawer, A. Towards a theory of ecosystems. *Strateg. Manag. J.* **2018**, *39*, 2255–2276. [CrossRef]
51. Azzam, J.E.; Ayerbe, C.; Dang, R.J.I. Using patents to orchestrate ecosystem stability: The case of a French aerospace company. *Int. J. Technol. Manag.* **2017**, *75*, 97–120. [CrossRef]
52. Adner, R.; Kapoor, R. Value creation in innovation ecosystems: How the structure of technological interdependence affects firm performance in new technology generations. *Strateg. Manag. J.* **2010**, *31*, 306–333. [CrossRef]
53. Alexy, O.; George, G.; Salter, A.J. Cui bono? The selective revealing of knowledge and its implications for innovative activity. *Acad. Manag. Rev.* **2013**, *38*, 270–291. [CrossRef]
54. Kapoor, R.; Lee, J.M. Coordinating and competing in ecosystems: How organizational forms shape new technology investments. *Strateg. Manag. J.* **2013**, *34*, 274–296. [CrossRef]
55. Brusoni, S.; Prencipe, A. The organization of innovation in ecosystems: Problem framing, problem solving, and patterns of coupling. *Adv. Strateg. Manag.* **2013**, *30*, 167–194. [CrossRef]
56. Etzkowitz, H. Is Silicon Valley a global model or unique anomaly? *Ind. High. Educ.* **2019**, *33*, 83–95. [CrossRef]
57. Feld, B. *Startup Communities: Building an Entrepreneurial Ecosystem in Your City*; John Wiley & Sons, Inc.: Hoboken, NJ, USA, 2012.

58. Etzkowitz, H.; Germain-Alamartine, E.; Keel, J.; Kumar, C.; Smith, K.N.; Albats, E. Entrepreneurial university dynamics: Structured ambivalence, relative deprivation and institution-formation in the Stanford innovation system. *Technol. Forecast. Soc. Chang.* **2019**, *141*, 159–171. [CrossRef]
59. Smith, H.L.; Romeo, S.; Bagchi-Sen, S. Oxfordshire biomedical university spin-offs: An evolving system. *Camb. J. Reg. Econ. Soc.* **2008**, *1*, 303–319. [CrossRef]
60. Mason, C.; Brown, R. Entrepreneurial ecosystems and growth oriented entrepreneurship. *Final. Rep. OECD Paris* **2014**, *30*, 77–102.
61. Boutillier, S.; Carré, D.; Levratto, N. *Entrepreneurial Ecosystems*; Wiley: Hoboken, NJ, USA, 2016; Volume 2.
62. Wasdani, K.P.; Manimala, M.J. *Entrepreneurial Ecosystem*; Springer: Berlin/Heidelberg, Germany, 2015.
63. Unceta, A.; Guerra, I.; Barandiaran, X. Integrating social innovation into the curriculum of higher education institutions in latin america: Insights from the students4change project. *Sustainability* **2021**, *13*, 5378. [CrossRef]
64. Siegel, D.S.; Lockett, A.; Wright, M. The rise of entrepreneurial activity at universities: Organizational and societal implications. *Ind. Corp. Chang.* **2007**, *16*, 489–504. [CrossRef]
65. Rice, M.P.; Fetters, M.L.; Greene, P.G. University-based entrepreneurship ecosystem: Key success factors and recommendations. In *The Development of University-Based Entrepreneurship Ecosystems: Global Practices*; Edward Elgar Publishing Limited: Cheltenham, UK, 2010; pp. 177–196.
66. Maritz, A.; Foley, D. Expanding Australian indigenous entrepreneurship education ecosystems. *Adm. Sci.* **2018**, *8*, 20. [CrossRef]
67. Maritz, A. Illuminating the black box of entrepreneurship education programmes: Part 2. *Educ. Train.* **2017**, *59*, 11. [CrossRef]
68. Regele, M.D.; Neck, H.M. The entrepreneurship education subecosystem in the United States: Opportunities to increase entrepreneurial activity. *J. Bus. Entrep.* **2012**, *23*, 25–47.
69. Ferrandiz, J.; Fidel, P.; Conchado, A. Promoting entrepreneurial intention through a higher education program integrated in an entrepreneurship ecosystem. *Int. J. Innov. Sci.* **2018**, *10*, 6–21. [CrossRef]
70. De Jager, H.J.; Mthembu, T.Z.; Ngowi, A.B.; Chipunza, C. Towards an innovation and entrepreneurship ecosystem: A case study of the central university of technology, free state. *Sci. Technol. Soc.* **2017**, *22*, 310–331. [CrossRef]
71. Mogollón, R.H.; Portillo, A.F.; Escobedo, M.C.S.; Pérez, J.L.C. The approach of the entrepreneur microecosystem for university entrepreneurial education: Model M2E EMFITUR. In *Economy, Business and Uncertainty: New Ideas for a Euro-Mediterranean Industrial Policy*; Springer: Cham, Switzerland, 2019; pp. 250–275.
72. Bischoff, K.; Volkmann, C.K.; Audretsch, D.B. Stakeholder collaboration in entrepreneurship education: An analysis of the entrepreneurial ecosystems of European higher educational institutions. *J. Technol. Transf.* **2018**, *43*, 20–46. [CrossRef]
73. Link, A.N.; Sarala, R.M. Advancing conceptualisation of university entrepreneurial ecosystems: The role of knowledge-intensive entrepreneurial firms. *Int. Small Bus. J. Res. Entrep.* **2019**, *37*, 289–310. [CrossRef]
74. Bedo, Z.; Erdos, K.; Pittaway, L. University-centred entrepreneurial ecosystems in resource-constrained contexts. *J. Small Bus. Enterp. Dev.* **2020**, *27*, 1149–1166. [CrossRef]
75. Kim, M.G.; Lee, J.-H.; Roh, T.; Son, H. Social entrepreneurship education as an innovation hub for building an entrepreneurial ecosystem: The case of the KAIST social entrepreneurship MBA program. *Sustainability* **2020**, *12*, 9736. [CrossRef]
76. Miller, D.J.; Acs, Z.J. The campus as entrepreneurial ecosystem: The University of Chicago. *Small Bus. Econ.* **2017**, *49*, 75–95. [CrossRef]
77. Jongbloed, B.; Enders, J.; Salerno, C. Higher education and its communities: Interconnections, interdependencies and a research agenda. *High. Educ.* **2008**, *56*, 303–324. [CrossRef]
78. Lahikainen, K.; Kolhinen, J.; Ruskovaara, E.; Pihkala, T. Challenges to the development of an entrepreneurial university ecosystem: The case of a Finnish university campus. *Ind. High. Educ.* **2018**, *33*, 96–107. [CrossRef]
79. Mukesh, H.V.; Pillai, K.R. Role of institutional ecosystem in entrepreneurship education: An empirical reiteration. *J. Entrep.* **2020**, *29*, 176–205. [CrossRef]
80. Cao, Z.; Zhou, M. Research on the innovation and entrepreneurship education mode in colleges and universities based on entrepreneurial ecosystem theory. *Educ. Sci. Theory Pract.* **2018**, *18*. [CrossRef]
81. Varano, M.; Kahkonen, E.; Aarnio, H.; Clavert, M.; Kaulio, M.; Thorén, K.; Haenen, C.; Petegem, W.; Colombelli, A.; Sansone, G.; et al. Entrepreneurship Education Ecosystems in Engineering and Technology. In Proceedings of the Annual Conference of the European Society for Engineering Education, Copenhagen, Denmark, 17–21 September 2018; pp. 1369–1378.
82. Antal, N.; Kingma, B.; Moore, D.; Streeter, D. University-Wide Entrepreneurship Education. In *Innovative Pathways for University Entrepreneurship in the 21st Century*; Emerald Publishing Limited: Bingley, UK, 2014; pp. 227–254.
83. Lyons, R.; Alshibani, S. *Entrepreneurship Education: A Tale of Two Countries*; International Council for Small Business: Washington, WA, USA, 2015; pp. 1–9.
84. Schmidt, J.J.; Molkentin, K.F. Building and maintaining a regional inter-university ecosystem for entrepreneurship: Entrepreneurship education consortium. *J. Entrep. Educ.* **2015**, *18*, 157–168.
85. Corbin, J.; Strauss, A. Strategies for Qualitative Data Analysis. In *Basics of qualitative Research: Techniques and Procedures for Developing Grounded Theory*; SAGE Publications, Inc.: Thousand Oaks, CA, USA, 2008; pp. 65–86.
86. Patton, M.Q. *Qualitative Research and Evaluation Methods*, 3rd ed.; Sage Publications: Thousand Oaks, CA, USA, 2002.
87. Maher, C.; Hadfield, M.; Hutchings, M.; de Eyto, A. Ensuring rigor in qualitative data analysis: A design research approach to coding combining nvivo with traditional material methods. *Int. J. Qual. Methods* **2018**, *17*, 1609406918786362. [CrossRef]

88. Galvao, A.R.; Marques, C.S.E.; Ferreira, J.J.; Braga, V. Stakeholders' role in entrepreneurship education and training programmes with impacts on regional development. *J. Rural Stud.* **2020**, *74*, 169–179. [CrossRef]
89. Hagebakken, G.; Reimers, C.; Solstad, E. Entrepreneurship education as a strategy to build regional sustainability. *Sustainability* **2021**, *13*, 2529. [CrossRef]
90. Mok, K.H.; Jiang, J. Towards corporatized collaborative governance: The multiple networks model and entrepreneurial universities in Hong Kong. *Stud. High. Educ.* **2020**, *45*, 2110–2120. [CrossRef]
91. Meyer, M.H.; Lee, C.; Kelley, D.; Collier, G. An assessment and planning methodology for university-based: Entrepreneurship ecosystems. *J. Entrep.* **2020**, *29*, 259–292. [CrossRef]
92. Wang, Q. Higher education institutions and entrepreneurship in underserved communities. *High. Educ.* **2020**, *81*, 1273–1291. [CrossRef]
93. Cai, Y.Z.; Ma, J.Y.; Chen, Q.Q. Higher education in innovation ecosystems. *Sustainability* **2020**, *12*, 4376. [CrossRef]
94. Fayolle, A.; Byrne, J. EM Lyon Business School. In *The Development of University-Based Entrepreneurship Ecosystems*; Edward Elgar Publishing Limited: Cheltenham, UK, 2010; pp. 45–75.
95. Fetters, M.L.; Greene, P.G.; Rice, M.P. Babson College. In *The Development of University-Based Entrepreneurship Ecosystems*; Edward Elgar Publishing Limited: Cheltenham, UK, 2010; pp. 15–44.
96. Nowiński, W.; Haddoud, M.Y.; Wach, K.; Schaefer, R. Perceived public support and entrepreneurship attitudes: A little reciprocity can go a long way! *J. Vocat. Behav.* **2020**, *121*, 103474. [CrossRef]

MDPI
St. Alban-Anlage 66
4052 Basel
Switzerland
Tel. +41 61 683 77 34
Fax +41 61 302 89 18
www.mdpi.com

Sustainability Editorial Office
E-mail: sustainability@mdpi.com
www.mdpi.com/journal/sustainability

www.ingramcontent.com/pod-product-compliance
Lightning Source LLC
LaVergne TN
LVHW070135170726
843515LV00007B/1801

* 9 7 8 3 0 3 6 5 6 2 4 1 4 *